# Synchronization
A universal concept in nonlinear sciences

First recognized in 1665 by Christiaan Huygens, synchronization phenomena
are abundant in science, nature, engineering, and social life. Systems as diverse
as clocks, singing crickets, cardiac pacemakers, firing neurons, and applauding
audiences exhibit a tendency to operate in synchrony. These phenomena are
universal and can be understood within a common framework based on modern
nonlinear dynamics. The first half of this book describes synchronization
without formulae, and is based on qualitative intuitive ideas. The main effects
are illustrated with experimental examples and figures; the historical
development is also outlined. The second half of the book presents the main
effects of synchronization in a rigorous and systematic manner, describing both
classical results on the synchronization of periodic oscillators and recent
developments in chaotic systems, large ensembles, and oscillatory media. This
comprehensive book will be of interest to a broad audience, from graduate
students to specialist researchers in physics, applied mathematics, engineering,
and natural sciences.

ARKADY PIKOVSKY was part of the Max-Planck research group on nonlinear
dynamics before becoming Professor of Statistical Physics and Theory of Chaos
at the University of Potsdam, Germany. A member of the German and American
Physical Societies, he is also part of the Editorial Board of *Physical Review* E,
for the term 2000–2002. Before this, he was a Humboldt fellow at the University
of Wuppertal. His PhD focused on the theory of chaos and nonlinear dynamics,
and was carried out at the Institute of Applied Physics of the USSR Academy of
Sciences. Arkady Pikovsky studied radiophysics and physics at Gorky State
University, and started to work on chaos in 1976, describing an electronic device
generating chaos in his Diploma thesis and later proving it experimentally.

MICHAEL ROSENBLUM has been a research associate in the Department of
Physics, University of Potsdam, since 1997. His main research interests are the
application of oscillation theory and nonlinear dynamics to biological systems
and time series analysis. He was a Humboldt fellow in the Max-Planck research
group on nonlinear dynamics, and a visiting scientist at Boston University.
Michael Rosenblum studied physics at Moscow Pedagogical University, and
went on to work in the Mechanical Engineering Research Institute of the USSR
Academy of Sciences, where he was awarded a PhD in physics and mathematics.

JÜRGEN KURTHS has been Professor of Nonlinear Dynamics at the University of Potsdam and director of the Interdisciplinary Centre for Dynamics of Complex Systems since 1994. He is a fellow of the American Physical Society and fellow of the Fraunhofer Society (Germany), and is currently vice-president of the European Geophysical Society. He is also a member of the Editorial Board of the *International Journal of Bifurcation and Chaos*. Professor Kurths was director of the group for nonlinear dynamics of the Max-Planck Society from 1992 to 1996. He studied mathematics at Rostock University, and then went on to work at the Solar–Terrestrial Physics Institute, and later the Astrophysical Institute of the GDR Academy of Sciences. He obtained his PhD in physics, and started to work on nonlinear data analysis and chaos in 1984. His main research interests are nonlinear dynamics and their application to geophysics and physiology and to time series analysis.

# Cambridge Nonlinear Science Series 12

## Editors

**Professor Boris Chirikov**
Budker Institute of Nuclear Physics,
Novosibirsk

**Professor Predrag Cvitanović**
Niels Bohr Institute, Copenhagen

**Professor Frank Moss**
University of Missouri, St Louis

**Professor Harry Swinney**
Center for Nonlinear Dynamics,
The University of Texas at Austin

The Cambridge Nonlinear Science Series contains books on all aspects of contemporary research in classical and quantum nonlinear dynamics, both deterministic and nondeterministic, at the level of graduate text and monograph. The intention is to have an approximately equal blend of experimental and theoretical works, with the emphasis in the latter on testable results. Specific subject areas suitable for consideration include: Hamiltonian and dissipative chaos; squeezed states and applications of quantum measurement theory; pattern selection; formation and recognition; networks and learning systems; complexity in low- and high-dimensional systems and random noise; cellular automata; fully developed, weak and phase turbulence; reaction–diffusion systems; bifurcation theory and applications; self-structured states leading to chaos; the physics of interfaces, including fractal and multifractal growth; and simulations used in studies of these topics.

## Titles in print in this series

# Synchronization

## A universal concept in nonlinear sciences

Arkady Pikovsky, Michael Rosenblum
and Jürgen Kurths
University of Potsdam, Germany

CAMBRIDGE
UNIVERSITY PRESS

CAMBRIDGE UNIVERSITY PRESS
Cambridge, New York, Melbourne, Madrid, Cape Town, Singapore, São Paulo

Cambridge University Press
The Edinburgh Building, Cambridge CB2 8RU, UK

Published in the United States of America by Cambridge University Press, New York

www.cambridge.org
Information on this title: www.cambridge.org/9780521592857

First published 2001
First paperback edition 2003
Reprinted 2003

*A catalogue record for this publication is available from the British Library*

*Library of Congress Cataloguing in Publication data*

Pikovsky, Arkady, 1956–
    Synchronization: a universal concept in nonlinear sciences / Arkady Pikovsky,
Michael Rosenblum, Jürgen Kurths.
        p.   cm. – (The Cambridge nonlinear science series; 12)
    Includes bibliographical references and index.
    ISBN 0 521 59285 2
        1. Synchronization.  2. Nonlinear theories.   I. Rosenblum, Michael, 1958–
II. Kurths, J. (Jürgen), 1953–   III. Title.   IV. Series.

Q172.5.S96 P54 2001.
003′.75–dc21      2001018104

ISBN 978-0-521-59285-7 hardback
ISBN 978-0-521-53352-2 paperback

Transferred to digital printing 2007

To my father Samuil    AP
To Sonya    MR
To my father Herbert    JK

# Contents

**Part III: Synchronization of chaotic systems**

## Appendices

## Appendix A1:  Discovery of synchronization by Christiaan Huygens   357

## Appendix A2:  Instantaneous phase and frequency of a signal   362

# Preface

The word "synchronous" is often encountered in both scientific and everyday language. Originating from the Greek words $\chi\rho\acute{o}\nu o\varsigma$ (*chronos*, meaning time) and $\sigma\acute{u}\nu$ (*syn*, meaning the same, common), in a direct translation "synchronous" means "sharing the common time", "occurring in the same time". This term, as well as the related words "synchronization" and "synchronized", refers to a variety of phenomena in almost all branches of natural sciences, engineering and social life, phenomena that appear to be rather different but nevertheless often obey universal laws.

A search in any scientific data base for publication titles containing the words with the root "synchro" produces many hundreds (if not thousands) of entries. Initially, this effect was found and investigated in different man-made devices, from pendulum clocks to musical instruments, electronic generators, electric power systems, and lasers. It has found numerous practical applications in electrical and mechanical engineering. Nowadays the "center of gravity" of the research has moved towards biological systems, where synchronization is encountered on different levels. Synchronous variation of cell nuclei, synchronous firing of neurons, adjustment of heart rate with respiration and/or locomotory rhythms, different forms of cooperative behavior of insects, animals and even humans – these are only some examples of the fundamental natural phenomenon that is the subject of this book.

Our surroundings are full of oscillating objects. Radio communication and electrical equipment, violins in an orchestra, fireflies emitting sequences of light pulses, crickets producing chirps, birds flapping their wings, chemical systems exhibiting oscillatory variation of the concentration of reagents, a neural center that controls the contraction of the human heart and the heart itself, a center of pathological activity that causes involuntary shaking of limbs as a consequence of Parkinson's disease – all these and many other systems have a common feature: they produce rhythms. Usually these objects are not isolated from their environment, but interact with other objects, in other words they are open systems. Indeed, biological clocks that govern daily (circadian) cycles are subject to the day–night and seasonal variations of illuminance and temperature, a violinist hears the tones played by her/his neighbors, a firefly is influenced by the light emission of the whole population, different centers of rhythmic brain activity may influence each other, etc. This interaction can be very weak, sometimes hardly perceptible, but nevertheless it often causes a qualitative transition: an object adjusts its rhythm in conformity with the rhythms of other objects. As a result, violinists play in unison, insects in a population emit acoustic or light pulses with a common rate, birds in a flock flap their wings simultaneously, the heart of a rapidly galloping horse contracts once per locomotory cycle.

**This adjustment of rhythms due to an interaction is the essence of synchronization**, the phenomenon that is systematically studied in this book.

The aim of the book is to address a broad readership: physicists, chemists, biologists, engineers, as well as other scientists conducting interdisciplinary research;[1] it is intended for both theoreticians and experimentalists. Therefore, the presentation of experimental facts, of the main principles, and of the mathematical tools is not uniform and sometimes repetitive. The diversity of the audience is reflected in the structure of the book.

The first part of this book, Synchronization without formulae, is aimed at readers with minimal mathematical background (pre-calculus), or at least it was written with this intention. Although Part I contains almost no equations, it describes and explains the main ideas and effects at a qualitative level.[2] Here we illustrate synchronization phenomena with experiments and observations from various fields. Part I can be skipped by theoretically oriented specialists in physics and nonlinear dynamics, or it may be useful for examples and applications.

Parts II and III cover the same ideas, but on a quantitative level; the reader of these parts is assumed to be acquainted with the basics of nonlinear dynamics. We hope that the bulk of the presentation will be comprehensible for graduate students. Here we review classical results on the synchronization of periodic oscillators, both with and without noisy perturbations; consider synchronization phenomena in ensembles of oscillators as well as in spatially distributed systems; present different effects that occur due to the interaction of chaotic systems; provide the reader with an extensive bibliography.

---

[1] As the authors are physicists, the book is inevitably biased towards the physical description of the natural phenomena.

[2] To simplify the presentation, we omit in Part I citations to the original works where these ideas were introduced; one can find the relevant references in the bibliographic section of the Introduction as well as in the bibliographic notes of Parts II and III.

We hope that this book bridges a gap in the literature. Indeed, although almost every book on oscillation theory (or, in modern terms, on nonlinear dynamics) treats synchronization among other nonlinear effects, only the books by Blekhman [1971, 1981], written in the "pre-chaotic" era, are devoted especially to the subject. These books mainly deal with mechanical and electromechanical systems, but they also contain extensive reviews on the theory, natural phenomena and applications in various fields. In writing our book we made an attempt to combine a description of classical theory and a comprehensive review of recent results, with an emphasis on interdisciplinary applications.

*Acknowledgments*
In the course of our studies of synchronization we enjoyed collaborations and discussions with V. S. Afraimovich, V. S. Anishchenko, B. Blasius, I. I. Blekhman, H. Chaté, U. Feudel, P. Glendinning, P. Grassberger, C. Grebogi, J. Hudson, S. P. Kuznetsov, P. S. Landa, A. Lichtenberg, R. Livi, Ph. Marcq, Yu. Maistrenko, E. Mosekilde, F. Moss, A. B. Neiman, G. V. Osipov, E.-H. Park, U. Parlitz, K. Piragas, A. Politi, O. Popovich, R. Roy, O. Rudzick, S. Ruffo, N. Rulkov, C. Schäfer, L. Schimansky-Geier, L. Stone, H. Swinney, P. Tass, E. Toledo, and A. Zaikin.

The comments of A. Nepomnyashchy, A. A. Pikovski, A. Politi, and C. Ziehmann, who read parts of the book, are highly appreciated.

O. Futer, N. B. Igosheva, and R. Mrowka patiently answered our numerous questions regarding medical and biological problems.

We would like to express our special gratitude to Michael Zaks, who supported our endeavor at all stages.

We also thank Philips International B.V., Company Archives, Eindhoven, the Netherlands for sending photographs and biography of Balthasar van der Pol and A. Kurths for her help in the preparation of the bibliography.

Finally, we acknowledge the kind assistance of the Cambridge University Press staff. We are especially thankful to S. Capelin for his encouragement and patience, and to F. Chapman for her excellent work on improving the manuscript.

*Book homepage*
We encourage all who wish to comment on the book to send e-mails to:
pikovsky@stat.physik.uni-potsdam.de;
mros@agnld.uni-potsdam.de;
jkurths@agnld.uni-potsdam.de.
All misprints and errors will be posted on the book homepage
(URL: http://www.agnld.uni-potsdam.de/~syn-book/).

# Chapter 1

## Introduction

## 1.1    Synchronization in historical perspective

The Dutch researcher Christiaan Huygens (Fig. 1.1), most famous for his studies in optics and the construction of telescopes and clocks, was probably the first scientist who observed and described the synchronization phenomenon as early as in the seventeenth century. He discovered that a couple of pendulum clocks hanging from a common support had synchronized, i.e., their oscillations coincided perfectly and the pendula moved always in opposite directions. This discovery was made during a sea trial of clocks intended for the determination of longitude. In fact, the invention and design of pendulum clocks was one of Huygens' most important achievements. It made a great impact on the technological and scientific developments of that time and increased the accuracy of time measurements enormously. In 1658, only two years after Huygens obtained a Dutch Patent for his invention, a clock-maker from Utrecht, Samuel Coster, built a church pendulum clock and guaranteed its weekly deviation to be less than eight minutes.

After this invention, Huygens continued his efforts to increase the precision and stability of such clocks. He paid special attention to the construction of clocks suitable for use on ships in the open sea. In his memoirs *Horologium Oscillatorium* (*The Pendulum Clock, or Geometrical Demonstrations Concerning the Motion of Pendula as Applied to Clocks*), where he summarized his theoretical and experimental achievements, Huygens [1673] gave a detailed description of such clocks.

> In these clocks the length of the pendulum was nine inches and its weight one-half pound. The wheels were rotated by the force of weights and were enclosed together with the weights in a case which was four feet long. At the

bottom of the case was added a lead weight of over one hundred pounds so that the instrument would better maintain a perpendicular orientation when suspended in the ship.

Although the motion of the clock was found to be very equal and constant in these experiments, nevertheless we made an effort to perfect it still further in another way as follows. ... the result is still greater equality of clocks than before.

Furthermore, Huygens shortly, but extremely precisely, described his observation of synchronization as follows.

... It is quite worth noting that when we suspended two clocks so constructed from two hooks imbedded in the same wooden beam, the motions of each

**Figure 1.1.** Christiaan Huygens (1629–1695), the famous Dutch mathematician, astronomer and physicist. Among his main achievements are the discovery of the first moon and the true shape of the rings of Saturn; the first printed work on the calculus of probabilities; the investigation of properties of curves; the formulation of a wave theory of light including what is well-known nowadays as the Huygens principle. In 1656 Christiaan Huygens patented the first pendulum clock, which greatly increased the accuracy of time measurement and helped him to tackle the longitude problem. During a sea trial, he observed synchronization of two such clocks (see also the introduction to the English translation of his book [Huygens 1673] for a historical survey). Photo credit: Rijksmuseum voor de Geschidenis der Natuuringtenschappen, courtesy American Institute of Physics Emilio Segrè Visual Archives.

pendulum in opposite swings were so much in agreement that they never receded the least bit from each other and the sound of each was always heard simultaneously. Further, if this agreement was disturbed by some interference, it reestablished itself in a short time. For a long time I was amazed at this unexpected result, but after a careful examination finally found that the cause of this is due to the motion of the beam, even though this is hardly perceptible. The cause is that the oscillations of the pendula, in proportion to their weight, communicate some motion to the clocks. This motion, impressed onto the beam, necessarily has the effect of making the pendula come to a state of exactly contrary swings if it happened that they moved otherwise at first, and from this finally the motion of the beam completely ceases. But this cause is not sufficiently powerful unless the opposite motions of the clocks are exactly equal and uniform.

The first mention of this discovery can be found in Huygens' letter to his father of 26 February 1665, reprinted in a collection of papers [Huygens 1967a] and reproduced in Appendix A1. According to this letter, the observation of synchronization was made while Huygens was sick and stayed in bed for a couple of days watching two clocks hanging on a wall (Fig. 1.2). Interestingly, in describing the discovered phenomenon, Huygens wrote about *"sympathy of two clocks"* (*le phénoméne de la sympathie, sympathie des horloges*).

Thus, Huygens had given not only an exact description, but also a brilliant qualitative explanation of this effect of **mutual synchronization**; he correctly understood that the conformity of the rhythms of two clocks had been caused by an imperceptible motion of the beam. In modern terminology this would mean that the clocks were synchronized in anti-phase due to **coupling** through the beam.

In the middle of the nineteenth century, in his famous treatise *The Theory of Sound*, William Strutt (Fig. 1.3) [Lord Rayleigh 1945] described the interesting phenomenon of synchronization in acoustical systems as follows.

> When two organ-pipes of the same pitch stand side by side, complications ensue which not unfrequently give trouble in practice. In extreme cases the pipes may

**Figure 1.2.** Original drawing of Christiaan Huygens illustrating his experiments with two pendulum clocks placed on a common support.

almost reduce one another to silence. Even when the mutual influence is more moderate, it may still go so far as to cause the pipes to speak in absolute unison, in spite of inevitable small differences.

Thus, Rayleigh observed not only mutual synchronization when two distinct but similar pipes begin to sound in unison, but also the effect of **quenching (oscillation death)** when the coupling results in suppression of oscillations of interacting systems.

A new stage in the investigation of synchronization was related to the development of electrical and radio engineering. On 17 February 1920 W. H. Eccles and J. H. Vincent applied for a British Patent confirming their discovery of the synchronization property of a triode generator – a rather simple electrical device based on a vacuum tube that produces a periodically alternating electrical current [Eccles and Vincent 1920]. The frequency of this current oscillation is determined by the parameters of the elements of the scheme, e.g., of the capacitance. In their

**Figure 1.3.** Sir John William Strutt, Lord Rayleigh (1842–1919). He studied at Trinity College, Cambridge University, graduating in 1864. His first paper in 1865 was on Maxwell's electromagnetic theory. He worked on the propagation of sound and, while on an excursion to Egypt taken for health reasons, Strutt wrote *Treatise on Sound* (1870–1871). In 1879 he wrote a paper on traveling waves, this theory has now developed into the theory of solitons. His theory of scattering (1871) was the first correct explanation of the blue color of the sky. In 1873 he succeeded to the title of Baron Rayleigh. From 1879 to 1884 he was the second Cavendish professor of experimental physics at Cambridge, succeeding Maxwell. Then in 1884 he became the secretary of the Royal Society. Rayleigh discovered the inert gas argon in 1895, the work which earned him a Nobel Prize in 1904. Photo credit: Photo Gen. Stab. Lit. Anst., courtesy AIP Emilio Segrè Visual Archives.

experiments, Eccles and Vincent coupled two generators which had slightly different frequencies and demonstrated that the coupling forced the systems to vibrate with a common frequency.

A few years later Edward Appleton (Fig. 1.4) and Balthasar van der Pol (Fig. 1.5) replicated and extended the experiments of Eccles and Vincent and made the first step in the theoretical study of this effect [Appleton 1922; van der Pol 1927]. Considering the simplest case, they showed that the frequency of a generator can be entrained, or synchronized, by a weak external signal of a slightly different frequency. These studies were of great practical importance because triode generators became the basic elements of radio communication systems. The synchronization phenomenon was used to stabilize the frequency of a powerful generator with the help of one which was weak but very precise.

Synchronization in living systems has also been known for centuries. In 1729 Jean-Jacques Dortous de Mairan, the French astronomer and mathematician, who

**Figure 1.4.** Sir Edward Victor Appleton (1892–1965). Educated at Cambridge University, he began research at the Cavendish Laboratory with W. L. Bragg. During the First World War he developed an interest in valves and "wireless" signals, which inspired his subsequent research career. He returned to the Cavendish Laboratory in 1919, continuing to work on valves and, with B. van der Pol, on nonlinearity, and on atmospherics. In 1924, in collaboration with M. F. Barnett, he performed a crucial experiment which enabled a reflecting layer in the atmosphere to be identified and measured. In 1936 he succeeded C. T. R. Wilson in the Jacksonian Chair of Natural Philosophy at Cambridge, where he continued collaborative research on many ionospheric problems. He was awarded the Nobel Prize for Physics in 1947 for his investigations of the ionosphere. Photo credit: AIP Emilio Segrè Visual Archives, E. Scott Barr Collection.

was later the Secretary of the Académie Royale des Sciences in Paris, reported on his experiments with a haricot bean. He noticed that the leaves of this plant moved up and down in accordance with the change of day into night. Having made this observation, de Mairan put the plant in a dark room and found that the motion of the leaves continued even without variations in the illuminance of the environment. Since that time these and much more complicated experiments have been replicated in different laboratories, and now it is well-known that all biological systems, from

**Figure 1.5.** Balthasar van der Pol (1889–1959). He studied physics and mathematics in Utrecht and then went to England, where he spent several years working at the Cavendish Laboratory in Cambridge. There he met E. Appleton, and they started to work together in radio science. In 1919 van der Pol returned to Holland and in 1920 he was awarded a doctorate from Utrecht University. In 1922 he accepted an offer from the Philips Company and began work at Philips Research Laboratories in Eindhoven; soon he became Director of Fundamental Radio Research. Van der Pol acquired his international reputation due to his pioneering work on the propagation of radio waves and nonlinear oscillations. His studies of oscillation in a triode circuit led to the derivation of the van der Pol equation, a paradigmatic model of oscillation theory and nonlinear dynamics (see Eq. (7.2)). Together with van der Mark he pioneered the application of oscillation theory to physiological systems. Their work on modeling and hardware simulation of the human heart by three coupled relaxation oscillators [van der Pol and van der Mark 1928] remains a masterpiece of biological physics. Photo credit: Philips International B.V., Company Archives, Eindhoven, The Netherlands (see Bremmer [1960/61] for details).

rather simple to highly organized ones, have internal biological clocks that provide their "owners" with information on the change between day and night. The origin of these clocks is still a challenging problem, but it is well established that they can adjust their circadian rhythms (from *circa* = about and *dies* = day) to external signals: if the system is completely isolated from the environment and is kept under controlled constant conditions (constant illuminance, temperature, pressure, parameters of electromagnetic fields, etc.), its internal cycle can essentially differ from a 24-hour cycle. Under natural conditions, biological clocks tune their rhythms in accordance with the 24-hour period of the Earth's daily cycle.

As the last historical example, we cite another Dutchman, the physician Engelbert Kaempfer [1727][1] who, after his voyage to Siam in 1680 wrote:

> The glowworms . . . represent another shew, which settle on some Trees, like a fiery cloud, with this surprising circumstance, that a whole swarm of these insects, having taken possession of one Tree, and spread themselves over its branches, sometimes hide their Light all at once, and a moment after make it appear again with the utmost regularity and exactness . . . .

To our knowledge, this is the first reported observation of synchronization in a large population of oscillating systems.

We end our historical excursus in the 1920s. Since then many interesting synchronization phenomena have been observed and reported in the literature; some of them are described in the following chapters. More importantly, it gradually became clear that diverse effects which at first sight have nothing in common, obey some universal laws. A great deal of research carried out by mathematicians, engineers, physicists, and scientists from other fields, has led to the development of an understanding that, say, the conformity of the sounds of organ-pipes or the songs of the snowy tree cricket is not occasional, but can be described by a unified theory. In the following chapters we intend to demonstrate that these and a variety of other seemingly different effects have common characteristic features and can be understood within a unified framework.

## 1.2    Synchronization: just a description

We have shown with a few introductory examples (and we will illustrate it with further examples below) that synchronization is encountered in various fields of science, in engineering and in social behavior. We do not intend to give any rigorous definition of this phenomenon now. Before we discuss this notion in detail, although without mathematical methods, in Part I, and before we present the theoretical description in Parts II and III, we just give here a simple qualitative description of the effect; this section can be skipped by readers with a basic knowledge of physics

[1] Citation taken from [Buck and Buck 1968].

and nonlinear dynamics. Using several characteristic examples, we explain what synchronization is, and outline the common properties of systems that allow this effect to occur. However, the answer to the question "Why does it take place?" is left to Chapter 2.

## 1.2.1   What is synchronization?

We understand synchronization as an **adjustment of rhythms of oscillating objects due to their weak interaction**. Except for rare cases when it is said explicitly otherwise, this concept is used throughout the book. To explain this concept in qualitative terms we will concentrate on the following four questions.

- What is an oscillating object?
- What do we understand by the notion "rhythm"?
- What is an interaction of oscillating systems?
- What is an adjustment of rhythms?

To illustrate this general definition we take the classical example – a pendulum clock.

### Self-sustained oscillator: a model of natural oscillating objects

Let us discuss how a clock works. Its mechanism transforms the potential energy of the lifted weight (or compressed spring, or electrical battery) into the oscillatory motion of the pendulum. In its turn, this oscillation is transferred into the rotation of the hands on the clock's face (Fig. 1.6a). We are not interested in the particular design of the mechanism; what is important is only that it takes energy from the source and maintains a steady oscillation of the pendulum, which continues without any change until the supply of energy expires. The next important property is that the exact form of the oscillatory motion is entirely determined by the internal parameters of the clock and does not depend on how the pendulum was put into motion. Moreover, after being slightly perturbed, following some transient process the pendulum restores its previous internal rhythm.

These features are typical not only of clocks, but also of many oscillating objects of diverse nature. The set of these features constitutes the answer to the first of the questions above. In physics such oscillatory objects are denoted **self-sustained oscillators**; below we discuss their properties in detail. Further on we often omit the word "self-sustained", but by default we describe only systems of this class. Here we briefly summarize the properties of self-sustained oscillatory systems.

- This oscillator is an **active system**. It contains an internal **source of energy** that is transformed into oscillatory movement. Being isolated, the oscillator continues to generate the same rhythm until the source of energy expires.

Mathematically, it is described by an **autonomous** (i.e., without explicit time dependence) dynamical system.

■ The form of oscillation is determined by the parameters of the system and does not depend on how the system was "switched on", i.e., on the transient to steady oscillation.

■ The oscillation is stable to (at least rather small) perturbations: being disturbed, the oscillation soon returns to its original shape.

Examples of self-sustained oscillatory systems are electronic circuits used for the generation of radio-frequency power, lasers, Belousov–Zhabotinsky and other oscillatory chemical reactions, pacemakers (sino-atrial nodes) of human hearts or artificial pacemakers that are used in cardiac pathologies, and many other natural and artificial systems. As we will see later, an outstanding common feature of such systems is their ability to be synchronized.

### Characterization of a rhythm: period and frequency

Self-sustained oscillators can exhibit rhythms of various shapes, from simple sine-like waveforms to a sequence of short pulses. Now we quantify such rhythms using our particular example – the pendulum clock. The oscillation of the pendulum is periodic (Fig. 1.6b), and the **period** $T$ is the main characteristic of the clock. Indeed, the mechanism that rotates the hands actually counts the number of pendulum oscillations, so that its period constitutes the base time unit.

Often it is convenient to characterize the rhythm by the number of oscillation cycles per time unit, or by the oscillation **cyclic frequency**

$$f = \frac{1}{T}.$$

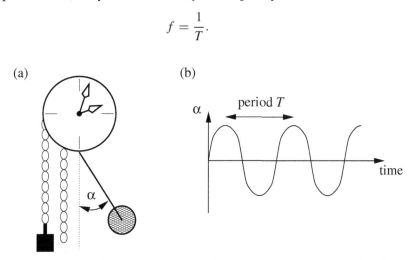

**Figure 1.6.** (a) An example of a self-sustained oscillator, the pendulum clock. The potential energy of the lifted weight is transformed into oscillatory motion of the pendulum and eventually in the rotation of hands. (b) The motion of the pendulum is periodic, i.e., its angle $\alpha$ with respect to the vertical varies in time with the period $T$.

In the theoretical treatment of synchronization, the **angular frequency** $\omega = 2\pi f = 2\pi/T$ is often more convenient; below we often omit the word "angular" and call it simply the frequency. Later on we will see that the frequency can be changed because of the external action on the oscillator, or due to its interaction with another system. To avoid ambiguity, we call the frequency of the autonomous (isolated) system the *natural frequency*.

## Coupling of oscillating objects

Now suppose that we have not one clock, but two. Even if they are of the same type or are made by the same fabricator, the clocks seem to be identical, but they are not. Some fine mechanical parameters always differ, probably by a tenth of a per cent, but this tiny difference causes a discrepancy in the oscillatory periods. Therefore, these two clocks show a slightly different time, and if we look at them at some instant of time, then typically we find the pendula in different positions (Fig. 1.7).

Let us now assume that these two nonidentical clocks are not independent, but interact weakly. There might be different forms of interaction, or **coupling**, between these two oscillators. Suppose that the two clocks are fixed on a common support, and let this be a not absolutely rigid beam (Fig. 1.8), as it was in the original observation of Huygens. This beam can bend, or it may vibrate slightly, moving from left to right, this does not matter much. What is really important is only that the motion of each pendulum is transmitted through the supporting structure to the other pendulum and, as a result, both clocks "feel" each other: they interact through the vibration of the common support. This vibration might be practically imperceptible; in order to detect and visualize it one has to perform high-precision mechanical measurements. However, in spite of its weakness, it may alter the rhythms of both clocks!

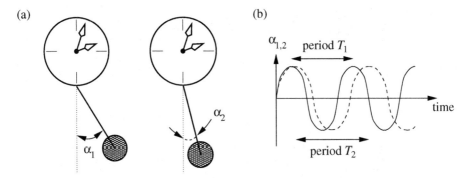

**Figure 1.7.** Two similar pendulum clocks (a) cannot be perfectly identical; due to a tiny parameter mismatch they have slightly different periods (here $T_2 > T_1$) (b). Therefore, if we look at them at some arbitrary moment of time, the pendula are, generally speaking, in different positions: $\alpha_1 \neq \alpha_2$.

## Adjustment of rhythms: frequency and phase locking

Experiments show that even a weak interaction can synchronize two clocks. That is, two nonidentical clocks which, if taken apart, have different oscillation periods, when coupled adjust their rhythms and start to oscillate with a common period. This phenomenon is often described in terms of coincidence of frequencies as **frequency entrainment** or **locking**:[2] if two nonidentical oscillators having their own frequencies $f_1$ and $f_2$ are coupled together, they may start to oscillate with a common frequency. Whether they synchronize or not depends on the following two factors.

### 1. Coupling strength

This describes how weak (or how strong) the interaction is. In an experimental situation it is not always clear how to measure this quantity. In the experiments described above, it depends in a complicated manner on the ability of the supporting beam to move. Indeed, if the beam is absolutely rigid, then the motions of pendula do not influence the support, and therefore there is no way for one clock to act on the other. If the clocks do not interact, the coupling strength is zero. If the beam is not rigid, but can vibrate longitudinally or bend, then an interaction takes place.

### 2. Frequency detuning

**Frequency detuning** or **mismatch** $\Delta f = f_1 - f_2$ quantifies how different the uncoupled oscillators are. In contrast to the coupling strength, in experiments with clocks detuning can be easily measured and varied. Indeed, one can tune the frequency of a clock by altering the pendulum length.[3] Thus, we can find out how the result of the interaction (i.e., whether the clocks synchronize or not) depends

---

[2] We use "entrainment" and "locking" as synonyms (see terminological remarks in Section 1.3.1).

[3] Mechanical clocks usually have a mechanism that easily allows one to do this. The process is used to force the clock to go faster if it is behind the exact time, and to force it to slow down if it is ahead.

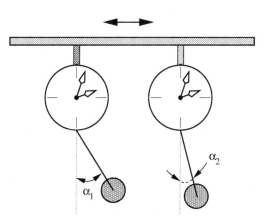

**Figure 1.8.** Two pendulum clocks coupled through a common support. The beam to which the clocks are fixed is not rigid, but can vibrate slightly, as indicated by the arrows at the top of the figure. This vibration is caused by the motions of both pendula; as a result the two clocks "feel" the presence of each other.

on the frequency mismatch. Imagine that we perform the following experiment. First we separate two clocks (e.g., we put them in different rooms) and measure their frequencies $f_1$ and $f_2$. Having done this, we put the clocks on a common support, and measure the frequencies $F_1$ and $F_2$ of the *coupled* systems. We can carry out these measurements for different values of the detuning to obtain the dependence of $\Delta F = F_1 - F_2$ on $\Delta f$. Plotting this dependence we get a curve as shown in Fig. 1.9, which is *typical for interacting oscillators*, independent of their nature (mechanical, chemical, electronic, etc.). Analyzing this curve we see that if the mismatch of autonomous systems is not very large, the frequencies of two clocks (two systems) become equal, or **entrained**, i.e., synchronization takes place. We emphasize that the frequencies $f_{1,2}$ and $F_{1,2}$ have to be measured for the same objects, but in different experimental conditions: $f_{1,2}$ characterize free (uncoupled, or autonomous) oscillators, whereas the frequencies $F_{1,2}$ are obtained in the presence of coupling. Generally, we expect the width of the synchronization region to increase with coupling strength.

A close examination of synchronous states reveals that the synchronization of two clocks can appear in different forms. It may happen that two pendula swing in a similar manner: for example, they both move to the left, nearly simultaneously attain the leftmost position and start to move to the right, nearly simultaneously cross the vertical line, and so on. The positions of the pendula then evolve in time in the way shown in Fig. 1.10a. Alternatively, one may find that two pendula always move in opposite directions: when the first pendulum attains, say, the leftmost position, the second pendulum attains the rightmost one; when they cross the vertical line, they move in opposite directions (Fig. 1.10b). To describe these two obviously distinct regimes, we introduce the key notion of synchronization theory, namely the **phase** of an oscillator.

We understand the phase as a quantity that increases by $2\pi$ within one oscillatory cycle, proportional to the fraction of the period (Fig. 1.11). The phase unambiguously determines the state of a periodic oscillator; like time, it parametrizes

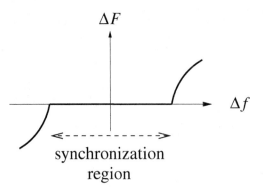

**Figure 1.9.** Frequency vs. detuning plot for a certain fixed strength of interaction. The difference of frequencies $\Delta F$ of two coupled oscillators is plotted vs. the detuning (frequency mismatch) $\Delta f$ of uncoupled systems. For a certain range of detuning the frequencies of coupled oscillators are identical ($\Delta F = 0$), indicating synchronization.

the waveform within the cycle. The phase seems to provide no new information about the system, but its advantage becomes evident if we consider the difference

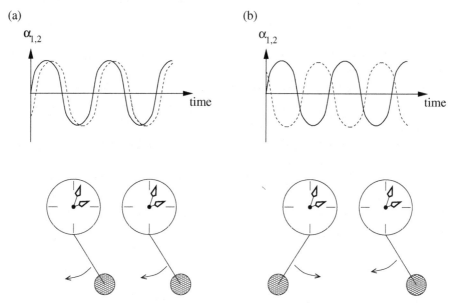

**Figure 1.10.** Possible synchronous regimes of two nearly identical clocks: they may be synchronized almost in-phase (a), i.e., with the phase difference $\phi_2 - \phi_1 \approx 0$, or in anti-phase (b), when $\phi_2 - \phi_1 \approx \pi$.

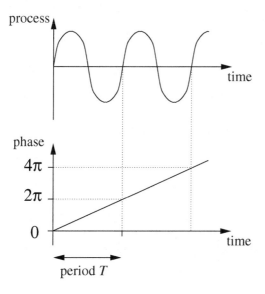

**Figure 1.11.** The definition of phase. The phase of a periodic oscillation grows uniformly in time and gains $2\pi$ at each period.

of the phases[4] of two clocks. This helps us to distinguish between two different synchronous regimes.

If two pendula move in the same direction and almost simultaneously attain, say, the rightmost position, then their phases $\phi_1$ and $\phi_2$ are close and this state is called **in-phase synchronization** (Fig. 1.10a). If we look at the motions of pendula precisely (we would probably need rather complicated equipment in order to do this), we can detect that the motions are not exactly simultaneous. One clock that was initially faster is a little bit ahead of the second one, so that one can speak of a **phase shift** between two oscillations. This phase shift may be very small, in the case of two clocks it may be not visible to the unaided eye, but it is always present if two systems have initially different oscillation periods, or different frequencies.

If the pendula of two synchronized clocks move in opposite directions then one speaks of synchronization in **anti-phase** (Fig. 1.10b). Exactly this synchronous state was observed and described by Christiaan Huygens. A recent reconstruction of this experiment carried out by I. I. Blekhman and co-workers and theoretical investigations demonstrate that both in- and anti-phase synchronous regimes are possible [Blekhman 1981], depending on the way the clocks are coupled (see Section 4.1.1 for details). Again, the oscillations of two clocks are shifted not exactly by half of the period – they are not exactly in anti-phase, but there exists an additional small phase shift.

The onset of a certain relationship between the phases of two synchronized self-sustained oscillators is often termed **phase locking**. Here we have described its simplest form; in the next chapters we consider more general forms of phase locking.

This imaginary experiment demonstrates the hallmark of synchronization, i.e., being coupled, two oscillators with initially different frequencies and independent phases adjust their rhythms and start to oscillate with a common frequency; this also implies a definite relation between the phases of systems. We would like to emphasize that this *exact identity* of frequencies holds within a certain range of initial frequency detuning (and not at one point as would happen for an occasional coincidence).

## 1.2.2    What is NOT synchronization?

The definition of synchronization just presented contains several constraints. We would like to emphasize them by illustrating the notion of synchronization with several counter-examples.

---

[4] Phase can be introduced in two different, although related, ways. One can reset it to zero at the beginning of each cycle, and thus consider it on the interval from 0 to $2\pi$; alternatively, one can sum up the gain in phase and, hence, let the phase grow infinitely. These two definitions are almost always equivalent because usually only the difference between the phases is important.

### There is no synchronization without oscillations in autonomous systems

First, we would like to stress the difference between synchronization and another well-known phenomenon in oscillating systems, namely resonance. For illustration we take a system that is to some extent similar to the clock, because its oscillatory element is also a pendulum. Let this pendulum rotate freely on a horizontal shaft and have a magnet at its free end (Fig. 1.12). The pendulum is not a self-sustained system and cannot oscillate continuously: being kicked, it starts to oscillate, but this free oscillation decays due to friction forces. The frequency $f_0$ of free oscillations (the eigenfrequency) is determined by the geometry of the pendulum and, for small oscillations, it does not depend on the amplitude. To achieve a nondecaying motion of the pendulum we place nearby an electromagnet fed by an alternating electrical current with frequency $f$. Correspondingly, the pendulum is forced by the periodical magnetic field. This oscillation is only visible if the frequency of the magnetic force $f$ is close to the eigenfrequency $f_0$, otherwise it is negligibly small; this is the well-known resonance phenomenon. After initial transience, the pendulum oscillates with the frequency of the magnetic field $f$. If this frequency were to vary, the frequency of the pendulum's oscillation would also vary. This seems to be the entrainment phenomenon described above, but it is not! This case cannot be considered as synchronization because one of two interacting systems, namely the pendulum, has no rhythm of its own. If we decouple the systems, e.g., if we place a metal plate between the electromagnet and the pendulum, the latter will stop after some transience. Hence, we cannot speak of an adjustment of rhythms here.

### Synchronous variation of two variables does not necessarily imply synchronization

Second, our understanding of synchronization implies that the object we observe can be separated into different subsystems that can (at least in principle, not necessarily in a particular experiment) generate independent signals. Thus we exclude the cases when two oscillating observables are just different coordinates of a single system. Consider, e.g., the velocity and the displacement of the pendulum in a clock; obviously these observables oscillate with a common frequency and a definite phase

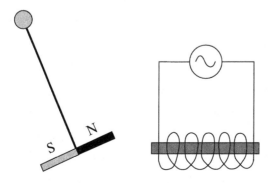

**Figure 1.12.** Resonance is not synchronization! The magnetic pendulum oscillates in the electromagnetic field with the frequency of the electric current. This is an example of a forced system that has no rhythm of its own: if the pendulum is shielded from the electromagnet then its oscillation decays

shift, but there cannot be any velocity without displacement and vice versa. Of course, in such a trivial example the nonrelevance of the notion of synchronization is evident, but in some complex cases it may not be obvious whether the observed signals should be attributed to one or to different systems.

We can illustrate this with the following example from population dynamics. The variation of population abundances of interacting species is a well-known ecological phenomenon; a good example is the hare–lynx cycle (Fig. 1.13). The numbers of predators and prey animals vary with the same period, $T \approx 10$ years, so that one can say that they vary synchronously with some phase shift. This plot resembles Fig. 1.10, but nevertheless we cannot speak of synchronization here because this ecological system cannot be decomposed into two oscillators. If the hares and lynxes were separated, there would be no oscillations at all: the preys would die out, and the number of predators would be limited by the amount of available food and other factors. Hares and lynxes represent two components of the same system. We cannot consider them as two subsystems having their own rhythms and, hence, we cannot speak about synchronization. On the contrary, it is very interesting to look for, and indeed find, synchronization of two (or many) hare–lynx populations which occupy adjacent regions and which are weakly interacting, e.g., via local migration of animals.

### Too strong coupling makes a system unified

Finally, we explain what we mean by "weak coupling", again using our "virtual" experiment. At the same time (we can also say "synchronously") we illustrate again

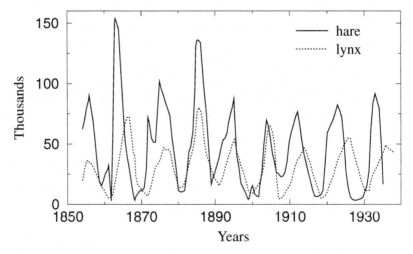

**Figure 1.13.** A classical set of data (taken from [Odum 1953]) for a predator–prey system: the Canadian lynx and snowshoe hare pelt-trading records of the Hudson Bay Company over almost a century. The notion of synchronization is not appropriate here because the lynxes and hares constitute a nondecomposable system.

the difference between the terms "synchronous motion" and "synchronization". Again, we take two clocks and mechanically connect the pendula with a rigid link (Fig. 1.14). Obviously, the clocks either stop, or the pendula move synchronously. We would not like to denote this trivial effect as synchronization: the coupling is not weak, it imposes too strong limitations on the motion of two systems, and therefore it is natural to consider the whole system as nondecomposable.

So, the possibility of separating several oscillating subsystems of a large system may depend on the parameters. Usually it is rather difficult, if at all possible, to determine strictly what can be considered as weak coupling, where the border between "weak" and "strong" lies, and, in turn, whether we are looking at a synchronization problem or should be studying the new unified system. In rather vague terms, we can say that the introduction of coupling should not qualitatively change the behavior of either one of the interacting systems and should not deprive the systems of their individuality. In particular, if one system ceases to oscillate, it should not prevent the second one keeping its own rhythm.

To summarize, if, in an experiment, we observe two variables that oscillate seemingly synchronously, it does not necessarily mean that we encounter synchronization. To call a phenomenon synchronization, we must be sure that:

- we analyse the behavior of two self-sustained oscillators, i.e., systems capable of generating their own rhythms;

- the systems adjust their rhythms due to weak interaction;

- the adjustment of rhythms occurs in a certain range of systems' mismatch; in particular, if the frequency of one oscillator is slowly varied, the second system follows this variation.

Correspondingly, a single observation is not sufficient to conclude on synchronization. Synchronization is **a complex dynamical process, not a state**.

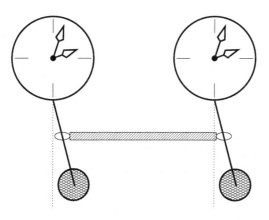

**Figure 1.14.** An example of what *cannot* be considered as weak coupling.

## 1.3    Synchronization: an overview of different cases

Thus far we have mentioned many examples of cooperative behavior of diverse oscillating objects, from simple mechanical or electrical devices to living systems. Now we want to list different forms of synchronization. In doing this we do not pay attention to the origin of the oscillations (i.e., whether they are produced by an electronic circuit or by a living cell) or the interaction (i.e., whether it takes place via a mechanical link or via some chemical mediator) but rather to certain general properties, e.g., whether the oscillations are periodic or irregular, or whether the coupling is bi- or unidirectional. This is not a full and rigorous classification, just a brief discussion of the main problems of synchronization theory that are addressed in this book.

The main example of the previous section – two interacting pendulum clocks – illustrates an important case that is denoted **mutual synchronization** (see also Chapters 4 and 8). Indeed, these two objects equally affect each other and mutually adjust their rhythms. It may also be that one oscillator is subject to an action that is completely independent of the oscillations of this driven system. We start with the description of such a case and then proceed to some more complex situations.

### Synchronization by an external force (Chapters 3 and 7)

Synchronization was discovered by Huygens as a side effect of his efforts to create high-precision clocks. Nowadays, this effect is employed to provide a precise and inexpensive measurement of time by means of radio-controlled clocks. Here, a weak broadcasted signal from a very precise central clock makes once-a-minute small adjustments to the rhythm of other clocks and watches, thus entraining them and making improvements in their quality superfluous. We emphasize the essential difference between radio-controlled and the "railway station clocks". The former are self-sustained oscillators and are therefore able to show the time (although not so precisely) even in the absence of synchronizing radio signals. On the contrary, railway station clocks are usually passive systems that are simply controlled by an electrical signal; they do not function if this signal disappears. Actually, these station clocks are nothing else than one (central) clock with many distant faces; hence, the notion of synchronization is not applicable here.

A similar synchronization scheme has been "implemented" by nature for the adjustment of biological clocks that regulate daily (circadian) and seasonal rhythms of living systems – from bacteria to humans. For most people the internal period of these clocks differs from 24 h, but it is entrained by environmental signals, e.g., illuminance, having the period of the Earth's rotation. Obviously, the action here is unidirectional: the revolution of a planet cannot be influenced by mankind (yet); thus, this is another example of synchronization by an external force. In usual circumstances this force is strong enough to ensure perfect entrainment; in order to desynchronize a biological clock one can either travel to polar regions or go caving. It is interesting that although normally the period of one's activity is exactly locked to that of the Earth's rotation, the phase shift between the internal clock and the external force varies from person to person: some people say that they are "early birds" whereas others call themselves "owls". Perturbation of the phase shift strongly violates normal activity. Every day many people perform such an experiment by rapidly changing their longitude (e.g., crossing the Atlantic) and experiencing jet lag. It can take up to several days to re-establish a proper phase relation to the force; in the language of nonlinear dynamics one can speak of different lengths of transients leading to the stable synchronous state.

### Ensembles of oscillators and oscillatory media (Chapters 4, 11 and 12)

In many natural situations more than two oscillating objects interact. If two oscillators can adjust their rhythms, we can expect that a large number of systems could do the same. One example has already been mentioned in Section 1.1: a large population of flashing fireflies constitutes what we can call an ensemble of mutually coupled oscillators, and can flash in synchrony. A very similar phenomenon, self-organization in a large applauding audience, has probably been experienced by every reader of this book, e.g., in a theater. Indeed, if the audience is large enough, then one can often hear a rather fast (several oscillatory periods) transition from noise to a rhythmic, nearly periodic, applause. This happens when the majority of the public applaud in unison, or synchronously.

A firefly communicates via light pulses with all other insects in the population, and a person in a theater hears every other member of the audience. In this case one speaks of **global** (all-to-all) coupling. There are other situations when oscillators are ordered into chains or lattices, where each element interacts only with its several neighbors. Such structures are common for man-made systems, examples are laser arrays and series of Josephson junctions, but may also be encountered in nature. So, mammalian intestinal smooth muscle may be electrically regarded as a series of loosely coupled pacemakers having intrinsic frequencies. Their activity triggers the muscle contraction. Experiments show that neighboring sources often adjust their frequencies and form synchronous clusters.

Quite often we cannot single out separate oscillating elements within a natural object. Instead, we have to consider the system as a continuous oscillatory medium, as it is in the case of the Belousov–Zhabotinsky chemical reaction. This can be conducted, e.g., in a thin membrane sandwiched between two reservoirs of reagents. Concentrations of chemicals vary locally, and collective oscillations that have a common frequency can be interpreted as synchronization in the medium.

### Phase and complete synchronization of chaotic oscillators (Chapters 5, 10 and Part III)

Nowadays it is well-known that self-sustained oscillators, e.g., nonlinear electronic devices, can generate rather complex, **chaotic** signals. Many oscillating natural systems also exhibit rather complex behavior. Recent studies have revealed that such systems, being coupled, are also able to undergo synchronization. Certainly, in this case we have to specify this notion more precisely because it is not obvious how to characterize the rhythm of a chaotic oscillator. It is helpful that sometimes chaotic waveforms are rather simple, like the example shown in Fig. 1.15. Such a signal is "almost periodic"; we can consider it as consisting of similar cycles with varying amplitude and period (which can be roughly defined as the time interval between the adjacent maxima). Fixing a large time interval $\tau$, we can count the number of cycles within this interval $N_\tau$, compute the **mean frequency**

$$\langle f \rangle = \frac{N_\tau}{\tau},$$

and take this as characterization of the chaotic oscillatory process.

With the help of mean frequencies we can describe the collective behavior of interacting chaotic systems in the same way as we did for periodic oscillators. If the coupling is large enough (e.g., in the case of resistively coupled electronic circuits this means that the resistor should be sufficiently small), the mean frequencies of two oscillators become equal and a plot such as that shown in Fig. 1.9 can be obtained. It is important that coincidence of mean frequencies does not imply that the signals coincide as well. It turns out that weak coupling does not affect the chaotic nature of both oscillators; the amplitudes remain irregular and uncorrelated whereas the frequencies are adjusted in a fashion that allows us to speak of the phase shift between

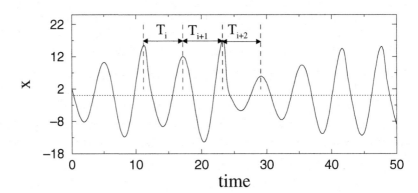

**Figure 1.15.** An example of the chaotic oscillation obtained by simulation of the Rössler system (it can be considered as a model of a generalized chemical reaction) [Rössler 1976]. (The Rössler system, as well as other dynamical models discussed in this book (e.g., the Lorenz and the van der Pol models), are usually written in a dimensionless form. Therefore, in Figs. 1.15, 1.16 and numerous other figures in this book, both the time and the time-dependent variables are dimensionless.) The interval between successive maxima irregularly varies from cycle to cycle, $T_i \neq T_{i+1} \neq T_{i+2}$, as well as the height of the maxima (the amplitude). Although the variability of $T_i$ is in this particular case barely seen, in general it can be rather large; therefore we characterize the rhythm via an average quantity, the mean frequency.

the signals (see Fig. 1.16c and compare it with Fig. 1.10a). This is denoted **phase synchronization** of chaotic systems.

Very strong coupling tends to make the states of both oscillators identical. It influences not only the mean frequencies but also the chaotic amplitudes. As a result, the signals coincide (or nearly coincide) and the regime of **complete synchronization** sets in (Fig. 1.16e, f).

Synchronization phenomena can be also observed in large ensembles of mutually coupled chaotic oscillators and in spatial structures formed by such systems. These effects are also discussed in the book.

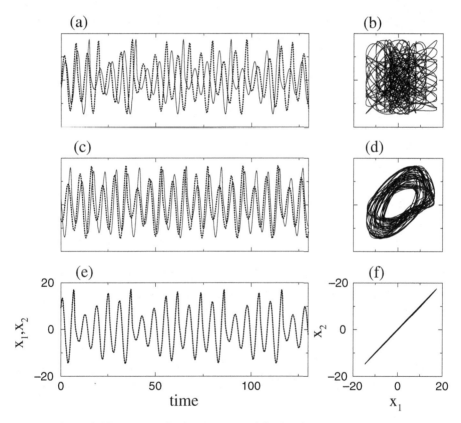

**Figure 1.16.** Two chaotic signals $x_1$, $x_2$ originating from nonidentical uncoupled systems are shown in (a). Within the time interval shown they have 21 and 22 maxima, respectively, hence the mean frequencies are different. Introduction of coupling between the oscillators adjusts the frequencies, although the amplitudes remain different (c). The plot of $x_1$ vs. $x_2$ (d) now shows some circular structure that is typical in the case of two signals with equal frequencies and a constant phase shift (compare with (b), where no structure is seen). Strong coupling makes the two signals nearly identical ((e) and (f)).

### What else is in the book

*Relaxation oscillators (Sections 2.4.2, 3.3 and 8.3)*
Quite often the oscillation waveforms are very far from sinusoidal. Many oscillators exhibit alternation of the epochs of "silence" and rapid activity; examples are contractions of the heart and spiking of neurons. Such systems are called **relaxation oscillators**, and a popular model is the **integrate-and-fire oscillator**. Understanding synchronization of such systems is important, e.g., in the context of neuroscience (neuronal ensembles) and cardiology (interaction of primary and secondary pacemakers of the heart).

*Rotators (Sections 4.1.8 and 7.4)*
Mechanical systems with rotating elements represent a special kind of objects that are able to synchronize. An electrical analog of rotators are the superconductive Josephson junctions. Synchronization of such systems is important for engineering applications.

*Noise (Section 3.4, Chapter 9)*
Periodic oscillators are idealized models of natural systems. Real systems cannot be considered perfectly isolated from the environment and are, therefore, always subject to different irregular perturbations. Besides, the internal parameters of oscillating objects slightly vary, e.g., due to thermal fluctuations. Thus, to be realistic, we have to study the properties of synchronization in the presence of noise.

*Inferring synchronization from data (Chapter 6)*
A stand-alone problem is an experimental investigation of possibly coupled oscillators. Quite often, especially in biological or geophysical applications, measured signals are much more complex than periodic motions of a clock's pendulum, and just an observation does not help here: special data analysis techniques are required to reveal synchronization. Moreover, sometimes we have no access to the parameters of the systems and coupling, but can only observe the oscillations. For example, the human organism contains a number of oscillators, such as the rhythmically contracting heart and respiratory system. Unlike the described "virtual" experiment with clocks, it is impossible (or, at least, very difficult) to tune these systems or to influence the coupling strength. The only way to detect the interaction is to analyze the oscillations registered under free-running conditions.

### 1.3.1   Terminological remarks

It appears very important to define the vocabulary of synchronization theory. Indeed, the understanding of such basic terms as *synchronization*, *locking* and *entrainment* differs, reflecting the background, individual viewpoints and taste of a researcher. In order to avoid ambiguity we describe here how *we* understand these terms.

We emphasize that we do not propose any general definition of *synchronization* that covers all the effects in interacting oscillatory systems. We understand synchronization as the adjustment of rhythms due to interaction and we specify this notion in particular cases, e.g., in considering noisy and chaotic oscillators. Generally, we do not restrict this phenomenon to a complete coincidence of signals, as is sometimes done.

We do not imply different meanings of the terms *locking* and *entrainment*; throughout the book these words are used as synonyms. We stress that we understand phase locking *not as the equality of phases*, but in a wider sense, that includes the constant phase shift and (small) fluctuations of the phase difference. That is, we say that the phases $\phi_1$ and $\phi_2$ are $n : m$ locked if the inequality $|n\phi_1 - m\phi_2| <$ constant holds.

In considering interacting chaotic systems we distinguish different stages of synchronization. In this context, the term *phase synchronization* is used to denote the state when only a relation between the phases of interacting systems sets in, whereas the amplitudes remain chaotic and can be nearly uncorrelated. This state can be described in terms of phase locking as well, and the words "phase synchronization" are used here to distinguish it from the *complete synchronization* when the chaotic processes become identical. The latter state is also frequently termed a full or an identical synchronization.

We emphasize that "oscillator", if not stated explicitly otherwise, means a self-sustained system. Such systems are sometimes denoted by the terms "self-oscillatory" ("auto-oscillatory") or "self-excited". By "limit cycle" we, according to the main object of our interest, mean here only the attractor of the self-sustained oscillator, not of the periodically forced system.

## 1.4     Main bibliography

Here we cite only some books and review articles. This list is definitely not complete, because descriptions of synchronization phenomena can be found in many other monographs and textbooks.

The only books solely devoted to synchronization problems are those by I. I. Blekhman [1971; 1981]; they primarily address mechanical oscillators, pendulum clocks in particular, systems with rotating elements, technological equipment, but also some electronic and quantum generators, chemical and biological systems.

A brief and popular introduction to synchronization can be found in [Strogatz and Stewart 1993]. An introduction to the synchronization theory illustrated by various biological examples is given in [Winfree 1980; Glass and Mackey 1988; Glass 2001].

The theory of synchronization of self-sustained oscillators by a harmonic force in the presence of noise was developed by R. L. Stratonovich [1963]. The influence of noise on mutual synchronization of two oscillators, synchronization by a force with fluctuating parameters and other problems are described by A. N. Malakhov [1968].

Different aspects of synchronization phenomena are studied in the monograph by P. S. Landa [1980]: synchronization of a self-sustained oscillator by an external force, mutual synchronization of two, three and many oscillators, impact of noise on the synchronization and entrainment of an oscillator by a narrow-band noise, synchronization of relaxation oscillators.

Y. Kuramoto [1984] developed the phase approximation approach that allows a universal description of weakly coupled oscillators. His book also presents a description of synchronization in large populations and synchronization of distributed systems (media). Some aspects of synchronization in spatially distributed systems, formation of synchronous clusters due to the effect of fluctuations, synchronization of globally (all-to-all) coupled oscillators with an emphasis on application to chemical and biological systems are discussed in [Romanovsky et al. 1975, 1984]. Synchronization effects in lasers are described in [Siegman 1986].

Synchronization of chaotic systems is addressed in chapters by Neimark and Landa [1992] and Anishchenko [1995]. We also mention a collection of papers and review articles [Schuster (ed.) 1999], as well as journal special issues on synchronization of chaotic systems [Pecora (ed.) 1997; Kurths (ed.) 2000].

We assume that the readers of Parts II and III of this book are acquainted with the basics of nonlinear science. If this is not the case, we can recommend the following books for an introductory reading on oscillation theory and nonlinear dynamics: [Andronov et al. 1937; Teodorchik 1952; Bogoliubov and Mitropolsky 1961; Hayashi 1964; Nayfeh and Mook 1979; Guckenheimer and Holmes 1986; Butenin et al. 1987; Rabinovich and Trubetskov 1989; Glendinning 1994; Strogatz 1994; Landa 1996]. In particular, [Rabinovich and Trubetskov 1989; Landa 1996] contain chapters highlighting the main problems of synchronization theory.

The reader wishing to know more about chaotic oscillations has many options, from introductory books [Moon 1987; Peitgen et al. 1992; Tufillaro et al. 1992; Lorenz 1993; Hilborn 1994; Strogatz 1994; Baker and Gollub 1996] to more advanced volumes [Guckenheimer and Holmes 1986; Schuster 1988; Wiggins 1988; Devaney 1989; Wiggins 1990; Lichtenberg and Lieberman 1992; Neimark and Landa 1992; Ott 1992; Argyris et al. 1994; Alligood et al. 1997].

# Part I

Synchronization without formulae

# Chapter 2

## Basic notions: the self-sustained oscillator and its phase

In this chapter we specify in more detail the notion of self-sustained oscillators that was sketched in the Introduction. We argue that such systems are ubiquitous in nature and engineering, and introduce their universal description in state space and their universal image – the limit cycle. Next, we discuss the notion and the properties of the phase, the variable that is of primary importance in the context of synchronization phenomena. Finally, we analyze several simple examples of self-sustained systems, as well as counter-examples. In this way we shall illustrate the features that make self-sustained oscillators distinct from forced and conservative systems; in the following chapters we will show that exactly these features allow synchronization to occur. Our presentation is not a systematic and complete introduction to the theory of self-sustained oscillation: we dwell only on the main aspects that are important for understanding synchronization phenomena.

The notion of **self-sustained oscillators** was introduced by Andronov and Vitt [Andronov *et al.* 1937]. Although Rayleigh had already distinguished between maintained and forced oscillations, and H. Poincaré had introduced the notion of the **limit cycle**, it was Andronov and Vitt and their disciples who combined rigorous mathematical methods with physical ideas. Self-sustained oscillators are a subset of the wider class of **dynamical systems**. The latter notion implies that we are dealing with a deterministic motion, i.e., if we know the state of a system at a certain instant in time then we can unambiguously determine its state in the future. Dynamical systems are idealized models that do not incorporate natural fluctuations of the system's parameters and other sources of noise that are inevitable in real world objects, as well as the quantum uncertainty of microscopic systems. In the 1930s only periodic self-oscillations were known. Nowadays, irregular, or **chaotic** self-sustained oscillators are also well-studied; we postpone consideration of such systems till Chapter 5.

## 2.1     Self-sustained oscillators: mathematical models of natural systems

### 2.1.1     Self-sustained oscillations are typical in nature

What is the common feature of such diverse oscillating objects as a vacuum tube radio generator, a pendulum clock, a firefly that emits light pulses, a contracting human heart and many other systems? The main universal property is that these are all active systems that taken apart, or being isolated, continue to oscillate in their own rhythms. This rhythm is entirely determined by the properties of the systems themselves; it is maintained due to an internal source of energy that compensates the *dissipation* in the system. Such oscillators are called *autonomous* and can be described within a class of *nonlinear* models that are known in physics and nonlinear dynamics as **self-sustained** or self-oscillatory (auto-oscillatory) systems.[1]

Quite often we can verify that an oscillator is self-sustained if we isolate it from the environment and check whether it still oscillates. So, one can isolate a firefly (or a cricket) from other insects, put it into a place with a constant temperature, illuminance, etc., and observe that, even being alone, it nevertheless produces rhythmic flashes (chirps). One can isolate a plant, an animal, or a human volunteer and establish that they still exhibit the rhythms of daily activity. Such experiments have been carried out by many researchers after de Mairan[2], and these studies clearly demonstrated that circadian rhythms exist in the absence of 24-hour external periodic perturbations. Thus, these rhythms are generated by an active, self-sustained system – a biological clock – in contrast to other process having the same periodicity, namely tidal waves. Tides are caused by the daily variation of gravitational forces due to the Moon. Although we cannot isolate oceans from this perturbation, the mechanism of the tides is well-understood and we know that the ocean is not a self-sustained oscillator; these oscillations would vanish in the absence of the periodic force, i.e., if there were no Moon.

Below we consider without any further discussion many natural rhythms, e.g., physiological, as self-sustained oscillations. The reason for this is that the systems which generate these rhythms are necessarily dissipative and therefore long-lasting rhythmical processes can only be maintained at the expense of some energy source. Hence, if it is evident that these systems are autonomous then we conclude that they are self-sustained.

---

[1]  In radiophysics and electrical engineering the term "generator" is traditionally used as a synonym of a self-sustained oscillatory system. In the following, if not said explicitly otherwise, we use the word "oscillator" as a short way of saying "self-sustained system". See also the terminological remarks in Section 1.3.1.

[2]  See the historical introduction in Section 1.1.

## 2.1.2   Geometrical image of periodic self-sustained oscillations: limit cycle

The universal phenomenon of self-sustained oscillation holds with a unified description that we introduce here. Suppose we observe an oscillator that generates a periodic process at its output; we denote the value of that process $x(t)$. In particular, $x(t)$ can refer to the angle of a clock's pendulum, or to the electrical current through a resistor in a vacuum tube generator, or to the intensity of light in a laser, or to any other oscillating quantity. Suppose we want to describe the state of the oscillator at some instant of time. It is not enough to know the value of $x$: indeed, for the same $x$ the process may either increase or decrease. Therefore, to determine unambiguously the state of the system we need more variables.[3] Quite often two variables, $x$ and $y$, are enough, and we proceed here with this simplest case. For a pendulum clock, these variables can be, e.g., the angle of the pendulum with respect to the vertical and its angular velocity. Thus, the behavior of the system can be completely described by the time evolution of a pair $(x, y)$ (Fig. 2.1). These variables are called coordinates in the **phase space** (state space) and the $y(t)$ vs. $x(t)$ plot is called the *phase portrait* of the system; the point with coordinates $(x, y)$ is often denoted the *phase point*. As the oscillation is periodic, i.e., it repeats itself after the period $T$, $x(t)$ corresponds to a *closed curve* in the phase plane, called the **limit cycle** (Fig. 2.2).

To understand the origin of the term "limit cycle" we have to determine how it differs from all other trajectories in the phase plane. For this purpose we consider the behavior of the trajectories in the vicinity of the cycle. In other words, we look at what happens if a phase point is pushed off the limit cycle. For an original physical (or mechanical, biological, etc.) system this would mean that we somehow perturb its periodic motion. Here we come to the essential feature of self-sustained oscillators: after perturbation of their oscillation, the original rhythm is restored, i.e., the phase

---

[3] The number of required variables depends on the particular dynamical system and is called its dimension.

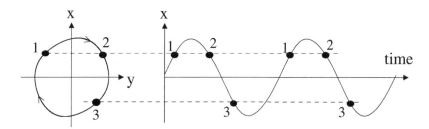

**Figure 2.1.** Periodic oscillation is represented by a closed curve in the phase space of the system: equivalent states $x(t)$ and $x(t + T)$ (denoted by a number on the time plot) correspond to the same point on that curve (denoted by the same number). On the contrary, the states with the same $x(t)$ are different if we take the second coordinate.

point returns to the limit cycle. This property also means that these oscillations do not depend on the initial conditions (at least in some range), or on how the system was initially put into motion. In the phase plane representation this corresponds to placing the initial phase point somewhere in the plane. We see from Fig. 2.2 that all trajectories tend to the cycle. Hence, after some transient time, the system performs steady state oscillations corresponding to the motion of the phase point along the limit cycle.[4]

The reason why we distinguish this curve from all others is thus that it attracts phase trajectories[5] and is therefore called an **attractor** of the dynamical system. The limit cycle is a simple attractor, in contrast to the notion of a **strange (chaotic) attractor**, to be encountered later.[6]

To conclude, self-sustained oscillations can be described by their image in the phase space – by the limit cycle. The form of the cycle, and, hence, the form of oscillation is entirely determined by the internal parameters of the system. If this oscillation is close in form to a sine wave, then the oscillator is called *quasilinear* (quasiharmonic). In this case the limit cycle can be represented as a circle. Strongly nonlinear systems can exhibit oscillation of a complicated form; in the following we analyze corresponding examples.

---

[4] Strictly speaking, the limit cycle may be not planar so that generally a high-dimensional phase space is required for the description of periodic oscillations. Nevertheless, very often this space can be reduced to the phase plane, i.e., two variables are sufficient. Analysis of the *transient* behavior, i.e., of the onset or perturbation of the limit cycle motion, may involve many, even infinitely many, variables. As the basic properties of self-sustained oscillations can be adequately illustrated in two dimensions, we restrict ourselves here to this case.

[5] At least from some neighborhood.

[6] A system can have more than one attractor having their own basins of attractions, e.g., several limit cycles, or a limit cycle and a stable equilibrium point. Different basins of attraction are separated by repelling sets (curves or surfaces), or *repellers*.

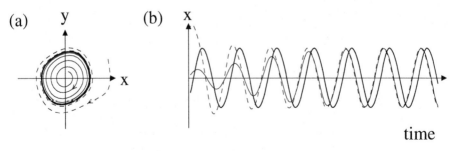

**Figure 2.2.** (a) The closed curve (bold curve) in the phase plane attracts all the trajectories from its neighborhood, and is therefore called the **limit cycle**. The same trajectories are shown in (b) as a time plot.

## 2.2    Phase: definition and properties

The notion of phase plays a key role in the theory of synchronization; we therefore discuss it in detail. We begin with the simple case of quasilinear oscillators, for which the notion of phase (and amplitude) can be easily illustrated. This example explains the general properties of the phase that are independent of the shape of the limit cycle. We complete this section with a demonstration of how phase can be defined for an arbitrary cycle.

### 2.2.1    Phase and amplitude of a quasilinear oscillator

The limit cycle of a quasilinear oscillator is nearly a circle, and the oscillation itself can be taken as the sine wave, $x(t) = A\sin(\omega_0 t + \phi_0)$. Here $\omega_0$ denotes the *angular frequency* which is related to the oscillation period by $\omega_0 = 2\pi/T$ and should be distinguished from the cyclic frequency of oscillation $f_0 = 1/T$. The intensity of oscillation is determined by its **amplitude** $A$, and the quantity $\phi(t) = \omega_0 t + \phi_0$ is called **phase** (see Fig. 1.11).

First of all we should note that the term "phase" is used in physics and nonlinear dynamics in different senses. For example, we have already encountered the notion of "phase space". The coordinates in this space are often denoted "phase point", and the time evolution of a system is described by the motion of this point. Here, as well as in the expression "phase transitions", the meaning of the word "phase" is completely different from what we understand by saying "the phase of oscillation". In this book we hope that the exact meaning will always be clear from the context.

The phase of the oscillation $\phi(t)$ increases without bound, but, as the sine is a periodic function, $\sin\phi = \sin(\phi + 2\pi)$, two phases that differ by $2\pi$ correspond to the same physical states. Sometimes, for convenience we consider a cyclic phase that is defined on a circle, i.e., varies from 0 to $2\pi$; we use the same notation for both cases and hope that it does not complicate the understanding.

We have not discussed yet the term $\phi_0$ that can be understood as the *initial phase*. We know already that when the self-sustained system is "switched on", then after some transient it achieves the limit cycle, irrespective of the initial state of the system. This means that the amplitude of oscillation does not depend on the initial conditions. On the contrary, the initial phase $\phi_0$ depends on the transient to the eventual state and can be arbitrary, i.e., all the values of $\phi_0$ are equivalent. Indeed, if we consider stationary oscillations only, then we can always change the initial phase by choosing some other starting point $t = 0$ for our observations.

We can easily interpret the notions of phase and amplitude, exploiting the picture of the limit cycle (Fig. 2.3a); this is nothing more than the representation of the point in the phase plane in polar coordinates.[7] This point rotates, e.g., counter-clockwise

---

[7] We remind the reader that the limit cycle of a quasilinear oscillator is nearly circular.

with angular velocity $\omega_0$, so that during the oscillation period $T$ it makes one rotation around the origin and the oscillatory phase $\phi(t)$ increases by $2\pi$.

## 2.2.2 Amplitude is stable, phase is free

Here we come to the most important point: the stability of the phase point on the limit cycle. In other words, we are interested in what happens to the oscillation if we slightly perturb the motion.

For convenience, we now analyze the system's behavior in a coordinate system rotating counter-clockwise with the same angular velocity $\omega_0$ as the phase point in the original coordinate system. From the viewpoint of an observer in this new reference frame, the phase point remains immobile, i.e., stationary oscillations correspond to the point that is at rest and has the coordinates $\phi(t) - \omega_0 t = \phi_0$ and $A$ (Fig. 2.3b).

Suppose now that the system is disturbed; we can describe this by a displacement of the point from the limit cycle (Fig. 2.3b). What is the evolution of this disturbance? As we know already, the perturbation of the amplitude of oscillation decays (see Fig. 2.2b). As to the phase, its perturbation *neither grows nor decays*: indeed, as all the values of $\phi_0$ are equivalent, then if the initial phase was changed from $\phi_0$ to $\phi_1$, it remains at this value unless the system is perturbed again.

We can illustrate the stability of the limit cycle oscillations by means of an analogy between the behavior of the phase point and that of a light particle placed in a viscous fluid. If some force acts on such a particle, it moves with a constant velocity. If the action is terminated, the particle stops and remains in that position.[8] The behavior of the amplitude can be then represented by a particle in a U-shaped potential (Fig. 2.4a).

---

[8]  If the mass of the particle is negligibly small, a constant force induces constant velocity, not constant acceleration. Such dynamics are often called overdamped.

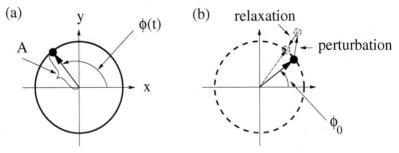

**Figure 2.3.** (a) A stationary self-sustained oscillation is described by the rotation of the phase point along the limit cycle; its polar coordinates correspond to the phase $\phi(t)$ and amplitude $A$ of the oscillation. (b) In the rotating coordinate system the stationary oscillations correspond to the resting point (shown by a filled circle). If this point is kicked off the limit cycle, the perturbation of the amplitude decays, while the perturbation of the phase remains; the perturbed state and the state after the perturbation decays are shown as small dashed circles.

The minimum of the potential corresponds to the amplitude $A$ of the unperturbed oscillation, so that with respect to radial perturbations the point on the limit cycle is in a state of stable equilibrium.

Consider now the variation of the phase, i.e., only the motion along the cycle. There is no preferred value of the phase, and therefore its dynamics can be illustrated by a particle on a plane (Fig. 2.4b,c). This particle is in a state of *neutral (indifferent) equilibrium*. It stays at rest until a perturbing force pushes it to a new position. It is very important that the particle can be shifted even by an infinitely small force, but, of course, in this case the velocity is also infinitely small. The main consequence of this fact is that *the phase can be very easily adjusted by an external action, and as a result the oscillator can be synchronized!*

### 2.2.3   General case: limit cycle of arbitrary shape

We now discuss how the phase can be introduced for a limit cycle of arbitrary, in general noncircular, shape (Fig. 2.5). Suppose that the periodic motion has period $T$. Starting at $t = t_0$ from an arbitrary point on the cycle, we define the phase to be

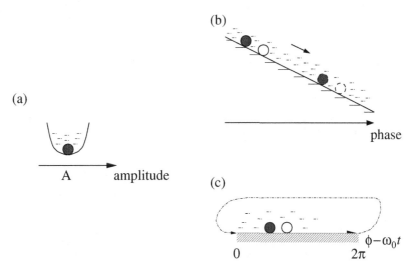

**Figure 2.4.** Stability of a point on the limit cycle illustrated by the dynamics of a light particle in a viscous fluid. (a) Transversely to the cycle, the point is in a state of stable equilibrium: perturbation of the amplitude rapidly decays to the stable value $A$. (b) Perturbation along the cycle, i.e., that of the phase, neither grows nor decays. Steady growth of the phase can be represented by a light particle sliding in a viscous fluid along an inclined plane. Two such particles (corresponding to the unperturbed and perturbed phases, filled and open circles) slide with a constant lag. (c) In a rotating (with velocity $\omega_0$) reference frame the phase is constant (see also the discussion in Section 3.1). This corresponds to a particle on a horizontal plane in the state of neutral (indifferent) equilibrium. The dashed line reminds us that the phase is a $2\pi$-periodic variable.

proportional to the fraction of the period, i.e.,

$$\phi(t) = \phi_0 + 2\pi \frac{t - t_0}{T}. \tag{2.1}$$

Then the phase increases monotonically along the trajectory, and each rotation of the point around the cycle corresponds to a $2\pi$ gain in phase. The amplitude can be then understood as a variable that characterizes the transversal deviation of the trajectory from the cycle. Note that the motion of the point along the cycle can be nonuniform. Moreover, it can happen that intervals of rather slow motion alternate with rapid jumps of the point along the cycle. Systems with such cycles are called *relaxation oscillators*; they are considered in Section 2.4.2. We emphasize here that although the motion of the point in the phase plane may be nonuniform, the growth of the phase in time is always uniform.

We stress that the main properties we have described for a quasilinear oscillator remain valid for general periodic self-sustained oscillations as well. Independently of its form, the limit cycle is stable in the transversal direction, whereas in the tangential direction the phase point is neither stable nor unstable.

If we consider two close trajectories in the phase space, i.e., the unperturbed and the perturbed ones, then we see that in the radial direction they converge with time, while in the direction along the cycle they neither converge nor diverge. In terms of nonlinear dynamics, the convergence/divergence properties of nearby trajectories are characterized by the **Lyapunov exponents**. Convergence of trajectories along some direction in the phase space corresponds to a negative Lyapunov exponent (Fig. 2.6). The absolute value of this exponent quantifies the rate of convergence. Similarly, the divergence of trajectories (we encounter this property later while studying chaotic

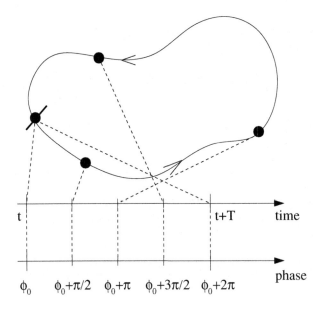

**Figure 2.5.** Definition of phase of a self-sustained oscillator with a limit cycle of arbitrary shape. The phase parameterizes the motion of the point along the cycle and can be understood as the fraction of the oscillation period $T$; it is a cyclic variable: the values $\phi$ and $\phi + 2\pi k$ are equivalent. The initial phase $\phi_0$ (shown by a short bold line orthogonal to the cycle) can be chosen arbitrarily. Note that the motion of the point along the cycle is generally nonuniform.

$\phi_0$     $\phi_0+\pi/2$     $\phi_0+\pi$     $\phi_0+3\pi/2$     $\phi_0+2\pi$

t                                                        t+T          time

phase

oscillators (Section 5.1)) is characterized by a positive exponent. Finally, the neutral direction (no divergence and no convergence) corresponds to a zero Lyapunov exponent. The most important conclusion that is used in the theoretical treatment in Part II is that **the phase of an oscillator can be considered as a variable that corresponds to the zero Lyapunov exponent**.

## 2.3 Self-sustained oscillators: main features

Self-sustained oscillations are nondecaying stable oscillations in autonomous dissipative systems. In this section we compare such systems with conservative and forced ones, and emphasize the distinctive features. The goal of this comparison is to show why the concept of self-sustained oscillations is so important for an adequate description of many natural phenomena, and of synchronization in particular. We also argue that self-sustained oscillations can be maintained only in nonlinear systems.

### 2.3.1 Dissipation, stability and nonlinearity

*Dissipation*

Macroscopic natural systems dissipate their energy. This happens, e.g., due to mechanical friction on the suspension of a pendulum, or due to electrical resistance, or due to other mechanisms of irreversible transformation of the system's energy into heat. Thus, excluding from our consideration examples like planet systems (where the dissipation seems to be negligible) or the vibration of molecules (that are described by the laws of quantum mechanics), we should always take into account the fact that, without a constant supply of energy into the system, its oscillations would decay. Therefore, self-sustained oscillators must possess an *internal energy source*. In the pendulum clock, oscillations are maintained due to the potential energy of the lifted

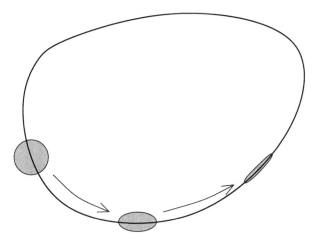

**Figure 2.6.** Convergence and divergence of trajectories is characterized by Lyapunov exponents. Suppose we consider a cloud of initial conditions around some point on the limit cycle (shaded circle). This phase volume decreases during the evolution, resulting in an elliptical form. This corresponds to a negative Lyapunov exponent in the direction transversal to the cycle, and to a zero Lyapunov exponent in the tangential direction.

weight or compressed spring. Modern clocks, as well as vacuum tubes and other electronic generators, obtain the required energy from some electrical power supply. Rhythmic contractions of the heart muscle or the emission of light pulses by fireflies occur due to the energy of chemical reactions in these systems.

*Stability*

The property of stability of oscillations with respect to perturbations also distinguishes self-sustained oscillators from conservative ones. Periodic motions in conservative systems can be described in the phase plane by a family of closed curves (Fig. 2.7a). Indeed, as these systems neither dissipate nor replenish their energy, then if the system were to be altered due to a perturbation, it would remain perturbed. Correspondingly, the altered value of the oscillation amplitude is preserved as well. In other words, conservative periodic systems do not "forget" the initial conditions. Thus, the periodic oscillations in such systems are dependent on how the system was brought into motion.[9]

As an illustration we mention here the Lotka–Volterra model that is well-known in population dynamics. It was proposed to explain the oscillations of population abundances in a predator–prey system (an example of such oscillations, the so-called hare–lynx cycle is shown in Fig. 1.13). This nonlinear model is conservative, and cannot therefore describe either stable oscillations or synchronization of several interacting populations. Modifications of the original Lotka–Volterra model that make it self-oscillatory help here.

---

[9] In mathematical language, the properties of dissipation and stability are described as the decrease of the initial phase volume. Indeed, take a set of initial conditions that corresponds to a cloud of points in the phase space. In a dissipative system, this cloud shrinks with time, and eventually becomes a point or a line (see Fig. 2.6). Conservative systems preserve the phase volume, and therefore they have no attractors.

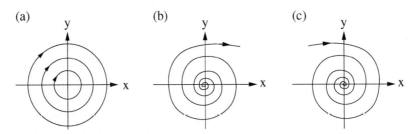

**Figure 2.7.** (a) Periodic motion in a conservative system is represented by a family of closed curves (they describe oscillations with different amplitudes and may be generally not circular). The phase portrait of a conservative linear system is similar to (a); alternatively, in dissipative linear oscillators only unstable (b) and stable (c) oscillatory motions are possible.

*Nonlinearity*

Nonlinearity is also essential for the maintenance of stable limit cycle oscillations. Stability means that a periodic motion exists only with a certain amplitude. This cannot be obtained within the class of models that are described by linear equations: if a solution $x(t)$ of a linear system is periodic, then for any factor $a$, $a \cdot x(t)$ is a periodic solution as well. Hence, the linear system exhibiting periodic oscillations is necessarily a conservative one (see Fig. 2.7a). Linear systems with dissipation (or a source of energy) admit infinitely growing or decaying solutions (Fig. 2.7b,c), but never a limit cycle motion.

The onset of stable oscillations is mathematically described as the attraction of a trajectory to the limit cycle. Physically, this attraction can be understood in the following way. Whatever the exact mechanism of dissipation and power supply is, the energy of some source, usually nonoscillating, is transformed into oscillatory motion. Typically, the larger the amplitude of oscillations, the more energy is taken from the source. From the other side, the amount of dissipated energy also depends on the amplitude.[10] The interplay between these two dependencies determines the amplitude of stationary oscillation, as illustrated in Fig. 2.8. If there is no oscillation (zero amplitude), there is no dissipation and no energy is taken from the source – therefore both these functions pass through the origin. The intersection of these two curves determines the amplitude $A$ of stationary oscillations, in this point the energy supplied to the system exactly compensates the dissipated energy. We note that these two curves can intersect in this way and simultaneously pass through the origin only if the system is *nonlinear*, i.e., is described by nonlinear differential equations.

[10] This consideration is valid for systems having the phase portrait as in Fig. 2.2, where the trajectory spirals towards the limit cycle, so that we can speak of oscillation with slowly varying amplitude.

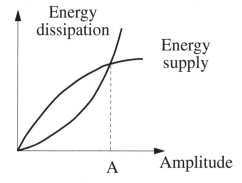

**Figure 2.8.** Interplay of the dissipation and supply of energy from a source determines the amplitude $A$ of the steady oscillations. In the example shown here this amplitude is stable: if it is increased due to some perturbation, the dissipation starts to prevail over the energy supply, and this leads to a decrease of the amplitude. Similarly, the occasional decrease of amplitude results in an excess of supply over the loss, and therefore the initial amplitude $A$ is re-established.

## 2.3.2  Autonomous and forced systems: phase of a forced system is not free!

Here, we emphasize again the difference between self-sustained and forced systems. Both the self-sustained system, a pendulum clock (Fig. 1.6), and the forced pendulum (Fig. 1.12) are represented by closed curves in the phase space that attract trajectories from their vicinity. Nevertheless, they have an essential difference: **the phase on a limit cycle is free, but the phase on a stable closed curve of the forced system is unambiguously related to the phase of the external force.** As we show in the next chapters, due to this crucial difference self-sustained oscillators are able to be synchronized, while forced systems are not. This difference can be revealed if we disturb the system, as illustrated in Fig. 2.9. Therefore, in the following, when we say "a limit cycle oscillator", we implicitly mean an autonomous, self-sustained system.

For a transparent example we take another very common system, namely the swing (Fig. 2.10a). The "algorithm" of how to put it into motion is known to everyone: after the swing is initially disturbed in some way from its equilibrium position, the person on it should sit down when the swing approaches the left- and rightmost positions, and stand up when it passes the vertical. By moving the center of gravity, the person pumps energy into the system; this energy compensates the dissipation (due to friction) and thus the stationary sway is maintained. Clearly, the source of energy here is muscular power, the oscillatory element is the pendulum with variable length, and the feedback is implemented by a person who moves the center of gravity (and hence,

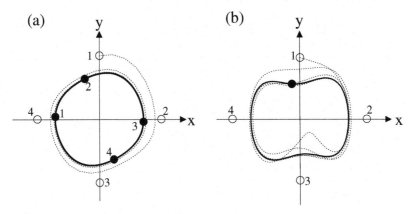

**Figure 2.9.** Phase is free for a self-sustained system (a) and not free for a forced one (b). The point is kicked off the attracting trajectory (bold curve); four different perturbed states are shown by circles. After some time the points return to the attractor (filled circles). In (a) the initial and final states are marked by the same numbers; in (b) the final states coincide; one actual trajectory is shown by the dotted curve. For the self-sustained oscillator the phase "remembers" its initial value, and can therefore be arbitrary. On the contrary, for the driven system the phase relaxes to a certain value that is determined by the external force and does not depend on the initial state.

changes the length of the pendulum) when appropriate. The crucial point here is that these movements occur not in accordance with some given periodic rhythm, but in accordance with the position of the swing. (Indeed, the period of free oscillation of the pendulum varies with the amplitude, and thus the period of the person's motion should vary with the amplitude as well.) This feature makes this system self-oscillatory and essentially different from the "mechanical (driven) swing" (Fig. 2.10b) that can be taken as an example of a *forced* system [Landa and Rosenblum 1993]. Suppose that the mechanical swing has a device that periodically changes the length of the rope. Then, if the swing is even slightly disturbed from the equilibrium, large oscillations arise (this is known as parametric excitation) such that the pendulum is in the vertical position when the rope is short, and in the left- or rightmost position when the rope is long. At first sight both systems, the self-oscillatory and the driven swing, seem to be very much alike. They both demonstrate excitation of oscillations, but there are essential distinctions between them. This difference can be seen if we disturb the systems, e.g., give impacts to the pendula or brake them, and observe them after some long time. In this case, we find the pendulum of the driven swing exactly in the same state as it would have been without perturbation, whereas the self-sustained swing could be found in any position, depending on the perturbation (cf. Fig. 2.9). In other words, the phase of the normal swing is free, but the phase of the mechanical swing is determined by the phase of the external force. Therefore, although oscillations of both systems can be described by a closed curve in the phase plane,[11] we use the term limit cycle only in the context of self-sustained systems like the usual swing with a person on it.

Below we show that it is exactly the freedom of the phase that makes self-sustained oscillators capable of being synchronized. The comparison of two swings illustrates the importance of the notion of self-oscillations. Having understood the correct mechanism that gives rise to oscillations, we can predict whether synchronization is possible

[11] More precisely, the closed curve for the forced system is obtained as an appropriate projection of the three-dimensional phase space of this system, where time is the third coordinate.

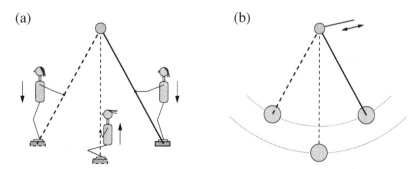

(a)                                              (b)

**Figure 2.10.** (a) A person on a swing represents an example of a self-sustained oscillator. (b) Mechanical swing whose length changes according to a prescribed function of time is a forced system.

or not. So, if the two self-sustained swings are fixed to a common support, i.e., are weakly coupled, then they can synchronize if their parameters are close. By contrast, the driven swings do not synchronize.

## 2.4    Self-sustained oscillators: further examples and discussion

This section contains an extended discussion of self-sustained oscillators. It can be skipped by the reader with a knowledge of this subject. On the other hand, in later chapters we refer to the models described in Section 2.4.2, so that it may be necessary to return to this section later.

### 2.4.1    Typical self-sustained system: internal feedback loop

Electronic generators are traditional objects for the study of synchronization phenomena; moreover, it was the appearance of radio communication and electronics that stimulated the rapid development of synchronization studies. Here we discuss a system that consists of an amplifier, a speaker and a microphone (Fig. 2.11). It is a commonly known effect that if a microphone is brought close to a speaker, self-excitation in the system may occur that results in a loud sound at a certain frequency. This system contains the components inherent in electronic generators: an amplifier, an oscillatory circuit, a feedback loop and a power supply. Any weak noise registered by a microphone is amplified and transmitted by the speaker, it returns to the input of the amplifier via the microphone, and is amplified again. This process goes on until saturation takes place due to the nonlinearity of the amplifier.

The internal feedback loop and other typical components of a self-sustained oscillator are not always easily identifiable in a particular system; some nondecomposable physical subsystems combine the function of, say, amplifier and oscillator. Nevertheless, the mechanisms of self-oscillations are universal and are common for such

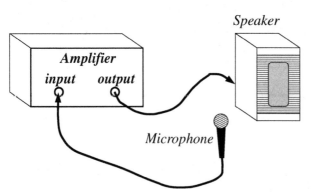

**Figure 2.11.** An amplifier, a speaker and a microphone constitute a self-sustained oscillator. If the microphone is close to the speaker, i.e., the feedback is strong enough, then any small perturbation can lead to unwanted self-excitation.

diverse systems as electronic generators and cellular biological clocks that generate circadian rhythms. It is now established that these clocks are based on a feedback loop in which the clock genes are rhythmically expressed, giving rise to cyclic levels of ribonucleic acid (RNA) and proteins [Andretic *et al.* 1999; Glossop *et al.* 1999; Moore 1999]. A feedback loop is an obligatory element of control systems. The presence of a time delay in a control loop often results in the excitation of self-sustained oscillations; one may find many interesting physiological and biological examples in [Winfree 1980; Glass and Mackey 1988].

## 2.4.2  Relaxation oscillators

In this section we introduce a particular class of self-sustained systems, known since the early works of van der Pol [1926] as **relaxation oscillators**. The essential feature of these systems is the presence of two time scales: within each cycle there are intervals of slow and fast motion. The form of the oscillation is therefore very far from the simple sine wave; rather, it resembles a sequence of pulses.[12]

We start our consideration of relaxation oscillators with a very simple mechanical model [Panovko and Gubanova 1964] sketched in Fig. 2.12. The main element of the system is a vessel being slowly filled with water. When the water level reaches a threshold value, the vessel empties quickly, and a new cycle begins; the energy of the pump that provides a constant water supply is transformed into oscillations of the water level in the vessel. As within each cycle water is slowly accumulated and then almost instantly removed, such systems are often called *integrate-and-fire*

---

[12] We note that sometimes the form of oscillation can be altered by a variation of parameters of the oscillator, so that the same generator can be relaxation or quasilinear, depending on, say, a resistance in the circuit.

(a)                                        (b)

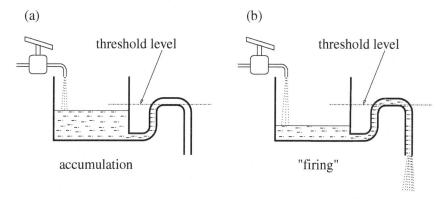

**Figure 2.12.** Mechanical model of a relaxation (integrate-and-fire) oscillator. (a) The water slowly fills the vessel until the threshold value is reached; this part of the cycle can be denoted "integration". (b) The water flows out through the trap and its level in the vessel quickly goes down: the oscillator "fires".

or *accumulate-and-fire* oscillators. If we plot the level of water and its flow vs. time (Fig. 2.13), the intervals of the slow and fast motions can be clearly seen.[13]

The oscillator described above is of course a toy model, but it captures the main features of real systems. The electronic generators studied by van der Pol in the 1920s [van der Pol 1926] operate in a very similar way. Such an oscillator consists of a battery, a capacitor, a resistance and a neon tube (Fig. 2.14). First, the capacitor is being slowly charged; the characteristic time of this process is determined by the capacitance and resistance. The growth of the voltage is similar to the increase of the water level in our toy model. When the voltage reaches a certain critical level, gas discharge is initiated in the tube and it begins to conduct electric current and to glow. As a result, the capacitor quickly discharges through the lamp, its voltage drops and the gas discharge stops. The lamp becomes nonconductive again, and the process repeats itself again and again. Like the flow of the water out of the vessel, the flow of electric current through the neon tube appears as a sequence of short pulses (and, accordingly, short flashes of the tube), whereas the voltage has a saw-tooth form. Generators of this kind are used in the circuitry of oscilloscopes, TV sets and computer displays. Indeed, an electron beam that forms an image on a screen should move horizontally across the screen with a constant speed and then almost instantly return; this is implemented by using a saw-tooth voltage to control the motion of the beam.

---

[13] We assume that the tap is only slightly opened so that the ratio of accumulate and firing intervals, $t_1/t_2$, is large. As we shall illustrate below, exactly such a regime is often interesting in practical situations.

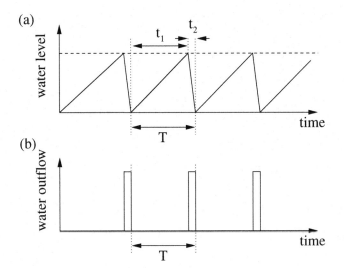

**Figure 2.13.** Time course of the mechanical integrate-and-fire (accumulate-and-fire) oscillator shown in Fig. 2.12. Water accumulates until it reaches the threshold level shown by a dashed line (a), then the water level is quickly reset to zero. The resetting corresponds to the pulse in the plot of the water outflow from the trap (b).

The general feature of relaxation oscillators – the slow growth (linear or not) of some quantity and its resetting at the threshold – is observed in many biological systems. For example, a mechanism that is very similar to pulse generation in the van der Pol relaxation oscillator is encountered in spontaneously firing neurons: if a current is injected into the cell, the electric potential (the voltage difference between the inside and outside of the neuron) is generally slowly changing, but occasionally it changes very rapidly producing spikes (action potentials) of about 2 ms duration (Fig. 2.15). Such a spike occurs every time the cell potential reaches a threshold $\approx -50$ mV, discharging the cell. After discharging, the cell resets to about $-70$ mV. The generation of spikes is due to the capacitance of the cell membrane and the nonlinear dependence of its conductivity on the voltage. When a constant current is injected into the cell, the action potentials are generated at a regular rate; slow variation of the current alters the firing rate of the neuron, just as the frequency of water level oscillation in the model in Fig. 2.12 can be adjusted by the tap. This is how sensor neurons work: the intensity of the stimulus is encoded by the firing rate. Understanding the interactions in neuronal ensembles that are often modeled as ensembles of integrate-and-fire oscillators is very important in various problems of neuroscience, e.g., in the binding problem.[14]

Another important example is the heartbeat. It is known that an isolated heart preserves the ability to rhythmic contraction *in vitro*: if the deinnervated heart is arterially perfused with a physiologic oxygenated solution, the activity of the primary pacemaker, the sino-atrial node, continues to trigger the beats. Certainly, this system is essentially different from the heart *in vivo*,[15] nevertheless, we can regard both systems as self-sustained ones. While speaking of the normal heartbeat we consider the whole cardiovascular system as one oscillator, which includes all the control loops of the

---

[14] Binding is the process of combining related but spatially distributed information in the brain, e.g. for pattern recognition.

[15] For example, the frequency of the heart *in vitro* is higher and the variability of the interbeat intervals is lower than those of the heart *in vivo*.

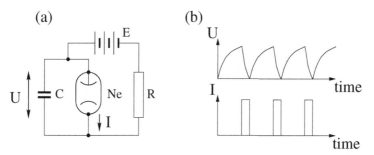

**Figure 2.14.** Schematic diagram of the van der Pol relaxation generator (a). The voltage at the capacitor increases until it reaches a threshold value at which the tube becomes conductive; then, the capacitor quickly discharges, the tube flashes, and a short pulse of current through the tube is observed. Each oscillatory cycle thus consists of epochs of accumulation and firing (b).

autonomic neural system and all the brain centers involved in the regulation of the heart rate. We understand that this system is affected by other physiological rhythms, primarily respiration, and it is impossible to isolate it completely. Nevertheless, it is clear that these other rhythms are not the cause of the heartbeat, but rather a perturbation to it. The heart is capable of producing its rhythm by itself in spite of these perturbations, and therefore we can call this system a self-sustained oscillator.

We note here that we can look at the functioning of the heart in more detail and find that this system consists of several, normally synchronized, self-sustained oscillators (primary and secondary pacemakers). Consideration at the lower level shows that each pacemaker, in its turn, represents an ensemble of rhythm-generating cells that synchronize their action potentials to trigger the contraction. Thus, synchronization phenomena are very important in functioning of the heart and we shall therefore return to this particular example later (see Sections 3.3.2 and 4.1.6).

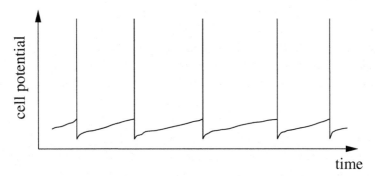

**Figure 2.15.** Intracellular potential in a neuron slowly increases towards the threshold level ($\approx -53$ mV in the particular example presented here) and then, after a short spike, resets. Schematical diagram, after [Hopfield 1994].

# Chapter 3

# Synchronization of a periodic oscillator by external force

This chapter is devoted to the simplest case of synchronization: entrainment of a self-sustained oscillator by an external force. We can consider this situation as a particular case of the experiments with interacting pendulum clocks described in Chapter 1. Let us assume that the coupling between clocks is unidirectional, so that only one clock influences another, and that is exactly the case of external forcing we are going to study in detail here. This is not only a simplification that makes the presentation easier; there are many real world effects that can be understood in this way. Probably, the best example in this context is not the pendulum clock, but a very modern device, the radio-controlled watch. Its stroke is corrected from time to time by a very precise clock, and therefore the watch is also able to keep its rhythm with very high precision. We have also previously mentioned examples of a living nature: the biological clocks that govern the circadian rhythms of cells and the organisms controlled by the periodic rhythms originating from the rotation of the Earth around its axis and around the Sun. Definitely, such actions are also unidirectional. We consider several other examples below.

We start by considering a harmonically driven quasilinear oscillator. With this example we explain entrainment by an external force and discuss in detail what happens to the phase and frequency of the driven system at the synchronization transition. Next, we introduce the technique of stroboscopic observation; in the following we use it to study synchronization of strongly nonlinear systems, entrainment of an oscillator by a pulse train, as well as synchronization of order $n : m$. We proceed with the particular features of the synchronization of relaxation oscillators and describe the effect of noise on the entrainment. We illustrate the presentation with various experimental examples. Finally, we discuss several effects related to synchronization, namely chaotization and the suppression of oscillations due to strong external forcing, as well as periodic stimulation of excitable systems and stochastic resonance phenomenon.

## 3.1    Weakly forced quasilinear oscillators

We have discussed autonomous self-sustained oscillators in detail. Now we investigate the case when such a system is subject to a *weak external action*. As an example we take a clock whose pendulum is made of a magnetic material and put it in the vicinity of an electromagnet fed by an alternating current. Then, the self-sustained oscillation of the clock is perturbed by a weak periodic magnetic field. Alternatively, we can vibrate the beam that suspends the clock or periodically kick the pendulum.

For simplicity of presentation we start with the case when an oscillator is quasi-linear, $x(t) = A\sin(\omega_0 t + \phi_0)$, with frequency $\omega_0$ and amplitude $A$, and the action is harmonic with the frequency $\omega$, i.e., the external force varies as $\varepsilon\cos(\omega t + \bar\phi_e)$, where $\phi_e(t) = \omega t + \bar\phi_e$ is the phase of the force and $\varepsilon$ is its amplitude. It is important that the frequency of the force $\omega$ is generally different from that of the oscillator $\omega_0$, the latter being termed the *natural frequency*. The difference between the frequencies $\omega - \omega_0$ is called *detuning*.

What is the consequence of this weak external action? Quite generally we can expect that the external force tries to change the amplitude as well as the phase of the oscillation. However, as already discussed in Section 2.2, the amplitude is stable, whereas the phase is neutral (it is neither stable nor unstable). Therefore a weak force can only influence the phase, not the amplitude (see Fig. 2.3b). Hence, we can mainly concentrate on the phase dynamics.

We emphasize here that we stick to the definition of the phase as it was introduced for an autonomous, nonforced oscillator. We do not redefine the phase when the forcing is applied. A relation between the phase and the position on the limit cycle remains (see Fig. 2.5). Therefore, under an external force this phase generally will not rotate uniformly, but in a more complex manner.

### 3.1.1    The autonomous oscillator and the force in the rotating reference frame

In this section we consider a quasilinear oscillator. As already described, its limit cycle is a circle and the phase point rotates around it uniformly with angular velocity $\omega_0$. It is convenient to study the phase dynamics of the forced system in a new reference frame that rotates in the same direction (let it be counter-clockwise) with the frequency of the external force $\omega$.

As the first step we show how the autonomous oscillator appears in the new coordinate frame. It means that we suppose for the moment that the force has zero amplitude ($\varepsilon = 0$). Obviously, depending on the relation between $\omega$ and $\omega_0$, the point in the new frame either continues to rotate counter-clockwise (for $\omega_0 > \omega$), or remains immobile (for $\omega_0 = \omega$), or rotates in the opposite direction (for $\omega_0 < \omega$), as illustrated in Fig. 3.1. We can characterize the position of the point by the phase difference $\phi - \phi_e$, which increases with a constant velocity $\omega_0 - \omega$, remains constant, or decreases with the velocity $\omega_0 - \omega$. Since we still assume that the force has zero amplitude, it seems to be senseless to introduce the quantity $\phi - \phi_e$. Nevertheless we do bear in mind that when we "switch on" the force, we are always interested in the phase difference $\phi - \phi_e$.

Let us now recall the analogy of phase dynamics with the motion of a light particle in a viscous fluid; this analogy was introduced in Section 2.2.2 and will be very helpful in the following. In the absence of detuning, $\omega_0 = \omega$, the phase point in the rotating frame is at rest, and we can picture this as a particle on a horizontal plane (see Fig. 2.4c). The uniform increase or decrease of the phase difference $\phi - \phi_e$, in the case of nonzero detuning, i.e., $\omega_0 \neq \omega$, can be represented by a particle sliding down along the tilted plane;[1] in this context one usually speaks of the motion of the particle in an inclined potential. This particle is shown in Fig. 3.2a for the case $\omega_0 > \omega$; for detuning with opposite sign the plane is tilted in such a way that the phase difference $\phi - \phi_e$ decreases with time.

---

[1] Under constant forcing such a particle moves with a constant velocity.

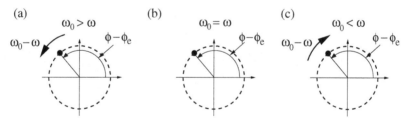

**Figure 3.1.** In the reference frame rotating with $\omega$ the limit cycle oscillation corresponds to a rotating (a and c) or to a resting point (b), depending on the detuning $\omega_0 - \omega$. The position of the point is characterized by the angle variable $\phi - \phi_e$, which increases (a), is constant (b) or decreases (c).

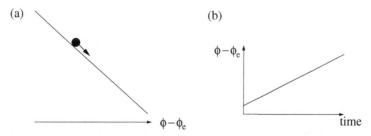

**Figure 3.2.** (a) A particle sliding with constant velocity along a tilted plane illustrates the phase dynamics of a limit cycle oscillator in a rotating frame; here it is shown for positive detuning ($\omega_0 > \omega$). The phase of the oscillator increases linearly with time (b).

Now we switch on the force, i.e., consider $\varepsilon \neq 0$. The oscillating force $\varepsilon \cos(\omega t + \bar{\phi}_e)$ is represented in the frame rotating with the same angular velocity $\omega$ by a constant vector[2] of length $\varepsilon$ acting at some angle $\phi^0$.[3] The effect of this force on the phase point on the cycle depends on the phase difference $\phi - \phi_e$ (Fig. 3.3a). Indeed, at points 1 and 2 the force is directed orthogonally to the trajectory, and cannot therefore shift the phase of the oscillator. At points 3 and 4 its effect is maximal, and at points 5–8 it is intermediate. Note that at some points the force tends to move the phase point in the clockwise direction, whereas at other points it acts counter-clockwise. It is easy to see that points 1 and 2 correspond to *stable and unstable equilibria*, respectively (Fig. 3.3a). Exploiting again the analogy with the motion of a light particle on a plane, we can say that the force creates a nonflat, curved surface (potential) out of this plane (Fig. 3.3b) with the maximum and the minimum corresponding to unstable and stable equilibria. In mathematical terminology, the stable equilibrium $\phi^0$ is asymptotically stable, whereas the phase in the unforced oscillator is marginally stable, but not asymptotically stable (see Fig. 2.4c).

---

[2] More precisely, the force in the rotating frame is represented by a constant vector and a vector that rotates in the clockwise direction with the frequency $2\omega$. As this rotation is much faster than the rotation of the phase point ($2\omega \gg |\omega_0 - \omega|$), we can neglect the influence of the second vector, because it acts alternatively in all directions and, thus, its action nearly cancels.

[3] This angle depends on the initial phase of the force $\bar{\phi}_e$ and on the way the force acts; typically, for quasilinear oscillators $\phi^0 = \bar{\phi}_e + \pi/2$, see Section 7.2.

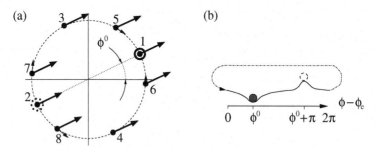

**Figure 3.3.** (a) A weak external force cannot influence the amplitude of the limit cycle but can shift the phase $\phi$ of the oscillator. The effect of the forcing depends on the phase difference $\phi - \phi_e$: at points 3 and 4 this effect is maximal, whereas at points 1 and 2 the force acts in the radial direction only, and cannot therefore shift the phase point along the cycle. Note that in the vicinity of point 2 the force propels the point away from point 2. On the contrary, in the vicinity of point 1 the phase point is pushed towards this equilibrium position. Hence, the external force creates a stable (open circle at point 1) and unstable (dotted bold circle at point 2) equilibrium positions on the cycle. These equilibrium positions are also shown in (b) (cf. Fig. 2.4c).

Intermediate summary

We have described how the effects of detuning and external forcing appear in the rotating reference frame. They can be summarized as follows.

- Detuning without a force corresponds to rotation of the phase point with an angular velocity $\omega_0 - \omega$. Detuning can be also envisaged as a particle sliding down a tilted plane whose slope increases with detuning.

- Force without detuning creates a stable and an unstable equilibrium position on the cycle. In another presentation (Fig. 3.3b), the force creates a curved periodic potential, with a minimum and a maximum.

We now consider the joint action of both factors, the force and the detuning, and determine the conditions when the force can synchronize the oscillator.

## 3.1.2 Phase and frequency locking

We start with the simplest case of zero detuning, i.e., the frequency of the force is the same as the frequency of natural oscillations. From Fig. 3.3 it is obvious what happens in this case: whatever the initial phase difference $\phi - \phi_e$,[4] the phase point moves towards the stable equilibrium (Fig. 3.3a), so that eventually $\phi = \phi_e + \phi^0$, i.e., the *phase of the oscillator is locked by the force*. We should emphasize that locking occurs even if the force is vanishingly small. Certainly, in this case one should wait a very long time until the synchronous regime sets in. This case is, of course, rather trivial, because the frequencies of the oscillator and of the force are equal from the very beginning, so that the synchronization manifests itself only in the appearance of a stable phase relation.

Now let the frequency of the force differ from the frequency of the autonomous oscillator; for definiteness we take $\omega_0 > \omega$. The effects of the force and detuning counteract: the force tries to make the phases $\phi$ and $\phi_e + \phi^0$ equal (the phase point tends to the minimum in the potential), whereas the detuning (rotation) drags the phases apart. Depending on the relation between the detuning $\omega_0 - \omega$ and the amplitude of the forcing $\varepsilon$, one of the factors wins. Consequently, two qualitatively different regimes are possible; we describe them below assuming that $\varepsilon$ is fixed, but detuning is varied.

Small detuning: synchronization

Here we consider the effect of an external force on a slowly rotating phase point (Fig. 3.4). As we already know, the effect of the drive depends on the difference $\phi - \phi_e$. Clearly, at some points the force promotes the rotation, whereas at other points it brakes it. At a certain value of the phase difference $\Delta\phi = \phi - \phi_e - \phi^0$ (see point 1 in Fig. 3.4) the force balances the rotation and stops the phase point. As a result, the frequency of the *driven oscillator* (we denote this by $\Omega$ and call it the *observed*

---

[4] We remember that the initial phase of the oscillator can be arbitrary.

*frequency*) becomes equal to the frequency of the drive, $\Omega = \omega$, and a stable relation between the phases is established. We call this motion **synchronous**.

We emphasize here that synchronization appears not as the equality of phases, but as the onset of a constant phase difference $\phi - \phi_e = \phi^0 + \Delta\phi$, where the angle $\Delta\phi$ is called the *phase shift*. Another representation of the entrainment is shown in Fig. 3.5: the force creates minima in the tilted potential, and the particle is trapped in one of them.[5] In fact, there exists a second point where the force compensates the rotation of the phase difference (point 2 in Fig. 3.4), but this regime is not stable; it corresponds to the position of the particle at the local maximum of the potential.

[5] We note that the potential is not $2\pi$ periodic any more, because the clockwise and counter-clockwise directions are not equivalent due to the rotation.

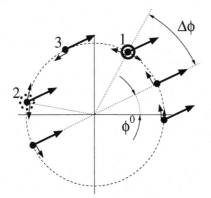

**Figure 3.4.** Small detuning. The rotation of the phase point is affected by the external force that accelerates or decelerates the rotation depending on the phase difference $\phi - \phi_e$. The rotation is pictured by arrows inside the cycle; here it is counter-clockwise, corresponding to $\omega_0 > \omega$. At point 1 the rotation is compensated by the force. This point then becomes the stable equilibrium position. Unstable equilibrium is then at point 2 (cf. Fig. 3.3a). The phase point is stopped by the force and a stable phase shift $\Delta\phi$ is maintained. Possible values of $\Delta\phi$ lie in the interval $-\pi/2 < \Delta\phi < \pi/2$.

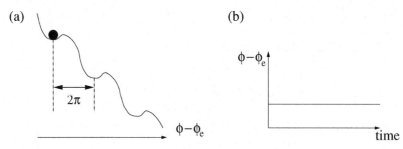

**Figure 3.5.** Small detuning. The external force creates minima in the tilted potential (a) and the particle rests in one of them. This corresponds to a stable constant phase difference, $\phi - \phi_e = \phi^0 + \Delta\phi$ (b).

### Large detuning: quasiperiodic motion

If the detuning exceeds some critical value, then the effect of the force is too weak to stop the rotation. Indeed, inspection of Fig. 3.4 shows that with an increase of detuning (i.e., of the rotation velocity), the equilibrium points shift towards point 3 where the braking effect of the force is maximal. Eventually, the stable and the unstable equilibria collide and disappear here, and the phase point begins to rotate with the so-called **beat frequency** $\Omega_b$. In terms of the motion of the particle, the force bends the potential, but does not create minima (Fig. 3.6a). Hence, the particle slides down, but its velocity (and, correspondingly, the phase growth) is not uniform (Fig. 3.6b).

Although for the given detuning the external force is too weak to synchronize the oscillator, it essentially affects its motion: the force makes the rotation nonuniform and on average decelerates it, so that $\Omega_b < \omega_0 - \omega$.[6] This deceleration happens because the particle nearly stops at the points where the slope of the potential is minimal.

Coming back to the original reference frame, we see that the oscillator has the frequency $\omega + \Omega_b < \omega_0$, and that its phase growth is modulated by the beat frequency $\Omega_b$. Such motion is characterized by two frequencies ($\omega + \Omega_b$ and $\Omega_b$) and is called **quasiperiodic**. More precisely, the motion is quasiperiodic if the ratio $(\omega + \Omega_b)/\Omega_b$ is irrational, a very typical case indeed.

To summarize, if the detuning is small, then even a very small force can create local minima of the potential and entrain the oscillator. The larger the detuning, the larger forcing is required for the onset of synchronization.

---

[6] We remind the reader that the autonomous oscillation is represented in the rotating reference frame by the point rotating with the frequency $\omega_0 - \omega$ (see Fig. 3.1) and that for the sake of definiteness we consider the case of positive detuning, $\omega_0 > \omega$.

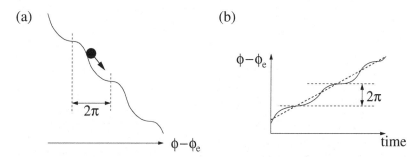

**Figure 3.6.** Large detuning. The force cannot entrain the particle: there are no local minima in the tilted potential and the particle slides down (a). Under-threshold force modulates the velocity of the particle, so that the phase difference grows nonuniformly (b). The average growth rate, shown by the dashed line, determines the *beat frequency*. From (a) one can see that the period of rotation (shown by two dashed lines) coincides with the period of modulation.

## Frequency locking. Synchronization region

We have seen that for a fixed value of the forcing amplitude $\varepsilon$ the frequency of the driven oscillator depends on the detuning $\omega_0 - \omega$. For sufficiently small detuning the external action entrains the oscillator, so that the frequency of the driven oscillator $\Omega$ becomes equal to $\omega$. For detuning exceeding a certain critical value, this equality breaks down, as shown in Fig. 3.7a.

The *identity of the frequencies that holds within a finite range of the detuning* is the hallmark of synchronization and is often called **frequency locking**.[7]

It is instructive to plot the whole family of curves $\Omega - \omega$ vs. $\omega$ (Fig. 3.7a) for different values of the forcing amplitude $\varepsilon$. These curves determine the region in the $(\omega, \varepsilon)$ plane that corresponds to the synchronized state of the oscillator (Fig. 3.7b,c); note that the parameters $\omega$ and $\varepsilon$ of the external force are usually varied in experiments. Within this region the frequency of the oscillator is equal to the frequency of the external action. Since the works of Appleton and van der Pol such a region has been designated the **synchronization region** or, more recently, the **Arnold tongue**.[8]

It is important to note that the synchronization region touches the $\omega$-axis. This means that for vanishing detuning the oscillator can be synchronized by an infinitesimal force (although in this case the transient to the synchronous state is infinitely long). This property is widely used, e.g., in electrical engineering, in particular, in

---

[7]  In the context of synchronization in lasers it is often termed *mode locking*.

[8]  For small $\varepsilon$ the borders of the tongue are straight lines; this is a general feature of weakly forced oscillators. For large $\varepsilon$ the form of the tongue depends on the particular properties of the oscillator and the force.

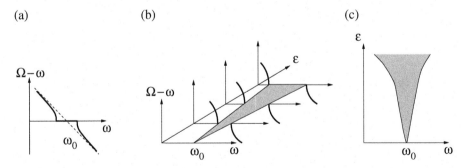

**Figure 3.7.** (a) Difference of the frequencies of the driven oscillator $\Omega$ and the external force $\omega$ as a function of $\omega$ for a fixed value of the forcing amplitude $\varepsilon$. In the vicinity of the frequency of the autonomous oscillator $\omega_0$, $\Omega - \omega$ is exactly zero; this is denoted **frequency locking**. With the breakdown of synchronization the frequency of the driven oscillator remains different from $\omega_0$ (the dashed line shows $\omega_0 - \omega$ vs. $\omega$): the force is too weak to entrain the oscillator, but it "pulls" the frequency of the system towards its own frequency. (b) The family of $\Omega - \omega$ vs. $\omega$ plots for different values of the driving amplitude $\varepsilon$ determines the domain where the frequency of the driven oscillator $\Omega$ is equal to that of the drive $\omega$. This domain, delineated by gray in (c), is known as the **synchronization region** or **Arnold tongue**.

radio communication systems, where the frequency of a powerful generator is often stabilized by a very weak but precise signal from an auxiliary high-quality reference generator.

### Phase locking: constant phase shift

Synchronization is also often described in terms of **phase locking**. As we have already seen, the nonsynchronous motion corresponds to an unbounded growth of the phase difference, whereas in the synchronous state *the phase difference is bounded*, and there exists a constant phase shift between the phases of the oscillator and of the force:

$$\phi(t) - \phi_e(t) = \text{constant}, \tag{3.1}$$

where the constant equals $\phi^0 + \Delta\phi$. The phase shift $\Delta\phi$ depends on the initial detuning. If the synchronization region is "crossed" along the line of constant $\varepsilon$, i.e., the amplitude of the force is kept constant whereas its frequency is varied, the phase shift varies by $\pi$; it is zero only if there is no detuning, i.e., in the middle of the synchronization region. It should be remembered that the constant angle $\phi^0$ depends on the initial phase of the force and on how it affects the oscillator. We emphasize here that the notion of phase locking implies that the phase difference remains bounded within a finite range of detuning, i.e., within the synchronization region.

## 3.1.3　Synchronization transition

We now describe the transition to synchronization, i.e., how the dynamics of the phase difference change at the border of the synchronization region. Actually, we already have all the required information; here we just summarize our knowledge.

### Phase slips and intermittent phase dynamics at the transition

Suppose that the amplitude of the forcing is fixed, whereas its frequency is varied. By drawing the Arnold tongue shown in Fig. 3.8a, we would like to know what occurs along a horizontal line in this diagram. Let us begin with zero initial detuning (point 1 in Fig. 3.8a), $\omega_0 = \omega$, and then gradually decrease the frequency of the external force $\omega$. In other words, we choose $\omega$ to correspond to the middle of the synchronization region and observe how synchronization is destroyed when the border of the tongue is crossed due to variation of the driving frequency (Fig. 3.8). We already know that if the detuning is zero then the phase difference is equal to $\phi^0$; for definiteness we take $\phi^0 = 0$ (see case 1 in Figs. 3.8a and b). With an increase in detuning, a nonzero phase shift appears; this is illustrated by the point and the line labeled 2. When the point crosses the border of the tongue, loss of synchronization occurs, and the phase difference grows infinitely (point 3 in (a) and curve 3 in (b)). However, this growth is not uniform! Indeed, there are epochs when the phase difference is nearly constant, and other, much shorter, epochs where the phase difference changes relatively rapidly by $2\pi$. Such a rapid change that appears practically as a jump is often called a *phase slip*.[9]

Alternation of epochs of almost constant phase difference with phase slips means that for some long intervals of time the system oscillates almost in synchrony with the external force, and then makes one additional cycle (or loses one cycle, if the point crosses the right border of the tongue). In the representation shown in Fig. 3.4 this would correspond to the following: the point that denotes the phase of the oscillator rotates with almost the same angular velocity as the phase of the external force, and then accelerates (decelerates) and makes one more (one less) revolution with respect to the drive.

In order to understand this behavior, we recall the analogy with the motion of a particle in a tilted potential (Fig. 3.6). The loss of synchronization corresponds to the disappearance of local minima in the potential, but nevertheless the potential remains curved. Hence, the particle slides very slowly along that small part of the potential where it is almost horizontal, it practically pauses there, and then moves relatively quickly where the potential is steeper. This corresponds to an almost synchronous epoch and a phase slip, respectively. These epochs continue to alternate so that one

---

[9] The rapidity of the phase jump depends on the amplitude of the drive. For very weak forcing the slip occurs within several, or even many, periods of the force.

(a)                                   (b)

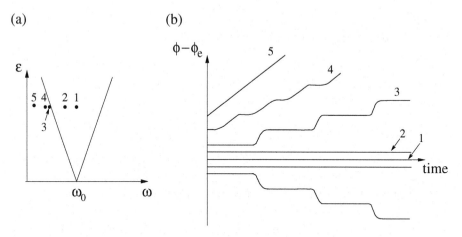

**Figure 3.8.** Dynamics of the phase at the synchronization transition. The phase difference is shown in (b) for different values of the frequency of the driving force; these values are indicated in (a) by points 1–5, within and outside the synchronization region. In the synchronous state (points 1 and 2) the phase difference is constant (lines 1 and 2 in (b)); it is zero in the very center of the tongue and nonzero otherwise. Just outside the tongue the dynamics of the phase are intermittent: the phase difference appears as a sequence of rapid jumps (slips) intermingled with epochs of almost synchronous behavior (point and curve 3). As one moves away from the border of the tongue the dynamics of the phase tend towards uniform growth (points and curves 4 and 5). The transition at the right border of the tongue occurs in a similar way, only the phase difference now decreases.

can say that the *dynamics of the phase difference are intermittent,* and that is what we see in Fig. 3.8b, curve 3.

With further increases in detuning, the duration of the epochs of almost synchronous behavior becomes smaller and smaller, and eventually the growth of the phase difference becomes almost uniform (Fig. 3.8b, curves 4 and 5). Naturally, this picture is symmetric with respect to detuning; an increase of the frequency of the external force corresponds to a negative slope of the potential.

### Synchronous vs. quasiperiodic motion: Lissajous figures

In terms of frequencies, the loss of synchronization is understood as a transition from motion with only one frequency $\omega$ to quasiperiodic motion, when two frequencies are present. This transition can be traced by a simple graphical presentation where an observable of the oscillator is plotted vs. the external force (Fig. 3.9). If the frequencies of both signals are identical (synchronous motion) then these plots are closed curves known as *Lissajous figures*. Otherwise, if the frequencies differ, the plot fills the whole region. This technique is quite convenient in experimental electronic setups: one can just feed the signals to the inputs of the oscilloscope (although nowadays the plots are usually done by means of computers).

Lissajous figures not only allow us to estimate the relation between the frequencies, but they also help us to estimate the phase shift in a synchronous state. For example,

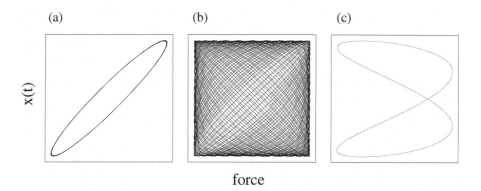

**Figure 3.9.** Lissajous figures indicate synchronization. An observable $x(t)$ of the forced oscillator is plotted vs. the force. (a) Synchronous state. The periods of oscillations along the force and $x$ axes are identical, therefore the plot is a closed curve. (b) Quasiperiodic state. The point never returns to the same coordinates and the plot fills the region. (c) The 8-shaped Lissajous figure corresponds to the case when the force performs exactly two oscillations within one cycle of the oscillator. This is an example of high-order (1 : 2) synchronization discussed in Section 3.2. (Note that the scales along the axes are chosen differently in order to obtain clearer presentation; in fact, the amplitude of the force is much smaller than the amplitude of the signal (weak forcing).)

in the case shown in Fig. 3.9a, the phase shift is close to zero; the curve shrinks to a line if $\phi(t) = \phi_e(t)$; a circular Lissajous figure corresponds to a phase shift of $\pm\pi/2$.

### 3.1.4   An example: entrainment of respiration by a mechanical ventilator

Not to be too abstract, we now consider a real situation. We illustrate the features of synchronization by external forcing with the results of experiments on entrainment of spontaneous breathing of anesthetized human subjects by a mechanical ventilator reported by Graves *et al.* [1986], see also [Glass and Mackey 1988]. The data were collected from eight subjects undergoing minor surgery requiring general anesthesia. The subjects were young adults of both sexes without history of cardiorespiratory or neurological diseases.

The respiratory rhythm originates in the breathing center in the brain stem. It has been known from animal experiments since the end of the 19th century that this rhythm is influenced by mechanical extension and compression of the lungs, known as the Hering–Breuer reflex.

Mechanical ventilation is a common clinical procedure, and the study of Graves *et al.* [1986], besides being very interesting from the viewpoint of nonlinear dynamics, is of practical importance. Indeed, for mechanical ventilation to be effective, the patient should not "struggle" with the apparatus. Therefore, for an effective treatment a certain phase relation should be achieved to ensure that mechanical inflation coincides with inspiration.

In the experiments we describe both the frequency and volume of mechanical ventilation were varied; in our notation these quantities are the frequency and the amplitude of the external force, respectively. Depending on these parameters, both synchronous and nonsynchronous states were observed (Fig. 3.10). One can see that, in the synchronous regime, each beat of the mechanical ventilator corresponds to one respiratory cycle. Let us look closely at the phase relations in such regimes. Graves *et al.* [1986] computed the instantaneous frequency and the phase difference for 24 consecutive cycles.[10] The plots of these quantities are shown in Fig. 3.11. Each subject was subjected to three different trials. In these trials the frequency of the ventilator was respectively below, equal to and higher than the mean "off pump" frequency, i.e., frequency of spontaneous breathing. The "off pump" frequency is indicated by a dashed line, whereas the frequency of the pump is indicated by a solid line. We see that the frequency of respiration is indeed entrained: when detuning is present, this frequency is shifted towards the value of the driving frequency. As to the phase relations, it is clear that the phase of the ventilator is ahead of the phase of the force,

---

[10] The frequency was estimated as the inverse of the period of each cycle. The phase difference was obtained by taking the delay from the beginning of mechanical inflation to the onset of spontaneous inspiration. In Chapter 6 we discuss thoroughly the methods of phase and frequency estimation from experimental data.

approximately coincides with it, or lags behind it depending on the detuning, as we expected.

Fig. 3.12 shows data for the case of relatively large detuning between the frequency of the ventilator and the mean "off pump" frequency, so that spontaneous breathing is not entrained by the ventilator. The decrease and growth of the phase difference then strongly resemble the theoretical curves sketched in Fig. 3.8.

Summarizing this example, we can say that the phase dynamics of oscillations in the forced complex living system, in locked as well as in nonlocked states, are in agreement with the main predictions of the theory. This example illustrates that the principal results are also valid for the case when a periodic force differs from a simple harmonic one. The only difference from our theoretical consideration is that the frequency of the entrained oscillator is not constant, and the phase difference

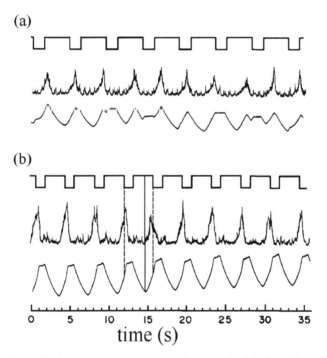

**Figure 3.10.** Nonsynchronous (a) and synchronous (b) breathing patterns. Top curves show mechanical ventilation (downwards corresponds to inflation). Middle curves are the preprocessed electromyograms of the diaphragm; these signals reflect the output of the central respiratory activity. Lower curves show the ventilation volume. Dashed and solid lines in (b) show the onset of inflation and inspiration, respectively. Note that entrainment of the spontaneous respiration by the external force (mechanical inflation) result in periodic variation of the ventilation volume (b). The loss of synchronization leads to counteraction of spontaneous respiration and inflation; as a result the amplitude of the volume oscillation decreases from time to time (a). Such behavior, typical for quasiperiodic motion, is often called a "waxing and waning" pattern in biological literature. From [Graves *et al.* 1986].

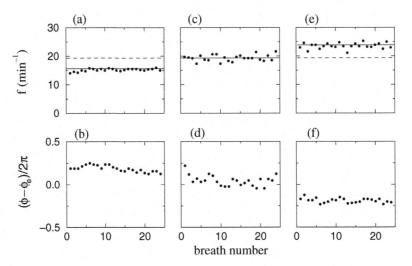

**Figure 3.11.** Frequency of respiration during mechanical ventilation (a, c, and e) and the difference between the phase of the ventilator and that of spontaneous breathing (b, d and f) computed for 24 consecutive breaths. The plots correspond to three locked states where the frequency of the external force (solid line) is smaller than (a, b), equal to (c, d) or larger than (e, f) the "off pump" frequency (dashed line). The frequency of respiration is entrained and fluctuates around the value of the external frequency. The average phase shift is respectively positive (b), close to zero (d) and negative (f). From [Graves *et al.* 1986].

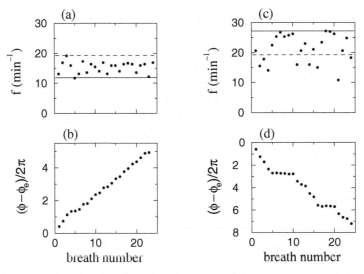

**Figure 3.12.** The same variables as in Fig. 3.11, but for two nonsynchronous states. The driving frequency is either smaller (a, b) or larger (c, d) than the "pump off" frequency. The phase difference grows almost uniformly (b) or in intermittent fashion (d), to be compared with curves 3 and 4 in Fig. 3.8b. (a), (c) from [Graves *et al.* 1986]. (b), (d) plotted using data from [Graves *et al.* 1986].

fluctuates in a rather irregular fashion. There are two reasons for this discrepancy. First, the living system is obviously not a perfectly periodic oscillator; its parameters vary with time and it is subject to some noise. These factors are treated in detail in Section 3.4. The second reason is that the time course of respiration deviates from a sine wave: inspiration takes less time than expiration. This issue will be discussed in the next section.

## 3.2 Synchronization by external force: extended discussion

We have described the simplest case of phase locking that can be well understood with a forced quasilinear oscillator. To consider more general situations, where also complex synchronization patterns can occur, we need to introduce a general method of stroboscopic observation. Next, we use this method to demonstrate entrainment of an oscillator by a sequence of pulses and to introduce a general case of $n : m$ synchronization. We complete this section with an extension of the notions of the phase and frequency locking.

### 3.2.1 Stroboscopic observation

Here we explain a technique that will be widely used in this book, both in theoretical and experimental studies; we call it the **stroboscopic technique**. The name of the method comes from the well-known optical devices that measure the frequency of rotation or oscillation of some mechanical object by shining a bright light at intervals so that the object appears stationary if the frequency of light flashes is equal to the measured one. If these frequencies are slightly different, then the pendulum (wheel) appears in the light to be slowly oscillating (rotating). For example, cinematography provides a stroboscopic observation with a fixed frequency of 24 pictures per second.

The method we introduce in this section is very similar to the stroboscope principle used in these optical devices. The only difference is that what we observe is not a real object in a physical space, but the position of a point in the phase space, i.e., its position on the attractor.[11] For periodic oscillators, the latter is unambiguously related to the phase of the oscillator. The idea of the technique is very simple: let us observe the phase of a periodically forced oscillator not continuously, but at the times $t_k = k \cdot T$, where $T$ is the period of the external force, and $k = 1, 2, \ldots$ . In other words, we observe the driven system at the moments when the phase of the external force attains some fixed value.

Before proceeding with some examples, we describe what we expect to obtain for synchronization of a nonquasilinear oscillator. The limit cycle of a strongly nonlinear

---

[11] In this section we deal with systems with limit cycles, but in the following we apply this technique to analyze systems with strange (chaotic) attractors as well.

oscillator is not necessarily circular and the motion of the point along the cycle is generally not uniform. This prevents us from using the rotating frame for the analysis of synchronization, as was described in the previous section; the stroboscopic technique helps here. If the oscillator is synchronized by an external periodic force, i.e., $\Omega = \omega$, then, obviously, if the limit cycle is "highlighted" with the forcing frequency $\omega$, a point on it is always found in the same position. If the oscillator is not entrained by the external force, and there is no relation between their phases, then we anticipate that at the moments of observation $\phi$ can attain an arbitrary value. If we observe the oscillator for a sufficiently long time, then we can compute the distribution of $\phi_k = \phi(t_k)$. In the case of synchronization this distribution is concentrated at a point, and in the nonsynchronized state this distribution is broad,[12] see Fig. 3.13.

[12] If the phase point rotates nonuniformly, then it is found more frequently at certain preferred positions, therefore the distribution is broad but not necessarily uniform. We note that even in the case of a quasilinear oscillator near the synchronization transition the distribution is nonuniform, because, due to the forcing, the phase point rotates nonuniformly and the phase difference grows nonlinearly (see Fig. 3.8).

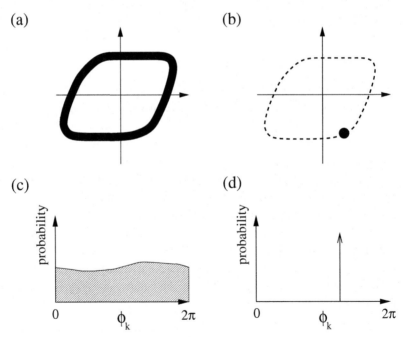

**Figure 3.13.** Stroboscopic observation of a point on a limit cycle. (a) If the frequency of the oscillator differs from the frequency of strobing (forcing), $\Omega \neq \omega$, then the point can be found in an arbitrary position on the cycle. (b) If the oscillator is entrained, $\Omega = \omega$, the phase $\phi_k$ of the oscillator corresponding to a fixed value of the phase of the force is always the same. (c, d) Distribution of the phase $\phi_k$ of the driven oscillator found at the times $t_k$ when the phase of the external force attains a fixed value, $\phi_e = $ constant. In the nonsynchronized state this distribution is broad (c), while in the synchronous state it is a $\delta$-function (d).

The stroboscopic technique is a particular case of the **Poincaré map**, well-known in oscillation theory and nonlinear dynamics. Its very important advantage is that it can be used to study arbitrary self-sustained oscillators, not only quasilinear, but also relaxation and even chaotic ones (see Chapter 5). The stroboscopic technique is equally applicable if the force is harmonic or can be described as a sequence of pulses; it is helpful in the case of noisy oscillators and in experimental studies of synchronization.

### 3.2.2 An example: periodically stimulated firefly

Male fireflies are known to emit rhythmic light pulses in order to attract females. They are able to synchronize their flashes with their neighbors (see Section 1.1 or the detailed analysis by Buck and Buck [1968]). To investigate the basics of this ability, Buck *et al.* [1981] (see also [Ermentrout and Rinzel 1984]) studied the flash response of a single firefly *Pteroptyx malaccae* to periodic stimulation by light pulses. The results are illustrated in Fig. 3.14. The interval of the pacing (the period of the external force) was switched from 0.77 to 0.75 s at $t \approx 22$ s. The result of this switch was transition from synchronization to desynchronization. Since this transition is not readily seen from Fig. 3.14, we illustrate it by applying the stroboscopic technique for the two time intervals, 0 s $< t <$ 22 s and 22 s $< t <$ 130 s (Fig. 3.15).

Note that both the force and the signal from the driven oscillator can be considered as sequences of events (instant pulses) or *point processes*. Naturally, an interval between the two pulses constitutes one cycle, and therefore the increase of the phase during such interval is $2\pi$. It is convenient to observe the phase of the firefly oscillation $\phi$ at the times of pacing; this phase $\phi_k$ can be computed as the fraction of the respective interflash interval. From Fig. 3.15 we find that in the case of stimulation with the period 0.77 s the stroboscopically observed phase $\phi_k$ is almost constant, whereas for the pacing interval 0.75 s it is scattered around the circle. This is an indication of synchronous and nonsynchronous states, respectively. (The phase $\phi_k$ is

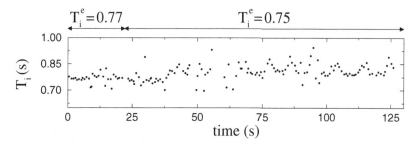

**Figure 3.14.** Flash response of a firefly to periodic external pacing. The intervals $T_i$ between the flashes of the firefly are plotted as a function of time. The period of pacing $T_i^e$ was altered from 0.77 to 0.75 s at $t \approx 22$ s. Plotted using data from [Ermentrout and Rinzel 1984].

not exactly constant in the synchronous state, because the firefly is, of course, a noisy oscillator.)

By plotting the stroboscopic picture we put the phases of the firefly flashes on a circle. This does not mean that we assume the oscillator to be quasilinear: we have arbitrarily chosen a constant value of the amplitude. We would like to attract the reader's attention to the fact that the amplitude, i.e., the shape of the cycle, is actually irrelevant. Synchronization appears as a relation between phases, and its onset can be established by means of stroboscopic plots and the distribution of the stroboscopically observed phase.

### 3.2.3  Entrainment by a pulse train

Quite often the external forcing action can be considered as a sequence of pulses. Therefore the oscillator is autonomous for almost the whole oscillation period, except for a short epoch when it is subject to the forcing. To illustrate this point, we again perform a "virtual" experiment with the pendulum clock. Suppose its pendulum is periodically kicked with a constant force and in a fixed direction. Sometimes the kicks tend to accelerate the pendulum (if it moves in the same direction at that instant), and sometimes they tend to slow down the motion of the pendulum. If the pendulum is initially slower, then the kicks more often accelerate it, and vice versa. As a result, synchronization can take place. This property is exploited to correct the time of modern radio-controlled clocks: they are periodically corrected by pulse signals transmitted from a central, very precise clock. Pulse-like forcing and interaction are also important in an understanding of the functioning of the biological systems, e.g., ensembles of firing neurons, heart pacemakers, etc. Such forcing is frequently used to

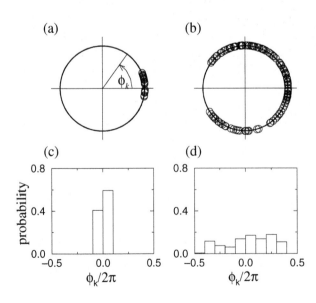

**Figure 3.15.** Stroboscopic observation of the phase of the flashes of a periodically stimulated firefly. The distribution of phases for the time interval 0 s < $t$ < 22 s is sharp (a and c), whereas the distribution for 22 s < $t$ < 130 s is broad (b and d), reflecting loss of synchronization at $t \approx$ 22 s (cf. Fig. 3.13). Plotted using data from [Ermentrout and Rinzel 1984].

study biological oscillators experimentally. We have already considered an example with periodic stimulation of a firefly, below we also describe the periodic stimulation of the heart by pacemaker cells. However, before we get to that example, we explain the effect of synchronization by a sequence of pulses using the stroboscopic technique (see also [Kharkevich 1962; Glass and Mackey 1988]).

### Phase resetting by a single pulse

As a preliminary step we consider the effect of a single pulse on the limit cycle oscillation. We exploit the property that perturbation of a point on the limit cycle in the radial direction (i.e., in amplitude) decays rapidly,[13] whereas the perturbation in phase remains (Fig. 3.16). Thus, the pulse resets the phase, $\phi \to \phi + \Delta$. Obviously, the sign and the absolute value of the phase resetting $\Delta$ depends on the phase at which the pulse was applied, and on the pulse strength.

### Periodic pulse train

Now we consider the effect of a $T$-periodic sequence of pulses on an oscillator with frequency $\omega_0$. For the sake of definiteness we assume that $\omega_0 > \omega = 2\pi/T$. Let us observe the oscillator stroboscopically at the instants of each pulse perturbation (Fig. 3.17). If there were no forcing, then, say, for three subsequent observations we would find the point in the positions denoted in Fig. 3.17a as 1, 2, and 3. As $\omega_0 > \omega$ then during the period between two observations the point on the cycle makes one complete rotation, plus a small advance that we denote $\nu$.

Suppose now that the pulse force is "switched on" when the point is found in position 2. If the pulse resets the phase exactly by the value $\nu$, then its action compensates the phase discrepancy due to detuning. Indeed, after resetting the point is found in position 1, and, until the next pulse comes, the point moves as if the oscillator were

---

[13] We note that in this consideration we do not assume, as we did before, that the amplitude is not influenced by the forcing. Nevertheless, we still assume that the amplitude is rather stable and, therefore, its relaxation to the unperturbed value is instant.

(a)                                              (b)

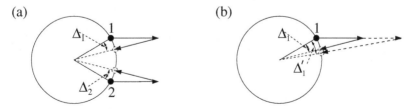

**Figure 3.16.** Resetting of the oscillator phase by a single pulse, $\phi \to \phi + \Delta$. (a) The effect of the pulse depends on the oscillation phase at the instant of stimulation. If the pulse is applied to the point at position 1 then the phase is delayed ($\Delta_1 < 0$); perturbation applied to the point 2 advances the phase ($\Delta_2 > 0$). (b) Effect of the strength of the pulse: the pulse of a larger amplitude (the vector shown by a dashed line) enlarges the resetting ($|\Delta_1'| > |\Delta_1|$).

autonomous. This means that, during the period $T$, the point evolves to position 2 and the next pulse resets it to point 1 again (Fig. 3.17b). Thus, in a stroboscopic observation with the period of forcing, we always find the point in the same position on the cycle, which indicates synchronization.

Obviously, synchronization is achieved only if the pulses have an appropriate amplitude and are applied at an appropriate phase. As we expected from our prior knowledge, large detuning requires a larger amplitude of forcing: if between two pulses the point evolves from position 1 to position 3, then the pulses should have an essentially larger amplitude in order to reset the phase by $2v$, and the stable phase difference between the force and the synchronized oscillator increases as well.

If the force is first applied in an arbitrary phase, then it generally cannot "stop" the phase point immediately, but it will do so after a transient time. Simple geometrical considerations show that if, say, the force is switched on when the point is in position 3 then the first pulse would put the point in between positions 2 and 1, the next one would put it closer to 1 and so on; eventually, synchronization sets in.

In a more formal way we can describe the synchronization of a periodically perturbed oscillator by writing the relation between the phases after the $n$th and $(n+1)$th pulses, $\phi_{n+1} = \phi_n + v + \Delta(\phi_n + v)$. Here the constant $v$ is the phase increase due to

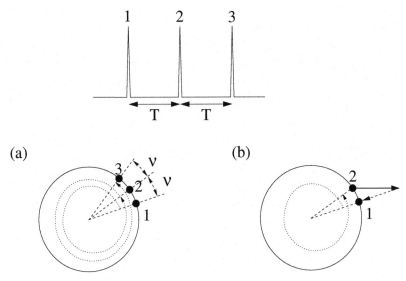

**Figure 3.17.** (a) Stroboscopic observation of an *autonomous oscillator* at the instants of pulse occurrence; when pulse 1 comes, the point is found in position 1, and so on. The frequency of the oscillator is chosen to be larger than the frequency of pulses, $\omega_0 > \omega = 2\pi/T$; therefore, the stroboscopically observed point moves along the cycle. (b) Synchronization by resetting. Periodic pulses applied to the point in position 2 reset it to position 1, thus compensating the phase discrepancy due to detuning. Between the pulses the phase evolves as in the autonomous system.

detuning,[14] and $\Delta(\phi_n + v)$ describes the resetting of the evolved phase by the $(n+1)$th pulse. The condition of synchronization is the equality of $\phi_{n+1}$ and $\phi_n$. The mapping $\phi_n \to \phi_{n+1}$ is known in nonlinear dynamics as the **circle map**. This powerful tool is discussed in detail in Section 7.3, and here we proceed with the qualitative discussion of synchronization.

### 3.2.4 Synchronization of higher order. Arnold tongues

Until now we have always assumed that the frequency of the autonomous oscillator is different from but close to the frequency of the external force, and synchronization was understood as the adjustment of these frequencies as a result of forcing, so that they become equal. Here we show that synchronization may also appear in a more complicated form. We continue to treat the periodically kicked oscillator, and suppose now that every second pulse is skipped. Consequently, the oscillator is autonomous within the time interval $2T$, and the point on the cycle evolves during this interval from position 1 to 3. From Fig. 3.18 it is clear that a pulse of sufficiently large amplitude can compensate the phase discrepancy $2v$ that was gained after the previous resetting.

Thus, an oscillator with frequency $\omega_0$ can be entrained by a force having a frequency close (but not necessarily equal!) to $\omega_0/2$, and synchronization then appears

---

[14] For an arbitrary limit cycle the phase grows linearly with time (see Section 2.2) and therefore $v$ is constant. Hence, our consideration is valid in a general case: it does not imply that the point rotates uniformly around the cycle. We also note that here $\phi + 2\pi$ is considered to be equivalent to $\phi$.

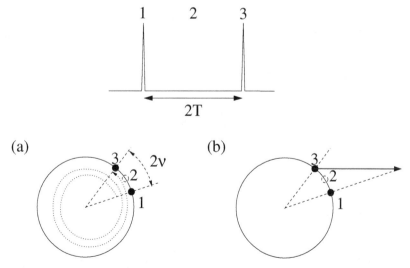

**Figure 3.18.** Synchronization of the oscillator by a $2T$-periodic pulse train.
(a) Between two pulses the point evolves from position 1 to position 3.
(b) Synchronization can be achieved even if every second pulse is skipped, but the amplitude of pulses must be larger (see Fig. 3.17b).

as the onset of the following relation between the frequencies: $2\omega = \Omega$. This regime
is called *synchronization of the order* 2 : 1. Obviously, entrainment by every third
pulse can be achieved as well, although it would require an even higher amplitude of
pulses for the same detuning. Generally, the synchronous regimes of arbitrary order
$n : m$ ($n$ pulses within $m$ oscillatory cycles) can be observed, and the whole family of
synchronization regions can be plotted (Fig. 3.19). These regions are now commonly
called **Arnold tongues**.

It is important to mention that high-order tongues are typically very narrow so that it
is very difficult (if not impossible) to observe them experimentally. We can explain this
by analyzing Figs. 3.17 and 3.18 again. We can see that, for the same value of detuning,
synchronization of order 2 : 1 requires an essentially larger amplitude of pulses. On
the contrary, if the amplitude is fixed, then resetting by, say, every second pulse can
compensate a smaller detuning than resetting by every pulse, meaning exactly that the
region of 2 : 1 frequency locking is narrower than the region of 1 : 1 locking.

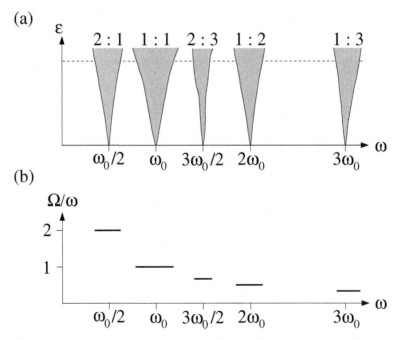

**Figure 3.19.** (a) Schematic representation of Arnold tongues, or regions of $n : m$
synchronization. The numbers on top of each tongue indicate the order of locking;
e.g., 2 : 3 means that the relation $2\omega = 3\Omega$ is fulfilled. (b) The $\Omega/\omega$ vs. $\omega$ plot for a
fixed amplitude of the force (shown by the dashed line in (a)) has a characteristic
shape, known as the *devil's staircase*, see Fig. 7.16 for full image (here the variation
of the frequency ratio between the steps is not shown).

### 3.2.5 An example: periodic stimulation of atrial pacemaker cells

We illustrate the theoretical considerations presented above with a description of experiments on the periodic stimulation of spontaneously beating aggregates of embryonic chick atrial heart cells by the Montreal group [Guevara *et al.* 1981, 1989; Glass and Mackey 1988; Zeng *et al.* 1990; Glass and Shrier 1991]. Figure 3.20 shows an experimental trace of transmembrane voltage as a function of time recorded with an intracellular microelectrode. A 20 ms current pulse is delivered via the same microelectrode giving rise to a large artifact; this pulse delays the phase, enlarging the oscillation period ($T' > T$). Periodic stimulation at a different rate results in the onset of stable phase locked states shown in Fig. 3.21.

### 3.2.6 Phase and frequency locking: general formulation

The synchronization properties we have described are general for weakly forced oscillators, and independent of the features of a particular system, i.e., whether it is a quasilinear or a relaxation oscillator. They are also independent of the form of the periodic forcing, whether it is harmonic (as assumed in the Section 3.1), rectangular (like in the experimental example in Fig. 3.10), or pulse-like (as in the example presented in Fig. 3.21). Generally, synchronization of order $n : m$ can be observed, with the Arnold tongues touching the $\omega$-axis; this means that synchronization can be achieved by an arbitrary small force (Fig. 3.19). For large $n$ and $m$ the

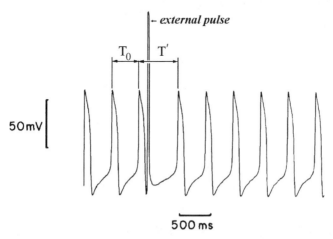

**Figure 3.20.** An experimental trace showing the transmembrane potential of a spontaneously beating cell (of embryonic chick heart) and the effect of injecting a current pulse. The pulse is delivered via the measuring electrode giving rise to a large artifact and resetting the phase of the oscillation. $T_0$ is the basic oscillation period of the preparation and $T'$ is the perturbed cycle length. From [Zeng *et al.* 1990].

synchronization regions are very narrow so that it is not always possible to observe them experimentally.

To incorporate high-order synchronization in the general framework, we reformulate the condition of frequency locking as

$$n\omega = m\Omega. \tag{3.2}$$

The condition of phase locking (Eq. (3.1)) should be reformulated for the case of $n : m$ synchronization as well. To this end, we first emphasize that the phase difference in the synchronous state is not necessarily constant but can oscillate around some value. For an example, let us again consider the $1 : 1$ synchronization of an oscillator by a pulse train (Fig. 3.17b) and plot both the phases of the oscillator and external force, as well as their difference (Fig. 3.22). We see that the phase difference is bounded but not constant. This is not inherent in the pulse perturbation, e.g., the phase difference generally oscillates if the motion of the phase point along the cycle (or growth of the phase of the force) is nonuniform. Now, taking into account the possibility of $n : m$ synchronization, we reformulate the condition of phase locking (3.1) in a more general form

$$|n\phi_e - m\phi| < \text{constant}. \tag{3.3}$$

Thus, the only crucial condition is that the phase difference $n\phi_e - m\phi$ remains *bounded*, which is equivalent to the condition of frequency locking (3.2).

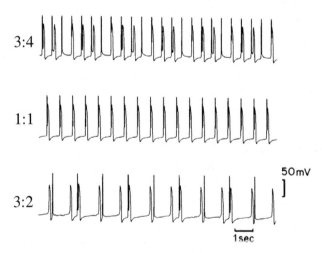

**Figure 3.21.** Synchronization of embryonic atrial chick heart cell aggregates by periodic sequence of pulses. The $3 : 4$, $1 : 1$ and $3 : 2$ synchronization regimes are shown here (stimulation period $T = 0.6$ s, $T = 0.78$ s and $T = 1.52$ s, respectively). All rhythms were obtained by stimulating for a long enough time to allow the transients to die out and the steady state rhythm to be established. From [Zeng *et al.* 1990]. Patterns of other orders $n : m$ are also reported there.

### 3.2.7  An example: synchronization of a laser

Simonet *et al.* [1994] experimentally and theoretically studied locking phenomena in a ruby nuclear magnetic resonance laser with a delayed feedback. This system is ideal for this study because of its excellent long-term stability and large signal-to-noise ratio, thus allowing observations of high-order synchronization. The unforced laser exhibited periodic oscillations of the light intensity with the cyclic frequency $f_0 \approx 40$ Hz. An external periodic voltage was added to the signal in the feedback loop. This voltage was either sinusoidal or square wave. In both cases synchronizations of different order were observed.

The locked states were identified by plotting the laser output vs. the forcing voltage $V$. When a stable Lissajous figure was found (Fig. 3.23, cf. Fig. 3.9), the system was considered to be locked. Arnold tongues for the forced laser are shown in Fig. 3.24. Note that high-order tongues are very narrow – within the resolution of the experiment they appear practically as lines.

The presented example illustrates once again that the synchronization properties of weakly forced systems are universal and do not depend on the features of the particular oscillator or form of the forcing. The latter can be a sequence of pulses, as in the preceding theoretical consideration, or sinusoidal or of square-wave form, as in the laser experiment described above. In any case, we observe a family of Arnold tongues touching the frequency axis. On the contrary, the properties of synchronization for moderate and large forcing are not universal. For example, in this laser experiment one finds that the 1 : 2 tongue is split into two (Fig. 3.24).

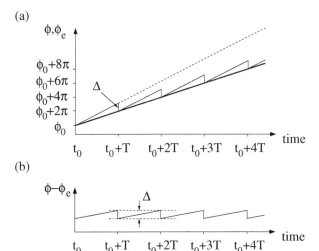

**Figure 3.22.** The phase difference in the synchronous state is not necessarily constant. (a) Phase of the external force $\phi_e$ increases linearly (bold line), whereas the phase $\phi$ (solid line) of the periodically kicked oscillator increases linearly between kicks, and is instantly reset, $\phi \to \phi - \Delta$, by each pulse. The phase of the autonomous oscillator would grow as shown by the dashed line. (b) The phase difference $\phi - \phi_e$ is bounded but oscillates.

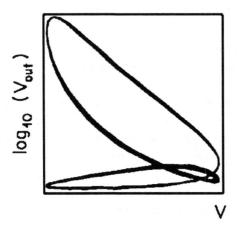

**Figure 3.23.** Intensity of the laser output log $V_{out}$ plotted vs. the forcing voltage $V$. The closed curve, known as a Lissajous figure, indicates that one period of the output oscillations exactly corresponds to two periods of the modulating signal (external force), i.e., 1 : 2 synchronization takes place (cf. Fig. 3.9). From Simonet *et al.*, *Physical Review* E, Vol. 50, 1994, pp. 3383–3391. Copyright 1994 by the American Physical Society.

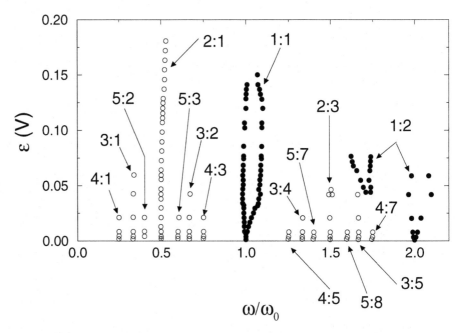

**Figure 3.24.** Synchronization tongues for the laser subject to external action. $\varepsilon$ and $\omega$ are the amplitude and the frequency of the force, respectively. The borders of 1 : 1 and 1 : 2 tongues are shown by filled circles. The high-order tongues are very narrow; in the precision of the experiment they appear as lines that are shown by open circles. From Simonet *et al.*, *Physical Review* E, Vol. 50, 1994, pp. 3383–3391. Copyright 1994 by the American Physical Society.

## 3.3   Synchronization of relaxation oscillators: special features

In this section we illustrate the synchronization properties of relaxation (integrate-and-fire) oscillators introduced in Section 2.4.2. We present three particular mechanisms of forcing of such oscillators: (i) resetting by pulses; (ii) variation of the threshold; and (iii) variation of the natural frequency, and present several experimental examples.

### 3.3.1   Resetting by external pulses. An example: the cardiac pacemaker

We again exploit the toy model from Section 2.4.2 (see Fig. 2.12), but now incorporate an external force as follows. Suppose that vessels filled with water are traveling on a conveyor belt and turn over into the oscillator tank (Fig. 3.25a). Thus, an amount of water is added to the system with some frequency $\omega$ which is determined by the velocity of the conveyor belt. Assuming that the water is added practically instantly, we can regard the force as a periodic sequence of pulses; the amplitude of these pulses corresponds to the volume of the added water. For simplification, we also make an assumption common in the study of integrate-and-fire oscillators    that the firing is instantaneous as well.

The effect of the pulses on the dynamics is obvious: they shorten the oscillation period and therefore directly influence the phase of the oscillator. Simple considerations (Fig. 3.25) yield the main properties of synchronization in this system as follows.

■ Additional supply of water completes the cycle earlier, but cannot make the period of the oscillator longer than it was without forcing.

■ Onset of synchronization is very fast, it occurs within a few cycles. Note that if the pulses are strong enough to reset the oscillator, i.e., their amplitude is larger than the threshold, then synchronization sets in with the first external pulse.

The mechanism of synchronization by resetting pulses has been implemented since the 1960s in devices that have saved many human lives – cardiac pacemakers. The first models were just constant rate pacers used to correct an abnormally low heart rate (bradycardia). In such devices electrical pulses from the implanted pacer stimulate the primary cardiac pacemaker (sino-atrial node) and thus trigger the heart contraction.[15] Similar devices are also used immediately after cardiac surgeries; they give pulses at a predefined frequency to induce the heart to beat at a higher rate than it would otherwise. Such a pacemaker is external and is connected to the heart by means of small electrodes. The electrodes are removed once the heart is healed.

---

[15] Modern pacemakers are programmable rate-adaptive devices that use complicated algorithms to monitor the cardiovascular system and alter the rate of pacing in accordance with daytime, physical activity, etc.

### 3.3.2    Electrical model of the heart by van der Pol and van der Mark

Another interesting example is the experiment conducted by van der Pol and van der Mark [1928]. Inspired by the idea, expressed by van der Pol in 1926, that heart beats are relaxation oscillations, they constructed an electrical model of the heart. Their device consisted of three relaxation oscillators of the type shown in Fig. 2.14, which were unidirectionally coupled (Fig. 3.26). Hence, one oscillator acted on the others as an external force.[16]

[16] It is known that contraction of the heart is normally caused by the primary pacemaker (sino-atrial node) having an intrinsic frequency of about 70 beats per minute. In case of its failure the heart beat is triggered by the secondary pacemaker (atrio-ventricular node); its frequency is $\approx$ 40–60 beats per minute. If the conduction through the atrio-ventricular node is completely blocked, the ventricules contract according to the rate of the third-order pacemakers ($\approx$ 30–40 beats per minute) [Schmidt and Thews 1983]. Thus, in normal functioning of the heart the pacemakers are synchronized by the sino-atrial node. As is typical for pulse coupled relaxation oscillators, the fastest one always dominates.

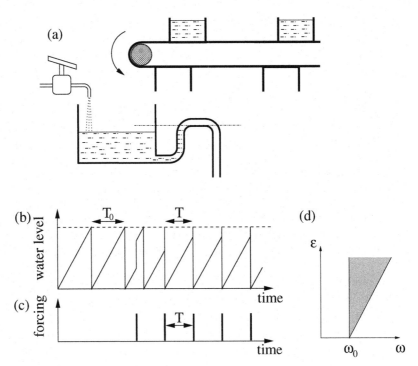

**Figure 3.25.** (a) Pulse forcing of the integrate-and-fire oscillator. An additional amount of water is periodically brought by a conveyor belt and instantly poured into the vessel. (b) The autonomous oscillator with the period $T_0$ is synchronized by the $T$-periodic sequence of pulses shown in (c). Each pulse results in an instant increase of the water level in the main vessel; hence the threshold value is reached earlier than in the absence of force. Note that synchronization sets in very quickly, within two cycles. (d) As the pulse force can only increase the frequency of the integrate-and-fire oscillator, the 1 : 1 synchronization region has a specific asymmetric shape.

Additionally, the device contained three keys so that short pulses could be applied to each generator, thus simulating extrasystols. By adding the current impulses from the atria and the ventricles (taken as P- and R-waves, respectively), the experimentalists simulated the electrocardiogram that was observed on an oscilloscope. This experiment is apparently the first application of nonlinear science to a biological problem.

Van der Pol and van der Mark varied the coupling between the second and the third generators (which imitated the conduction between the atria and the ventricles), starting with a sufficiently strong one. Indeed, in a normal contraction of the heart all three generators must be synchronized; every P-wave is followed by an R-wave. By reducing the coupling it was possible to simulate a certain cardiac pathology, the atrio-ventricular block, i.e., violation of normal conduction of stimuli from atria to ventricles. In such a pathology, termed nowadays a second-degree atrio-ventricular block, $m$ contractions of atria are followed by $n$ ventrical contractions ($m > n$). In the electrical model atrio-ventricular block corresponded to the $n : m$ synchronization of the second and the third generators.

Van der Pol and van der Mark noted that a decrease of coupling may be understood as a reduction of the amplitude of the stimulus that arrives at the ventricules. If this amplitude is not large enough to reset the third generator, then $1 : 1$ synchronization breaks down. Nevertheless, the pulse reduces the period of the third generator, and synchronization with $m > n$ becomes possible.[17]

### 3.3.3  Variation of the threshold. An example: the electronic relaxation oscillator

Another possibility to affect the integrate-and-fire oscillator is to alter the value of the threshold. The basic action in this case is that if the threshold is increased (decreased)

---

[17]  Readers interested in modern approaches to modeling different pathological cardiac rhythms are advised to consult [Glass and Mackey 1988; Guevara 1991; Yehia *et al.* 1999].

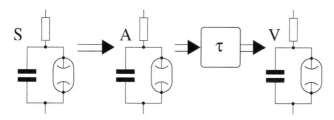

**Figure 3.26.** Electrical model of the heart by van der Pol and van der Mark. Three unidirectionally coupled relaxation generators represent the sino-atrial node (S), atria (A) and ventricles (V). The retardation system between the second and third oscillators imitates the time delay $\tau$ required for a stimulus to be transmitted from the atria to the ventricles. Schematically drawn after [van der Pol and van der Mark 1928].

then the accumulation epoch will be longer (shorter). A periodic variation (modulation) of the threshold results in synchronization of the relaxation oscillator.

Here we describe the mechanism that was first studied in the classical experiments of van der Pol and van der Mark [1927]. In their setup (shown in Fig. 3.27) a neon tube is nonconductive until the voltage applied to it exceeds some threshold value $u_{thresh}$. Hence, the voltage $u$ at the capacitor slowly increases towards this threshold, then the capacitor quickly discharges through the ignited tube and a new cycle starts. If an alternating voltage is introduced into the circuit, then the tube discharges when $u = u_{thresh} - \varepsilon \sin \omega t$; that is, the value of the threshold is periodically varied. Figure 3.28 explains why the variation of the threshold leads to synchronization: if the period of variation is slightly shorter than that of the autonomous system, then the threshold is reached earlier, thus making the frequency of the oscillator equal to the frequency of action, and vice versa. It is obvious that synchronization by a force with frequency $\omega \approx m\omega_0$ can also be achieved.

The remarkable synchronization property of the relaxation oscillator was used by van der Pol and van der Mark [1927] to implement frequency demultiplication (division). In their experiments the frequency $\omega$ of the applied harmonic voltage was chosen equal to the frequency of autonomous oscillation $\omega_0$ and kept constant, while the capacitance in the circuit was gradually increased. They observed that the frequency of oscillation in the system first remained constant and equal to $\omega$, and then suddenly dropped to $\omega/2$. With a further increase of the capacitance the frequency jumped to $\omega/3$ and so on up to $\omega/40$ (Fig. 3.29). These jumps correspond to the transitions between synchronous states of order $1 : n$ and $1 : (n + 1)$. In later experiments the authors achieved demultiplication to a factor of 200.

It is very interesting to emphasize that van der Pol and van der Mark noted that "often an irregular noise is heard in the telephone receivers before the frequency jumps to the next lower value"; that noise was probably the first experimental observation of chaotic dynamics (see Chapter 5).

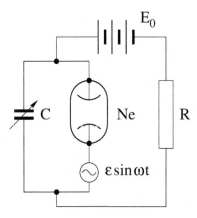

**Figure 3.27.** The setup of the relaxation oscillator experiment by van der Pol and van der Mark [1927]. The source of alternating voltage varies the threshold at which the neon tube flashes and becomes conductive, thus discharging the capacitor.

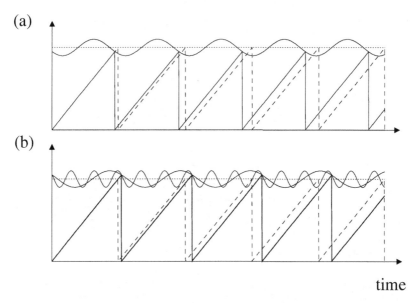

**Figure 3.28.** Synchronization of a relaxation oscillator via variation of the threshold. The threshold of the unforced system is shown by the dotted line; the time course of autonomous oscillation is shown by the dashed line. Periodic variation of the threshold can force the system to oscillate with the frequency of variation; this action can both decrease (a) or increase (b) the frequency of oscillation. As is clearly seen in (b), synchronization of higher order (here it is 1 : 3) can also be easily achieved.

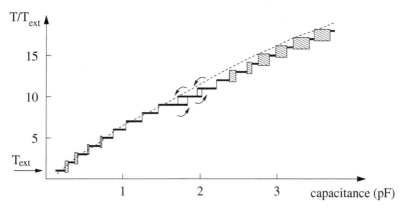

**Figure 3.29.** Frequency demultiplication implemented via synchronization of the neon tube relaxation generator (Fig. 3.27). The horizontal plateaux correspond to 1 : $m$ frequency locking; the shadowed regions denote those settings of the capacitor where an irregular noise was heard. The dashed line indicates the frequency with which the system oscillates in the absence of the external force. The curvy arrows show the hysteresis in the transition between neighboring synchronous states. From [van der Pol and van der Mark 1927].

Another very interesting experimental finding of van der Pol and van der Mark was the presence of hysteresis at the transition between neighboring synchronous states. That is, they found that the frequency jumps from $\omega/9$ to $\omega/10$ and from $\omega/10$ to $\omega/11$ occur at different values of the capacitance depending on whether it is being increased or decreased in the course of the experiments. This can be explained by the overlap of synchronization regions (see Section 7.3.4).

## 3.3.4  Variation of the natural frequency

Another basic mechanism of synchronization is related to the direct action of the forcing on the natural frequency $\omega_0$ of the oscillator, via its variation. In our integrate-and-fire model this can be represented as a periodic variation of the slope that characterizes the integration epoch. Indeed, between two fire events at $t_i$ and $t_{i+1} = t_i + T_0$ the level of water in the vessel or the voltage across the capacitor grows linearly in time as

$$x = x_{\text{thresh}} \cdot \frac{t - t_i}{T_0} = x_{\text{thresh}} \cdot \frac{\omega_0 \cdot (t - t_i)}{2\pi},$$

so that the slope is just directly proportional to the firing frequency.[18] It is easy to understand that the force, which immediately changes the frequency, entrains the oscillator. Suppose for simplicity that the slope between two firings is constant and is determined by the value of the force at $t_i$, i.e., when the new integration starts

$$x = x_{\text{thresh}} \cdot \frac{(\omega_0 + \varepsilon \sin(\omega t_i + \bar{\phi}_{\text{e}})) \cdot (t - t_i)}{2\pi}.$$

Depending on the phase $\omega t_i + \bar{\phi}_{\text{e}}$ of the force at $t_i$, the firing rate is either increased or decreased, and therefore synchronization can set in.[19]

An interesting regime appears in the case of $n : 1$ locking (Fig. 3.30). In this case the period of the forced oscillator varies from cycle to cycle (this is also called modulation of the frequency), and the spikes (the firings) are not equidistant.

This synchronization mechanism may be important for an understanding of the firing of sensory neurons. These neurons are known to respond to a slowly varying stimulus (i.e., to external forcing) by variation of the firing rate; this is called rate coding or frequency modulation.

Another physiological example where this mechanism of forcing is relevant is the cardiovascular system. Indeed, the autonomic neural system alters the slope of the integrate-and-fire oscillators that constitute a pacemaker (sino-atrial node) and thus affects the heart rate. We shall return to these examples in Chapter 6.

---

[18] Note that integrate-and-fire models can be understood in the following way: instead of accumulation of some variable $x$ until it reaches $x_{\text{thresh}}$, we can consider "accumulation" of the phase until it reaches $2\pi$. Alternatively, we can say that the instantaneous frequency $d\phi/dt$ is being integrated. In this way these models can be reduced to *phase oscillator* models.

[19] Simple consideration of $\omega = \omega_0 + \varepsilon \sin(\omega t_i + \bar{\phi}_{\text{e}})$ gives the condition of $1 : 1$ entrainment $|\omega_0 - \omega| \leq \varepsilon$ and the value of the constant phase shift in the locked state $\sin^{-1}((\omega - \omega_0)/\varepsilon)$.

### 3.3.5   Modulation vs. synchronization

Here we emphasize the important differences between synchronization and modulation. We stress that in the case of $n : 1$ locking the effect of the forcing is twofold:

(i)  it causes the *modulation* of the period of the oscillator that occurs with the period of the forcing;

(ii)  the force adjusts the average period of oscillation so that $m \cdot \langle T_i \rangle = T$ (*synchronization*).

These phenomena are distinct, although they can overlap. Indeed, there can be synchronization without modulation (e.g., $1 : 1$ phase locking), modulation without synchronization (described below), or a combination of both effects, as in Fig. 3.30.

Generally, modulation without synchronization is observed when a force affects oscillations, but cannot adjust their frequency. Such a situation appears, e.g., when a force affects not the oscillator itself, but the channel transmitting the oscillating signal. As an (somewhat artificial) example let us consider the contraction of the heart. It can be characterized by the pulse waves measured on a finger. If the velocity of pulse propagation along the blood vessels is varied (e.g., by changing the pressure) then some pulses propagate faster and some are delayed, and as a result we observe modulation in the pulse train. Definitely, the number of pulses, i.e., the average contraction frequency, cannot be changed in this way and synchronization due to this action is not possible. We illustrate the difference between synchronization and phase modulation in Fig. 3.31.

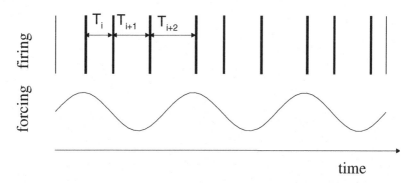

**Figure 3.30.** Synchronization of the integrate-and-fire oscillator via variation of its natural frequency. A regime of $3 : 1$ locking is shown here: three spikes occur within one period of forcing. Note that the period of the entrained oscillator is varied (modulated) by the force, $T_i < T_{i+1} < T_{i+2}$, and the spikes are not evenly spaced in time.

### 3.3.6   An example: synchronization of the songs of snowy tree crickets

In a particular experiment it is not always possible to reveal the exact mechanism of forcing. Generally, the action occurs in some combined form. Thus, in the following example (as well as in other examples to be presented in the next section) we do not focus on the mechanism of forcing, but rather dwell on the general features of synchronization.

The snowy tree cricket *Oecanthus fultoni* is a common dooryard species throughout most of the United States. These insects are able to synchronize their chirps by responding to the preceding chirps of their neighbors. The song of the snowy tree cricket is a long-continuing sequence of chirps produced when the male elevates his specialized forewings and rubs them together. Each chirp consists of 2 to 11 pulses that correspond to wing closures. The chirp rhythm is highly regular in choruses, and it is usually so for solitary singers [Walker 1969].

To investigate the mechanisms of acoustic synchrony in crickets, Walker [1969] performed the following experiments. He pre-recorded one chirp and then played it continuously to males singing on perches in individual glass cylinders. The test sounds were broadcast at approximately natural intensity and different rates, and the acoustic response of the test cricket was tape-recorded for subsequent analysis. Thus, a test cricket was subject to a periodic external action.

The main result of this study is shown schematically in Fig. 3.32. We can see the onset of synchronization following the onset of external chirps. For the diagram shown in Fig. 3.32a the rate of the chirps was 242 per minute, and the chirp rate of the cricket prior to broadcast was 185. As expected, the phase of the rhythm that was initially slower lags behind the phase of the faster artificial rhythm. Accordingly, when the initial chirp rate was 192 per minute to be compared with the rate of the broadcast of 166 per minute, the cricket was ahead of the phase of the synchronizing signal

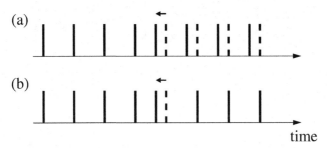

**Figure 3.31.** Synchronization vs. modulation. If the next pulse is generated with a constant lag to the previous one, then shifting one pulse results in changing the times of all subsequent pulses (a). Here the force can influence the frequency of the pulse train and synchronize it. If, on the contrary, shifting one pulse does not influence the times of subsequent pulses (b), then the force cannot change the frequency of the train. The pulses can be modulated by the force, but not synchronized.

(Fig. 3.32b). Finally, 1 : 2 synchronization (two artificial chirps per one natural) is also possible, as can be seen from Fig. 3.32c.

As is typical for relaxation oscillators, a cricket is able to adjust its song to any other song like its own very quickly. By either lengthening or shortening its own period in response to the preceding chirp, the insect can achieve synchrony within two cycles.

## 3.4 Synchronization in the presence of noise

Self-sustained systems are idealized models of natural oscillators. These models neglect the fact that the macroscopic parameters of natural oscillators fluctuate, and that the oscillators cannot be considered perfectly isolated from the environment. Indeed, real systems are always subject to thermal fluctuations; often they are disturbed by some weak external sources. Usually all these factors cannot be exactly taken into account and therefore the effect of natural fluctuations, as well as the influence of the environment, is modeled by some random process, or noise.

In this section we study the entrainment of a limit cycle oscillator by an external force in the presence of noisy perturbations. For this purpose we first discuss how noise affects an autonomous oscillator, and the main effect here is phase diffusion.

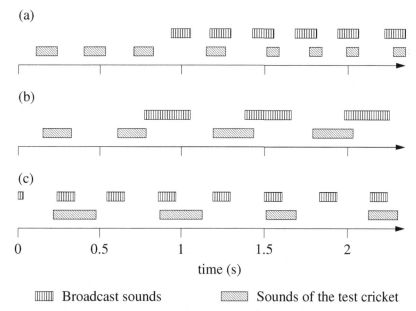

**Figure 3.32.** Synchronization of the songs of a snowy tree cricket. The left-hand side of panels (a) and (b) shows free songs. The sounds of the test cricket are entrained by a faster (a) or slower (b) artificial rhythm. An example of 1 : 2 locking is shown in (c). Schematically drawn using data from [Walker 1969].

### 3.4.1   Phase diffusion in a noisy oscillator

As in Section 3.1, we again consider a phase point in a rotating reference frame and model the dynamics of the phase by those of a weightless particle in a viscous fluid. If there were no noise, the phase point would rest at some arbitrary position on the cycle (at the moment we suppose that there is no external forcing!), and we know that with respect to the shifts along the cycle, the phase point is in a state of neutral (indifferent) equilibrium (see Fig. 2.3 and Fig. 2.4c). Hence, even weak noise affects the phase, shifting it back and forth. Thus, the motion of the phase point in the rotating reference frame can be regarded as a *random walk*.

It is very important to remember that even very small phase perturbations are accumulated, because phase disturbances neither grow nor decay. As a result, the phase can display large random deviations from the linear growth ($\phi = \omega_0 t + \phi_0$) that would be observed if there were no noise. Due to the analogy to the motion of a diffusing Brownian particle randomly kicked by molecules, one often speaks of **phase diffusion**.[20] We can illustrate noisy phase dynamics by plotting the phase of human respiration (Fig. 3.33).

Because of the influence of noise, the period of oscillation is not constant. Nevertheless, on average the random deviations in the clockwise and counter-clockwise directions compensate, and the average period $\langle T_0 \rangle$, computed by counting the number

---

[20]  In contrast to the walk of the Brownian particle, the walk of the phase is one-dimensional (along the cycle).

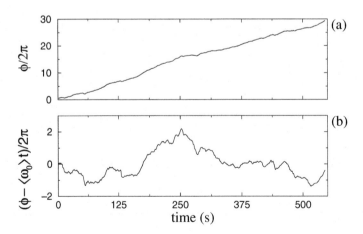

**Figure 3.33.** Random walk of the phase of respiration. Human respiration is governed by a rhythm generator situated in the brain stem. This system definitely cannot be regarded as an ideal, noise-free oscillator. The phase of respiration is computed from a record of the air flow measured by a thermistor at the nose. Due to noise, the increase of the phase is not linear (a) but can be regarded as a random walk. It is more illustrative to plot the deviation of the phase from linear growth $\langle \omega_0 \rangle t$ (b). Note that this deviation is not small if compared with $2\pi$.

of cycles within a very long time interval, coincides with the period of noise-free oscillations $T_0$.[21] Correspondingly the average frequency is $\omega_0$.

## 3.4.2 Forced noisy oscillators. Phase slips

Consider first a weak external forcing with frequency $\omega = \omega_0$ (i.e., there is no detuning). As a result of the driving, the particle walks not in a line, but in a $2\pi$-periodic potential (Fig. 3.34, cf. Fig. 3.3b), and we see two counteracting tendencies: the force tries to entrain the particle and to hold it at an equilibrium position, while the noise kicks the particle out of it. The outcome of this counter-play depends on the intensity and the distribution of the noise. If the noise is weak and *bounded* so that it can never push the particle over the potential barrier, then the particle just fluctuates around the stable equilibrium point in a minimum of the potential. If the noise influence is strong enough, or if the noise is *unbounded* (e.g., Gaussian), then the particle from time to time jumps over the potential barrier and rather quickly goes to the neighboring equilibrium $\phi^0 \pm 2\pi$. Physically, this means that the phase point of the oscillator makes an additional rotation with respect to the phase of the external force (or is delayed by one rotation). These relatively rapid changes of the phase difference by $2\pi$ are called *phase slips*.[22]

The situation changes qualitatively if the frequency of the force differs from that of the autonomous oscillator. In this case the potential is not horizontal, but tilted (cf. Fig. 3.5a). Again, we distinguish two cases. If the slips are not possible (i.e., the noise is bounded and weak), the phase fluctuates around a constant value (Fig. 3.35). Otherwise, if the noise is sufficiently strong or unbounded, phase slips occur. However, these phase slips now have different probabilities of jumping to the left and to the right: naturally, the particle more frequently jumps down than up (Fig. 3.36). Thus, although most of the time the particle fluctuates around an equilibrium point, on average it slides

[21] We suppose here that the noise is symmetric, i.e., it kicks the phase point clockwise and counter-clockwise with equal probability so that the random walk is unbiased. Otherwise the noise shifts the value of the average frequency $\langle \omega_0 \rangle$.

[22] Actually, if the noise is rather strong, then several jumps can immediately follow one another, so that the slip can be also $\pm 4\pi$, and so on.

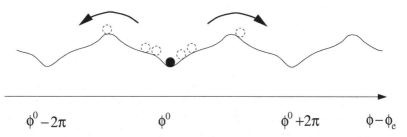

**Figure 3.34.** Phase diffusion in the oscillator that is forced without detuning: the phase point oscillates around the stable equilibrium point $\phi - \phi_e = \phi^0$ and sometimes jumps to physically equivalent states $\phi^0 \pm 2\pi k$.

down the potential. The phase dynamics are therefore very inhomogeneous (at least, for weak noise): long synchronous epochs are intermingled with phase slips.

### Characterizing synchronization of a noisy oscillator

We now discuss how synchronization of noisy systems can be characterized and how/whether the notions of locking can also be used for such systems. Indeed, we have already presented several experimental examples, and we have noted that the observations do not perfectly agree with the theory because the systems under study are noisy. To treat this imperfection we distinguish between two cases: weak bounded noise and unbounded or strong bounded noise.

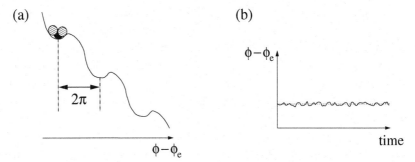

**Figure 3.35.** Phase dynamics of a forced oscillator perturbed by weak bounded noise. (a) The particle oscillates around a stable equilibrium point, but cannot escape from the minimum of the potential. Correspondingly, the phase difference fluctuates around a constant value (dotted line) that it would have in the absence of noise (b).

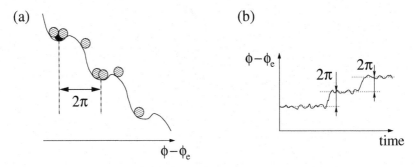

**Figure 3.36.** Phase dynamics of a forced oscillator subject to unbounded noise. (a) The particle oscillates around the stable equilibrium point $\phi^0$ and sometimes jumps to physically equivalent states $\phi^0 \pm 2\pi k$. Although jumps in both directions are possible, the particle more frequently jumps over the small barrier, i.e., downwards. These jumps (phase slips) are clearly seen in (b). The time course of the phase difference resembles the phase dynamics of a noise-free oscillator at the desynchronization transition (see curve 3 in Fig. 3.8b), but in the noisy case the slips appear irregularly.

*Weak bounded noise*

If the detuning is small, this noise never causes phase slips. In other words, the noise cannot move the particle between the wells of the potential (Fig. 3.35). Then, due to noisy perturbations, the phase difference fluctuates in a random manner, but remains bounded, so that the condition of phase locking (3.3) continues to be valid. The (average) frequency $\Omega$ of the noisy oscillator is also locked to the frequency of the force. For some larger detuning the intensity of the noise becomes sufficient in order to overcome the potential barrier, and the particle starts to slide downwards. Note that the transition occurs for smaller detuning than would be the case without noise (Fig. 3.37a). We emphasize here that, in the considered case, the synchronization region does not touch the $\omega$-axis: small-amplitude forcing cannot keep the particle in the potential well.

*Unbounded or strong bounded noise*

If the noise is not bounded (e.g., Gaussian), or bounded but strong, phase slips occur. If the potential is tilted then the jumps downwards happen more frequently, and on average the phase difference grows for arbitrary small detuning. We come to an important point: strictly speaking, unbounded noise destroys synchronization and neither phase locking nor frequency locking conditions, formulated for the deterministic case, are fulfilled. Nevertheless, at least for a weak noise we can say that the frequencies are approximately equal in some range of detuning (Fig. 3.37b). With an increase of noise

(a)                                    (b)

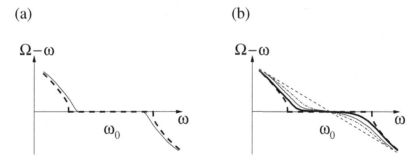

**Figure 3.37.** Frequency–detuning curves for noisy oscillators. (a) For weak bounded noise there exists a region of detuning where the frequencies of the oscillator $\Omega$ and the force $\omega$ are exactly equal. This region (solid line) is smaller than in the noise-free case (bold dashed line). (b) For unbounded noise the frequencies $\Omega$ and $\omega$ are equal only at one point, not in an interval. If the noise is very weak (bold line) then we can speak of approximate equality of frequencies in some range of detuning $\omega - \omega_0$. If the noise is stronger (solid and dashed-dotted lines) then this range shrinks to a point. Strictly speaking, synchronization appears only as a tendency: the external force "pulls" the frequency of the oscillator towards its own, but the noise prevents the entrainment. Very strong noise completely destroys synchronization (dashed line). The bold dashed line shows again the synchronization region for a noise-free oscillator.

intensity this range shrinks, and synchronization appears as a weak "attraction" of the observed frequency $\Omega$ by the frequency of the external force $\omega$. In other words, the frequencies tend to be adjusted, but (except for one point) there is no perfect coincidence.

We now discuss the relationship between phases of the oscillator and the force. We have already noted that the phase difference $\phi - \phi_e$ can be arbitrarily large due to phase diffusion (unless the noise is weak and bounded). Thus, generally, for a noisy oscillator we cannot speak of phase locking because the phase difference is not limited. On the other hand, the particle tends to stay near the minima of the potential,[23] and therefore certain values of $\phi - \phi_e$ are more likely to be observed. If we take the phase difference on a circle $[0, 2\pi]$,[24] and plot the distribution of the phase, then we find that it is not uniform, but has a well-expressed maximum (Fig. 3.38). The position of this maximum corresponds to the value of the phase difference that would be observed in the absence of noise. Noise broadens the peak, but the maximum is preserved. We can interpret this preference of a certain value of $\phi - \phi_e$ as some statistical analogy of phase locking. Note, however, that a similar distribution of the phase appears in the noise-free oscillator not far from the desynchronization transition, when the phase dynamics are intermittent (see curves 3 and 4 in Fig. 3.8), as well as in case of modulation without synchronization (see Section 3.3.4).

[23] The force tries to keep the particle at certain positions and, hence, suppresses phase diffusion; in the case of weak bounded noise the diffusion is eliminated.

[24] In other words, we consider different equilibrium positions as equivalent, and take the phase difference $(\phi - \phi_e)$ mod $2\pi$.

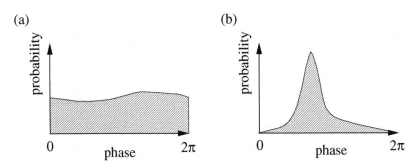

**Figure 3.38.** Broad (a) and unimodal (b) distributions of the phase difference in a noisy oscillator, without (a) and with (b) external forcing. If the forcing is absent, then the phase difference attains any value with almost the same probability, due to phase diffusion (a). The external force, creating minima in the potential, makes a certain value more probable (b). Similar distributions are obtained if we observe the phase of the oscillator stroboscopically, with the period of the external force. By such an observation, the phase of an entrained noise-free oscillator is always found in the same position (see. Fig. 3.13); the noise broadens the distribution but the maximum in it is preserved, indicating synchronization. (See also the experimental example shown in Fig. 3.15.) A distribution similar to (b) is also observed in the absence of synchronization, when the phase of a process is modulated.

### 3.4.3  An example: entrainment of respiration by mechanical ventilation

For illustration we again return to the experiments of Graves *et al.* [1986] (see Section 3.1.4), where the synchronization region for the case of mechanical ventilation was computed (Fig. 3.39). There is no distinct border of synchronization. Moreover, for very low amplitudes of the driving force synchronization is impossible: the forcing is too weak to overcome the destructive role of the noise.

### 3.4.4  An example: entrainment of the cardiac rhythm by weak external stimuli

Another example of synchronization of noisy oscillators by periodic external force is reported by Anishchenko *et al.* [2000], who studied the entrainment of a human cardiac rhythm by a weak periodic stimulation. In these experiments the subjects were asked to sit at rest in front of a monitor, while a computer generated a periodic sequence of acoustic and visual stimuli, i.e., the appearance of color rectangles on the screen was simultaneously accompanied by sound pulses.

To characterize the response to stimulation, the electrocardiogram was registered by a standard technique. It is well known that every normal cardiocycle contains a very sharp and distinct peak, denoted the R-peak (for illustration see Fig. 6.1); the interval between two neighboring R-peaks is usually taken as the interbeat interval. Naturally, an intact heart is far from being a periodic oscillator,[25] and one can see that the interbeat intervals vary essentially with time; this variability is a well-known

---

[25] There exist controversial viewpoints on whether the irregularity of the heart rate is caused by noise or by chaotic dynamics in the presence of noise; in the following we show that this issue does not affect the synchronization properties.

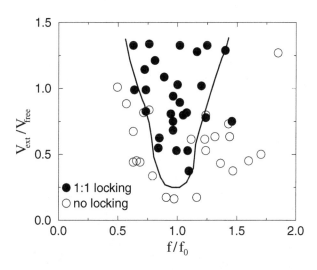

**Figure 3.39.** A region of 1 : 1 phase locking of spontaneous breathing by a mechanical ventilator obtained for seven subjects. The strength of the forcing is characterized by the volume $V_{ex}$ of the inflation. Frequency and volume axes have been normalized with respect to the mean "off pump" frequency and tidal volume of separate experiments, respectively. The boundary of the synchronous regime has been drawn arbitrary. From Graves *et al.* [1986].

physiological phenomenon. Therefore, we can only speak about the average period and frequency of cardiac oscillations. To estimate this mean cyclic frequency $F$, one can just count the number $n$ of R-peaks within a fixed time interval $\tau$ and compute $F = n/\tau$.

In the experiments we describe, first the cyclic frequency $f_0$ of the cardiac rhythm of a subject *without stimulation* was estimated. Then, the subjects were exposed to external stimulation with a different frequency $f$. Within each 10 minute long trial $f$ was kept constant, but it varied in the range $0.75 f_0$–$1.25 f_0$ from trial to trial. The resulting frequency vs. detuning curve for one subject is shown in Fig. 3.40. A plateau around $f \approx f_0$, although not perfectly horizontal, indicates frequency locking. We note that the stimulation in these experiments was indeed weak: it was verified that stimulation without detuning (i.e., with $f = f_0$) did not alter the mean heart rate or important parameters such as the blood pressure and the stroke volume.

## 3.5      Diverse examples

In this section we present several examples of synchronization in living systems.

### 3.5.1   Circadian rhythms

The behavior of humans and animals is characterized by a precise 24-hour cycle of rest and activity, sleep and wakefulness. This cycle (termed circadian rhythm) represents the fundamental adaptation of organisms to an environmental stimulus, the daily cycle of light and dark (see [Aschoff *et al.* 1982; Moore 1999; Sassone-Corsi 1999] and references therein). This behavioral cycle is accompanied by a daily oscillation in hormone secretion, core body temperature, lymphocyte number and other important physiological functions.

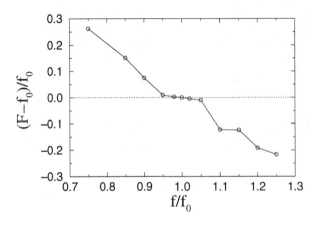

**Figure 3.40.** The observed mean frequency $F$ of the heart rhythm of a human as a function of the frequency $f$ of the weak external stimulation. This experimental curve is consistent with the corresponding frequency vs. detuning curves for a periodically driven noisy oscillator (cf. Fig. 3.37). From [Anishchenko *et al.* 2000].

It is a well-established fact that in the absence of a light–dark cycle the period of the circadian rhythm deviates from 24 hours; it can be either longer or shorter. Under normal conditions, the cycle is entrained by the daily variation of the illuminance. This entrainment was studied in numerous experiments within which the subject was isolated from the normal light–dark cycle and deprived of all other time cues. The results of these experiments can be schematically represented in the way shown in Fig. 3.41. This is a direct indication of the endogenous character of the circadian rhythm, i.e., that there exists a self-sustained clock that governs the rhythm. All available evidence indicates that there is one principal circadian pacemaker in mammals, namely the suprachiasmatic nucleus of the hypothalamus. This nucleus receives entraining information from retinal ganglion cells [Moore 1999].

Quantification of the circadian period in humans has yielded inconsistent results [Czeisler *et al.* 1999; Sassone-Corsi 1999; Moore 1999]: the average was controversially estimated to be 25 or 23 h, and individual variation from 13 to 65 h was reported in normal subjects. This large variation may be responsible for the behavioral patterns of "early birds" and "owls". Indeed, from general properties of synchronization we expect that the subjects with an intrinsic period $T_0 < 24$ h should be ahead in phase with respect to the external force, whereas those with a period $T_0 > 24$ h should lag in phase. Similarly, an age-related shortening of the intrinsic period has been

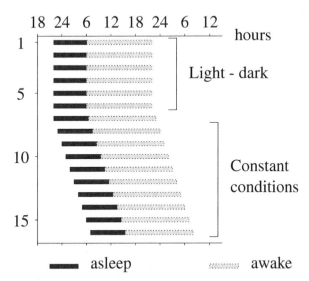

**Figure 3.41.** Schematic digram of the behavioral sleep–wake rhythm. Here the circadian rhythm is shown entrained for five days by the environmental light–dark cycle and autonomous for the rest of the experiment when the subject is placed under constant light conditions. The intrinsic period of the circadian oscillator is in this particular case greater than 24 hours. Correspondingly, the phase difference between the sleep–wake cycle and daily cycle increases: the internal "day" begins later and later. Such plots are typically observed in experiments with both animals and humans [Aschoff *et al.* 1982; Czeisler *et al.* 1986; Moore 1999].

hypothesized to account for the circadian phase advance and early-morning awakening frequently observed in the elderly (see [Czeisler *et al.* 1999] and references therein). We note that the phase shift depends not only on the initial detuning but also on whether the oscillator is a quasilinear or a relaxation oscillator. It is quite possible that an age-related change in circadian pacemaker activity can be understood in this way.

In recent experiments, Czeisler *et al.* [1999] attempted a precise determination of the intrinsic period of the human circadian pacemaker. To this end, the experimentalists used the so-called "forced desynchrony protocol": the sleep–wake time of each subject was scheduled to a 28-hour "day". (Czeisler *et al.* [1999] suppose that their method is superior to previously used ones because isolation under constant illuminance conditions influences the normal pacemaker activity.) In this way they ensured a quasiperiodic regime in the system: the 4-hour detuning is too large and the internal pacemaker is not entrained by the external force. In this regime, a time series of physiological parameters (core body temperature, plasma melatonin and plasma cortisol) is a mixture of two oscillations with the 28-hour period of the force and with the unknown period $T$. From this mixture, the latter was precisely determined by means of a special spectral technique and interpreted as the internal period of the pacemaker. This interpretation appears to be doubtful. Indeed, outside the synchronization region the observed period of the oscillation tends to the period of the autonomous system, but their difference remains finite (see Fig. 3.7a). It is not evident that this difference is negligibly small.

We note that in contrast to some other examples, the forcing of the circadian oscillator is not exactly periodic: some days are sunny, and some are not, so that the light–dark cycle differs from day to day. Thus, the force, although it has a strong periodic component, should be considered noisy. As one can see, it does not prevent synchronization.

We would like to complete this discussion of circadian rhythms with a remark on *chronotherapeutics*. The physiology and the biochemistry of a human vary essentially during a 24-hour cycle. Moreover, the circadian rhythms in critical bioprocesses give rise to a significant day–night variability in the severity of many human diseases and their symptoms. The effect of medication often depends strongly on the time it is taken [Smolensky 1997]. This fact (clearly very important for practical medicine) is quite natural from the viewpoint of nonlinear science: as we know, the effect of a stimulus on an oscillator depends crucially on the phase at which it is applied.

### 3.5.2  The menstrual cycle

The menstrual cycle can be also regarded as a noisy oscillation of the concentration of several hormones. Regulation of hormone production occurs via positive and negative feedback loops (remembering that the presence of feedback is a typical feature of self-sustained oscillators). The period of oscillation is about 28 days but it may fluctuate considerably (Fig. 3.42). It is natural to expect that a periodical addition of hormones could entrain the cycle; this indeed happens under treatment with hormonal pills (Fig. 3.42). The external forcing suppresses the phase diffusion and makes the oscillation period exactly 28 days.

A plausible model of forced oscillation is an integrate-and-fire oscillator with a varying threshold level: due to medication the hormonal regulation is changed in such a way that the threshold is kept very high and then dropped low enough to ensure firing (this is obvious from the fact that the interruption of treatment causes almost immediate onset of menstruation). Probably, in this case the forcing cannot be considered weak.

### 3.5.3   Entrainment of pulsatile insulin secretion by oscillatory glucose infusion

The main function of the hormone insulin is to regulate the uptake of glucose by muscle and other cells, thereby controlling the blood glucose concentration. The most important regulator of pancreatic insulin secretion is glucose, and thus insulin and glucose are two main components of a feedback system. In normal humans, ultradian oscillations with an approximate 2-hour period can be observed in pancreatic insulin secretion and blood glucose concentration (see [Sturis et al. 1991, 1995] and references therein).

Experiments performed by Sturis et al. [1991] have shown that this ultradian oscillation persists during constant intravenous glucose infusion, and can be entrained to the pattern of an oscillatory infusion. When glucose is infused as a sine wave with a relative amplitude of 33% and a period either 20% below or 20% above the period occurring during constant glucose infusion, 1 : 1 synchronization takes place (Fig. 3.43). Infusion with an ultraslow period of 320 min resulted in two pulses of glucose and insulin secretion for each cycle of the infusion, i.e., in the 2 : 1 synchronization.

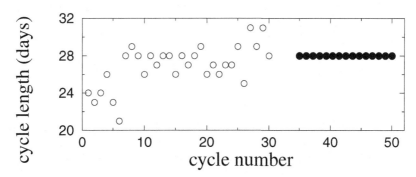

**Figure 3.42.** The length of the menstrual cycle normally fluctuates but becomes constant under hormonal medication. Circles and filled circles denote the length of the cycle prior to and under hormonal treatment, respectively.

### 3.5.4    Synchronization in protoplasmic strands of *Physarum*

Plasmodia of the acellular slime mold *Physarum* are one of the simplest organisms. They are differentiated into a frontal zone consisting of a more or less continuous sheet of protoplasm and into a rear region consisting of protoplasmic strands in which the protoplasm is transported in the form of shuttle streaming. The force for this mass transport is generated by oscillating contraction activities of the strands (see [Achenbach and Wohlfarth-Bottermann 1980, 1981] and references therein).

The stream of the protoplasm rhythmically changes its velocity and direction with a period of about 1–3 min. The period and amplitude of this oscillation depend, in particular, on the temperature of the environment. Kolin'ko *et al.* [1985] studied synchronization of the protoplasmic shuttle streaming by periodic variation of the temperature gradient. The latter is believed to cause the variation of the intracellular pressure gradient. The velocity of the plasmodic flow was measured by means of a laser Doppler anemometer. First, the period of autonomous oscillation $T_0$ was determined for a constant temperature of 19 °C. Next, the temperature gradient was harmonically varied with an amplitude of 0.5 °C; the period $T_e$ of this oscillation was changed in steps, starting from $T_e \approx T_0$. For a certain range of the period $T_e$ of the forcing, the period of the protoplasmic oscillation was entrained (Fig. 3.44).

## 3.6    Phenomena around synchronization

In this section we discuss several phenomena that are related to synchronization, but can be hardly described as a locking of a self-sustained oscillator by a periodic force.

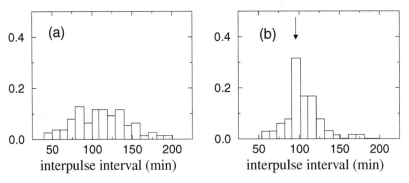

**Figure 3.43.** Distribution of the interpulse intervals (cycle length) of the insulin secretion rate for the case of constant (a) and oscillatory (b) glucose infusion. As expected for noisy systems, the period of oscillation is not constant. Periodic forcing makes the oscillation "more periodic", reducing the fluctuation of the cycle length. This is reflected by the appearance of the peak in the distribution in (b). The position of the peak corresponds to the period of the forcing (shown by the arrow), i.e., the average period of oscillation is locked to that of the force. From [Sturis *et al.* 1991].

### 3.6.1 Related effects at strong external forcing

We have discussed in detail and illustrated with several examples that an external force can entrain an oscillator, and the larger the mismatch between the frequencies of the autonomous system and the force, the larger must be the amplitude of the drive to ensure synchronization; see the sketch of a synchronization region in Fig. 3.7. From this sketch one can get the spurious impression that very strong forcing makes synchronization possible even for very strong detuning. We should remember that our explanation of synchronization implies that the force is weak; indeed, we assumed that the force only influences the phase of the oscillator and does not affect the shape of the limit cycle (i.e., the amplitude). If the force is not weak, the form of the Arnold tongues is no longer triangular, and the picture of synchronization is not universal – it depends now on the particular properties of the oscillator and the force. Moreover, strong forcing can cause qualitatively new effects.

#### Chaotization of oscillation

The motion of the forced oscillator can become irregular, or chaotic (see Chapter 5 for an introduction to chaotic oscillation and Section 7.3.4 in Part II for a discussion of transition to chaos by strong forcing). Such regimes were found in experiments with periodic stimulation of embryonic chick atrial heart cell aggregates [Guevara

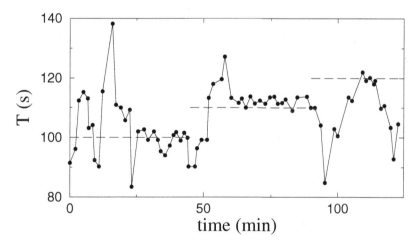

**Figure 3.44.** Synchronization of the protoplasm flow oscillation in the plasmodia of the slime mold *Physarum* by the harmonically alternating temperature gradient. The period of the protoplasm oscillation $T$ is shown as a function of time; the period $T_e$ of the forcing was changed stepwise (dashed line). For $T_e = 100$ s and $T_e = 110$ s synchronization sets in after some transient time; in the synchronous regime $T$ fluctuates around $T_e$. With further increase of $T$ synchronization is destroyed and a low-frequency modulation of $T$ is observed. From [Kolin'ko *et al.* 1985]. Copyright Overseas Publishers Association N.V., with permission from Gordon and Breach Publishers.

*et al.* 1981; Zeng *et al.* 1990]; see Fig. 3.45 and compare it with Fig. 3.21. These
authors have also observed synchronous patterns where two stimuli correspond to two
cycles. The difference with the simple regime of 1 : 1 locking is that two subsequent
stimuli occur at two different phases. Such behavior (period doubling) is typical for
the transition to chaos.

### Suppression of oscillations

Sometimes, a sufficiently strong external force, or even a single pulse, can suppress
oscillations. This may happen, e.g., if in the phase plane of a self-sustained system
a limit cycle coexists with a stable equilibrium point.[26] Suppression occurs if the
stimulus is applied in a "vulnerable phase" (Fig. 3.46). This effect was observed in
experiments with squid axons [Guttman *et al.* 1980] and pacemaker cells from the
sino-atrial node of cat heart [Jalife and Antzelevitch 1979] (see also [Winfree 1980;
Glass and Mackey 1988]).

It is also clear that a periodic sequence of pulses can essentially decrease the
amplitude of oscillation even if the equilibrium point is unstable: the first pulse kicks
the point towards the equilibrium and subsequent pulses prevent its return to the
limit cycle. An important possible application of these ideas is the elimination of
pathological brain activity (tremor rhythm) in Parkinsonian patients (see [Tass 1999]
and references therein).

---

[26] Such systems are called oscillators with hard excitation: in order to put them into motion one
should apply a finite perturbation that places the phase point into the basin of attraction of the
limit cycle; this mechanism of excitation is also called subcritical Hopf bifurcation.

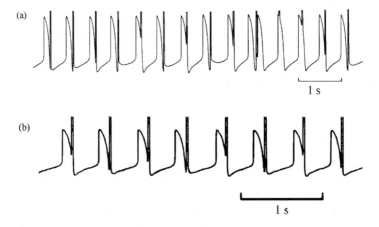

(a)

1 s

(b)

1 s

**Figure 3.45.** (a) Chaotic oscillation of periodically stimulated aggregates of
embryonic chick heart cells. From [Glass and Shrier 1991], Fig. 12.5a, Copyright
Springer-Verlag. (b) The 2 : 2 pattern (period doubling). From [Glass and Mackey
1988]. Copyright ©1988 by PUP. Reprinted by permission of Princeton University
Press.

## 3.6.2    Stimulation of excitable systems

Excitable systems are intrinsically quiescent, and therefore they do not conform to our understanding of synchronization. Nevertheless, their response to periodic stimulation strongly resembles synchronization of active systems, and therefore we briefly discuss them here. The main properties of excitable systems are the following.

- In response to a superthreshold stimulus they generate a spike called the action potential. Subthreshold stimuli do not cause any response, or the response is negligibly small.

- Immediately after the action potential they are refractory, i.e., they do not respond to stimuli of any amplitude.

Periodic stimulation of such systems gives rise to various patterns, depending on the frequency of stimulation.

Excitability is a typical feature of neural and muscle cells, and these synchronization-like phenomena have been observed in experiments with both tissue types. Matsumoto *et al.* [1987] studied the response of the giant squid axon to periodic pulse stimulation. They found different regular and irregular (possibly chaotic) rhythms. The intervals of the stimulation rate that caused a certain $n : m$ response rhythm were identified and found to resemble the phase locking regions (Fig. 3.47, cf. Fig. 3.19b).

The majority of cardiac cells are not spontaneously active. They excite, or generate an action potential, being driven by stimulation originating in a pacemaker region. Normally, the pacemaker imposes a $1 : 1$ rhythm on the excitable cells. However, the $1 : 1$ response may be lost when the excitability of the paced cells is decreased, or when the heart rate is increased. As a result, a variety of abnormal cardiac arrhythmias can arise [Glass and Shrier 1991; Guevara 1991; Yehia *et al.* 1999; Hall *et al.* 1999]. In particular, it may happen that one of $n$ stimuli is applied at the time when the cell is

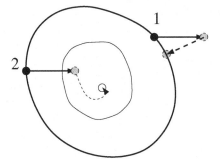

**Figure 3.46.** The effect of a single pulse on an oscillator with a coexisting limit cycle and stable equilibrium point depends on both amplitude and phase of stimulation. A pulse applied when the point is in position 1 just resets the phase; the pulse of the same amplitude applied at position 2 puts the phase point inside the basin of attraction of the stable equilibrium point (unfilled circle) thus quenching the oscillation; the basin boundary is shown by the solid curve.

refractory, and is therefore blocked. Hence, an $n : (n - 1)$ rhythm is observed. Because excitable systems are not self-oscillatory, we prefer to classify these regimes not as synchronization but as a *complex nonlinear resonance*.

### 3.6.3 Stochastic resonance from the synchronization viewpoint

Except for the excitable systems discussed above, we have always considered external forcing of systems that exhibit oscillations in the absence of forcing. To this end we distinguished these self-sustained systems from driven oscillators. An autonomous self-sustained system has an arbitrary neutrally stable phase, which can be easily adjusted by an external action or due to coupling with another oscillator. Contrary to this, in driven systems the phase of oscillations is unambiguously related to the phase of forcing, and cannot be easily shifted due to external action or coupling.[27] Accordingly, driven systems cannot be synchronized.

In this subsection we describe a class of oscillators which, in a certain sense, are in between self-sustained and driven systems. Namely, we study *noise-induced* oscillations, excited by rapidly fluctuating forces. Contrary to the case of noise-driven periodic oscillators discussed in Section 3.4, here the noise plays a crucial role in the dynamics: without fluctuations there are no oscillations at all. On the other hand, such a system exhibits oscillations without external periodic forcing, and that makes it different from an excitable system. In this sense these systems resemble the active, self-sustained ones, and we can therefore expect to observe synchronization-like phenomena. As is well-known, periodic forcing of a large class of noise-driven systems demonstrates the effect of *stochastic resonance*, which has been observed in many experiments (see the review by Gammaitoni *et al.* [1998]). We first describe the

---

[27] In the language of the theory of dynamical systems, the difference can be represented as the existence of zero Lyapunov exponent in the autonomous case, and the nonexistence of such an exponent in the forced case.

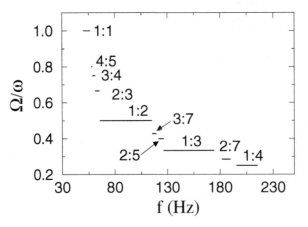

**Figure 3.47.** Devil's staircase picture for the stimulated giant squid axon. Plotted using data from [Matsumoto *et al.* 1987].

common phenomenological features of the stochastic resonance and synchronization, and then discuss the physical reasons for this analogy.

## Threshold systems

We start with a model that resembles an integrate-and-fire oscillator. Consider a system where a variable does not grow monotonically in a regular fashion, but fluctuates randomly (Fig. 3.48a) and from time to time crosses the threshold.[28] Let us count each such crossing and denote it as an event, or spike. Such models are believed to describe the functioning of some sensory neurons that in the absence of stimulation fire randomly.

If the threshold level is varied periodically (Fig. 3.48b), the system exhibits stochastic resonance; its manifestation is a typical structure in the distribution of interspike intervals [Moss *et al.* 1994, 1993; Wiesenfeld and Moss 1995; Gammaitoni *et al.* 1998], see Fig. 3.48c. Indeed, the system fires with high probability when the threshold is at its lowest, and therefore the interval between the spikes is most likely to be equal

---

[28] These models are sometimes called leaky integrate-and-fire models [Gammaitoni *et al.* 1998].

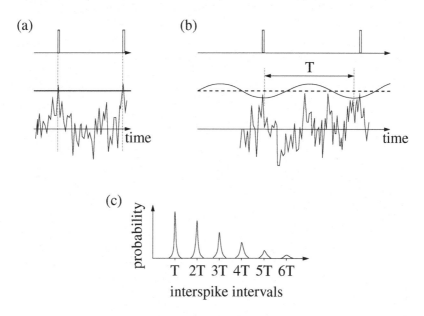

**Figure 3.48.** (a) Schematic representation of a noisy threshold system. This system produces a spike each time the noise crosses the threshold. (b) If the threshold is varied by a weak external signal, the firing predominantly occurs at a certain phase of the force, when the threshold is low. This results in the characteristic structure of the distribution of interspike intervals shown in (c). Such distributions have been observed in experiments with periodic stimulation of the primary auditory nerve of a squirrel monkey [Rose *et al.* 1967] and a cat [Longtin *et al.* 1994], as well as in periodically stimulated crayfish hair cells [Petracchi *et al.* 1995].

to the period of external force or its multiple. Thus, there is a correlation between the temporal sequence of neural discharge and the time dependence of the stimulus. In the neurophysiological literature this correlation is termed phase locking (see, e.g., [Rose *et al.* 1967; Longtin and Chialvo 1998]).

### Bistable systems

Another popular model is a system with two stable equilibria, where the noise induces transitions from one state to the other. It is appropriate to represent this as motion of a particle in a potential that has two minima (Fig. 3.49a). There are many physical and biological phenomena that can be described by such a model. We have chosen for illustration an experiment carried out by Barbay *et al.* [2000] with a laser that can emit two modes with different polarization, corresponding to two steady states. Due to noise, the laser radiation switches irregularly from one state to the other; for simplicity of presentation we continue speaking of a "particle" in a bistable potential. This potential, reconstructed from the measurements, is shown in Fig. 3.49b.

The rate of transition of the particle between the two wells depends on the noise intensity. Small fluctuations induce rare jumps; with increasing noise intensity the transition rate grows. Finally, if the noise is very strong, the particle does not "feel" the potential minima, and moves back and forth in an erratic manner. A weak external force modifies the bistable potential,[29] making one state more stable than the other (Fig. 3.49a). Inter-well jumps are more likely when the barrier is low; consequently, for

---

[29] The weakness of the force means that it cannot cause any transition if noise is absent.

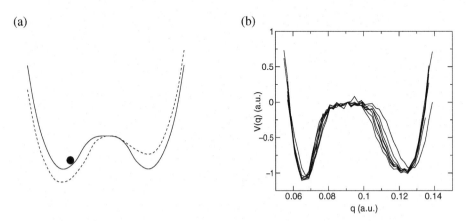

**Figure 3.49.** (a) Motion of a particle in a double-well potential. Noise induces irregular transitions between two stable states. A weak periodic force acting on the system modifies the potential so that the height of the wells is varied periodically (one snapshot is shown by a dashed line). (b) The effective bistable potential $V(q)$ obtained from the data of a laser experiment (the spatial coordinate and the potential height are in arbitrary units). Part (b) from Barbay *et al.*, *Physical Review* E, Vol. 61, 2000, pp. 157–166. Copyright 2000 by the American Physical Society.

a certain range of noise intensities, noise-induced oscillations appear in approximate synchrony with the force (Fig. 3.50).[30] This manifestation of *stochastic resonance* has been observed in many experiments (see, e.g., [Gammaitoni *et al.* 1998]). In a sense, stochastic resonance can be interpreted as synchronization of noise-induced oscillations [Simon and Libchaber 1992; Shulgin *et al.* 1995]. Recently, the notion of phase was also used to characterize this phenomenon [Neiman *et al.* 1998, 1999c]; in this context one jump of the particle corresponds to an increase in phase by $\pi$.

A similarity of the synchronization features of such different systems as self-sustained oscillators and stochastic bistable systems can be understood in the following way. We have thoroughly discussed the fact that the ability of the former to be synchronized is related to the fact that their phase is free (and therefore it can be easily shifted), whereas the phase of the periodically driven system is not (see Fig. 2.9 and Section 2.3.2). This follows from the fact that for autonomous self-sustained oscillators all the moments of time are equivalent, while for the driven systems they differ

[30] Note the similarity in the dynamics of periodically driven bistable and threshold stochastic systems: in both cases the force modulates the barrier, thus promoting transitions at certain phases of the force.

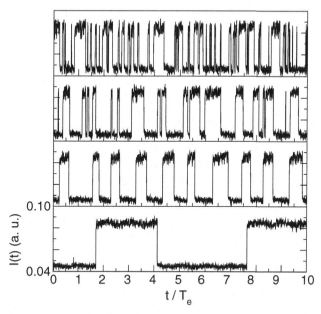

**Figure 3.50.** Time series of the polarized output laser intensity in the experiments of Barbay *et al.* [2000] in the presence of an external periodic force of period $T_e$. Different curves correspond to different noise intensities (increasing from bottom to top). There is no synchronization for weak and strong noise, but for medium noise (the second curve from the bottom) the transitions are nearly perfectly synchronized to the external force (the reader can simply correspond the external force to the tick marks). From Barbay *et al.*, *Physical Review* E, Vol. 61, 2000, pp. 157–166. Copyright 2000 by the American Physical Society.

because the force varies with time. The latter is certainly also true for noise-driven bistable systems: the position of the particle (and therefore the phase) is determined by the fluctuating force.

The key observation allowing one to describe stochastic resonance as synchronization is the existence of two time scales. One (microscopic) scale is related to the correlation time of the noise; it is small. The other one (macroscopic) is the characteristic time between the macroscopic events (spikes in a threshold system, jumps in a bistable system); it is much larger than the correlation time of the noise. We are interested in the jumps (spikes) and the times when they occur. The difference between the two time scales makes it possible for characteristic macroscopic events to occur at any time. For example, between two crossings of the threshold in Fig. 3.48a the process many times approaches the threshold value but does not reach it. By slightly changing the threshold at some instant of time, we can thus produce a spike at this time. This means that the phase of the macroscopic process can be shifted by weak action, and this is exactly the property yielding synchronization.

### 3.6.4    Entrainment of several oscillators by a common drive

In this section we consider a rather simple consequence of the effect of synchronization by external forcing: the appearance of coherent oscillations in an ensemble of oscillators driven by a common force (field).

**Coherent summation of oscillations. An example: injection locking of a laser array**

Suppose that there is a number of similar self-sustained oscillators influenced by a common external force (Fig. 3.51). Let these oscillators be different but have close frequencies. Then, a periodic external force can synchronize all or almost all oscillators in the ensemble. As a result, the oscillators will move coherently, with the same frequency (but, probably, with a phase shift).

This simple fact suggests a practical application of the synchronization effect: it can be used to achieve a coherent summation of the outputs of different generators. For example, this is a way to obtain a high-intensity laser beam. Modern technology allows the manufacture of arrays of semiconductor lasers situated on a common crystal, these

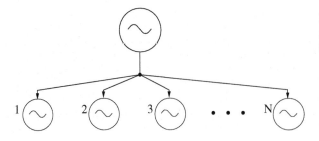

**Figure 3.51.** Ensemble of similar oscillators influenced by a common external force.

devices being comparatively cheap, but not powerful. Just a summation of the outputs of a number of lasers does not help in increasing the power. Indeed, even if the lasers were completely identical and lased with exactly the same frequencies, their oscillation would have different phases.[31] Then, at a certain instant of time, some oscillations would be positive and some would be negative, so that generally they annihilate each other. The difference in frequencies facilitates that the sum of a large number of oscillations remains relatively small.[32] Hence, to obtain a high-intensity output one has to achieve lasing not only with a common frequency, but also with close phases, i.e., the lasers should be synchronous. This can be implemented if one laser is used in order to synchronize all others. Such a mechanism, known in optics as *injection locking*, is quite effective (see [Buczek *et al.* 1973; Kurokawa 1973] for details). Goldberg *et al.* [1985] reported experiments in which they used less than 3 mW of injected power to generate 105 mW output power from a ten-element laser diode array (Fig. 3.52).

This mechanism of coherent summation of oscillations can be regarded as a technique to detect a weak signal. Indeed, the array of lasers can be considered an "amplifier" that provides a powerful output at the frequency of the weak input. Certainly, this amplifier responds only at signals with a certain frequency (the detuning of the majority of the ensemble elements should be not too large).

It is probable that a similar principle is realized by living organisms to increase the sensitivity of sensory organs. Such organs are known to consist of a large number of neurons firing at different rates. In the case of external stimulation, the group of neurons with frequencies close to that of the stimulus can be entrained and therefore

---

[31] Remember that the phase of a self-sustained oscillator is arbitrary; it depends on how the oscillator was "switched on", i.e., on the initial conditions. Therefore, the oscillators in an ensemble would generally have different phases.

[32] In a noncoherent summation of oscillations the sum is proportional to the square root of a number of oscillations, while for a coherent summation it is proportional to this number.

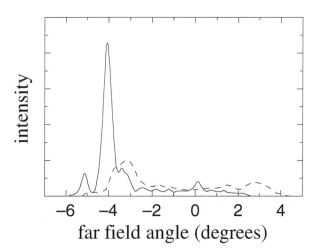

**Figure 3.52.** Injection locking of a laser array. Summation of the outputs of independent lasers provides a rather low intensity of the light field (dashed curve). If the array is illuminated by a master laser beam, a high-intensity output is achieved due to synchronization of individual lasers (solid curve). From [Goldberg *et al.* 1985].

provide the brain with a strong signal. The other groups respond to stimulation at other frequencies, and hence a weak signal can be revealed.

## An example: circadian oscillations in cells

External forcing of an ensemble of nonidentical and noninteracting oscillators results in the appearance of a macroscopic oscillation. Clearly, if the force ceases to exist, this oscillation dies out because the elements now oscillate with their own frequencies – after some time their phases diverge and the signals cancel each other. Consider the experiment described by Whitmore *et al.* [2000] that supports the evidence that the circadian system in vertebrates exists as a decentralized collection of peripheral clocks, and that these peripheral oscillators can be directly influenced by light. The observable here is naturally the sum of the oscillations produced by different cells.

Whitmore *et al.* [2000] studied *Clock* gene expression in the peripheral organs (heart and kidney) of zebrafish. They placed two groups of organs into culture, one group in constant darkness and the other in a light–dark cycle. In both groups the *Clock* expression continued to oscillate, as it was *in vivo*. In the organs exposed to the light–dark cycle, however, the amplitude and robustness of *Clock* oscillation were greater than those of the group in the dark. Exposure to the reversed cycle (i.e., phase shift of $\pi$ of the force) resulted in the entrainment to the new phase.

Circadian oscillations have also been reported in immortalized cell lines. One zebrafish embryonic cell line showed a constant level of *Clock* expression in the darkness. However, when these cells were exposed to a light–dark cycle, an oscillation in *Clock* level was apparent on the first day of the regime, with a relatively broad and low-amplitude peak (i.e., the maximum was spread over several hours). By the second day of the cycle, the peak of expression had consolidated at a certain phase of the force, and the amplitude of the rhythm had increased. When these cells were returned to darkness the oscillation continued for two cycles, but again with a reduced amplitude and broadening of expression. By the third day the oscillation was no longer apparent. The continuation of rhythmicity in the darkness following exposure to a light–dark cycle supports the hypothesis that these cells contain a clock that is entrained by light.

Whitmore *et al.* [2000] concluded that the absence of rhythmic expression in the dark indicates that the light–dark cycle either synchronizes single-cell oscillators with a random phase distribution or initiates circadian oscillation that was not functional before light exposure. Single-cell imaging of cells expressing a fluorescent reporter gene under the control of a clock-regulated promoter may help to resolve this issue.

## An example: synchronization of the mitotic cycle in acute leukaemia

A peculiar mechanism of synchronization was implemented by Lampkin *et al.* [1969] in a treatment of a five-year old boy with acute lymphoblastic leukaemia. The idea is that most chemotheurapeutic agents effective in this disease act principally on one phase of the mitotic cycle. Hence, therapy might be more effective if leukaemic cells were synchronized in a sensitive phase to produce the most susceptible cells for another mitotic cycle-dependent drug.

Normally, the proliferating cells of the leukaemic population are found to be randomly distributed throughout the mitotic cycle. Synchronization was achieved by injecting a certain drug (cytosine arabinoside) that slows or stops DNA synthesis. Cells newly entering the DNA synthesis phase are affected by the drug, and so they are trapped in this part of the mitotic cycle. Thus there is progressive accumulation of cells in the S phase of the mitotic cycle. As the effect of the drug wore off, DNA synthesis resumed in all these cells and they were synchronized. Then, another drug (vincrestine) was administrated.

Such a double-step treatment was repeated three times with a one-week interval, and after that the patient was in complete remission; at the time of writing the paper [Lampkin *et al.* 1969] he was receiving maintenance therapy and was still in remission.

Recent investigations reveal that this mechanism of synchronization plays an important role in mitosis (nuclear division). This process can be likened to a symphony in which many instruments, working individually, are coordinated by the conductor. These conductors, called "checkpoints" control the transitions between different mitotic stages and prevent errors in chromosome segregation that can lead to diseases such as Down's syndrome and cancer [Cortez and Elledge 2000; Scolnick and Halazonetis 2000].

# Chapter 4

# Synchronization of two and many oscillators

In Chapter 3 we studied in detail synchronization of an oscillator by an external force. Here we extend these ideas to more complicated situations when two or several oscillators are interrelated.

We start with two mutually coupled oscillators. This case covers the classical experiments of Huygens, Rayleigh and Appleton, as well as many other experiments and natural phenomena. We describe frequency and phase locking effects in these interacting systems, as well as in the presence of noise. Further, we illustrate some particular features of synchronization of relaxation oscillators and briefly discuss the case when several oscillators interact. Here we also discuss synchronization properties of a special class of systems, namely rotators.

This chapter also covers synchronization phenomena in large ordered ensembles of systems (chains and lattices), as well as in continuous oscillatory media. An interesting effect in these systems is the formation of synchronous clusters.

We proceed with a description and qualitative explanation of self-synchronization in large populations of all-to-all (globally) coupled oscillators. An example of this phenomenon – synchronous flashing in a population of fireflies – was described in Chapter 1; further examples are presented in this chapter. We conclude this chapter by presenting diverse examples.

## 4.1    Mutual synchronization of self-sustained oscillators

In this section we discuss synchronization of mutually coupled oscillators. This effect is quite similar to the case of external forcing that we described in detail in Chapter 3. Nevertheless, there are some specific features, and we consider them below. We also briefly mention the case when several oscillators interact.

## 4.1.1  Two interacting oscillators

Synchronization was first discovered in two mutually coupled oscillators. We have already described in Chapter 1 the observation of interacting pendulum clocks by Christiaan Huygens and of organ-pipes by Lord Rayleigh. Here, we explain these and other experiments using the ideas and notions introduced in the previous chapter. First, we discuss the adjustment of frequencies.

### Frequency locking

Generally, the interaction between two systems is nonsymmetrical: either one oscillator is more powerful than the other, or they influence each other to different extents, or both. If the action in one direction is essentially stronger than in another one, then we can return to the particular case of external forcing. We know that in this case the frequency of the driven system is pulled towards the frequency of the drive. The main point in a bidirectional interaction is that the frequencies of both oscillators change. Let us denote the frequencies of the autonomous systems (often called partial frequencies) as $\omega_1$ and $\omega_2$, and let $\omega_1 < \omega_2$; the observed frequencies of interacting oscillators we denote as $\Omega_{1,2}$. Then, if the coupling is sufficiently strong, frequency locking appears as the mutual adjustment of frequencies, so that $\Omega_1 = \Omega_2 = \Omega$, where typically $\omega_1 < \Omega < \omega_2$, see Fig. 4.1.[1]

### Phase locking

Frequency locking implies a certain relation between the phases that depends not only on the frequency detuning and coupling strength, but also on the way in which the systems are interacting. In the Introduction we mentioned the experiments with pendulum clocks made by Blekhman and co-workers [Blekhman 1981] who reported both anti-phase (phase difference nearly $\pi$) and in-phase (phase difference nearly zero) synchronization of clocks.[2] We remind the reader that the discoverer of synchronization, Christiaan Huygens, observed synchronization in anti-phase.

Consider two nearly identical, symmetrically coupled oscillators. If the interaction is weak, then, in full analogy to the case of external forcing, we can assume that it influences only the phases, shifting the points along the limit cycles, but not the amplitudes. The interaction depends in some way on the two phases, and the two simplest cases are when coupling either brings the phases together (Fig. 4.2a), or moves them apart (Fig. 4.2b). Clearly, the phase-attractive interaction leads to in-phase synchronization, whereas the phase-repulsive one results in anti-phase (out-of-phase)

---

[1]  Two oscillators may be coupled in a rather complicated way. For example, two electronic generators can be connected via a resistor and additionally interact via overlapping magnetic fields of the coils. Thus, generally, the coupling is characterized by several parameters. In the case of such complex coupling the frequency in the synchronous state can also lie outside the frequency range $[\omega_1, \omega_2]$.

[2]  Two nearly identical clocks synchronize in anti-phase if the natural vibration frequency of the beam is not much different from the frequencies of the clocks; otherwise, both regimes are possible (see [Blekhman 1971, 1981; Landa 1980] where this problem is treated analytically).

synchronization. Using the same argument as for the case of an externally forced oscillator (see Section 3.1), one comes to the conclusion that detuning makes the phase difference not exactly zero (not exactly $\pi$).

### High-order synchronization

Generally, when the frequencies of the uncoupled systems obey the relation $n\omega_1 \approx m\omega_2$, synchronization of order $n : m$ arises for sufficiently strong coupling. The frequencies of interacting systems become locked, $n\Omega_1 = m\Omega_2$, and the phases are also related. The condition of phase locking can be formulated as

$$|n\phi_1 - m\phi_2| < \text{constant}, \qquad (4.1)$$

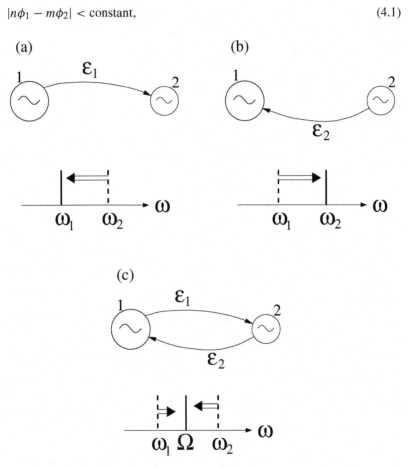

**Figure 4.1.** Adjustment of frequencies of two interacting oscillators ($\omega_1$ and $\omega_2$ are their natural frequencies). If the coupling goes in one direction (a and b), the frequency of the driven system (dashed vertical bar) is pulled towards the frequency of the drive. This is equivalent to the case of external forcing. If the interaction is bidirectional ($\varepsilon_{1,2} \neq 0$) then the frequencies of both systems change (c); the common frequency $\Omega$ of synchronized oscillators is typically in between $\omega_1$ and $\omega_2$. The frequency diagrams can be regarded as schematic drawings of the *power spectra* of oscillations.

(cf. Eq. (3.3) for the case of an externally driven oscillator). A phase shift between the oscillators depends on the initial detuning of interacting systems and the type and parameters of the coupling.

## 4.1.2   An example: synchronization of triode generators

E. V. Appleton [1922] systematically studied synchronization properties of triode generators in a specially designed experiment. He investigated both synchronization by an external force, and mutual synchronization of two coupled nonidentical systems. The setup of the latter experiment is sketched in Fig. 4.3. Each generator consists of an amplifier (triode vacuum tube), an oscillatory $LC$-circuit and a feedback implemented by means of the second inductance. This coil, submitting a signal proportional to the oscillation in the $LC$-circuit to the grid, therefore connects the output and input of the amplifier.

There are several ways to couple two triode generators. For example, they can be coupled via a resistor. In his experiments, Appleton placed the coils nearby so that their magnetic fields overlapped, and, hence, the currents in the $LC$-circuits influenced each other.

The experiment was carried out with oscillators having low frequencies of $\approx 400$ Hz. The frequency of one system was varied by tuning a capacitor. The effect of detuning was followed in two ways. First, the Lissajous figure was observed on the screen of the oscilloscope indicating the equality of frequency for a certain range of detuning. The phase shift between the synchronized generators was estimated from the

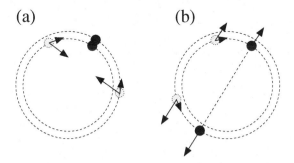

**Figure 4.2.** Two mutually coupled oscillators with phase-attractive (a) and phase-repulsive (b) interaction. The sketch shows the phases of two systems with equal natural frequencies; they are shown in a frame rotating with frequency $\omega = \omega_{1,2}$. Attraction (repulsion) of phases (shown by arrows) results in in-phase (anti-phase) synchronization. The components of the forces that push the phase points along the limit cycle are also shown by arrows. In the case of nonzero detuning (i.e., for nonidentical oscillators), the phase difference in the locked state is not exactly zero (or not exactly $\pi$); an additional phase shift arises, so that the interaction compensates the divergence of phases due to the difference in frequencies (cf. Fig. 3.4).

Lissajous figures. Second, the beat frequency was measured in a rather simple way: the beats were so slow that Appleton was able to count them by ear. The beat frequency ($|\Omega_1 - \Omega_2|$) is depicted in Fig. 4.4 as a function of the readings of the tuning capacitor (arbitrary units), i.e., as a function of detuning.

We note that the phase difference across the synchronization region varied from 0 to $\pi$, attaining $\pi/2$ for the vanishing detuning. A possible explanation for this is

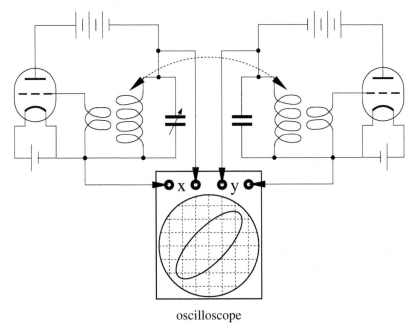

oscilloscope

**Figure 4.3.** Setup of the triode generator experiment by E. V. Appleton [1922]. The dashed arc indicates that the coils were placed in such a way that their magnetic fields overlapped, thus coupling the generators.

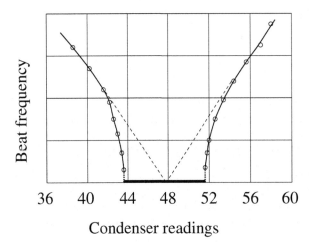

**Figure 4.4.** Results of the experiment with coupled triode generators. If no synchronization effects are taken into account, the theoretical change of the beat frequency with capacity is indicated by the dotted lines. The continuous lines are drawn through the experimental values. Synchronization region is shown by the horizontal bar. From [Appleton 1922].

that the generators in the experiments by Appleton were not only detuned, but also had different amplitudes, so that one generator dominated. Therefore, the features of synchronization in this system are very much like the case of synchronization by external force.

### 4.1.3   An example: respiratory and wing beat frequency of free-flying barnacle geese

The relation between heart rate, respiratory frequency and wing beat frequency of free-flying barnacle geese *Branta leucopsis* was studied by Butler and Woakes [1980]. Two barnacle geese were successfully imprinted on a human and trained to fly behind a truck containing their foster parent. A two-channel radio-transmitter was implanted into these geese so that heart rate and respiratory frequency could be recorded before, during and after flights of relatively long duration. Air flow in the trachea and an electrocardiogram were acquired and transmitted to the receiver, and ciné film was taken. The wing beat frequency was obtained from this film.

Butler and Woakes [1980] found no relationship between any of the measured frequencies and flight velocity of the birds, or between heart rate and respiration. There was, however, a 3 : 1 correspondence between the wing beat frequency and the respiratory frequency and a tight phase locking between the two. 1 : 3 and possibly 1 : 2 frequency locking is indicated by the distribution of the respiratory frequency shown in Fig. 4.5.

During the majority of the respiratory cycles, there were three wing beats per cycle with the wing always being fully elevated at the transition between expiration and inspiration. The wing beat was tightly locked to fixed phases of the respiratory cycle and, on average, the wing was fully elevated at $\approx 6$, 40.5 and 74%

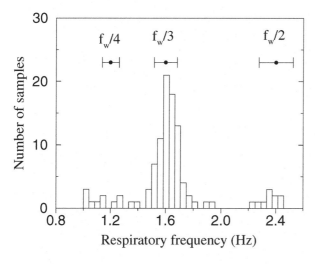

**Figure 4.5.** Histogram showing the distribution of values of respiratory frequency of barnacle geese during flight. Each individual measurement was the average frequency of six successive cycles. At the top of the plot the mean values $\pm$ standard deviation of the wing beat frequency $f_{\mathrm{w}}$ are shown, divided by integers corresponding to different locking ratios. From [Butler and Woakes 1980] with the permission of Company of Biologists Ltd.

of the respiratory cycle (Fig. 4.6). Phase relationships are present during flights of rather long duration and can be maintained even during transient changes in frequency of one of the processes. This corresponds to the results of other studies cited by Butler and Woakes [1980] where 1 : 1 and even as high as 1 : 5 locking was found for other species of birds. It was also found that the wings were fully elevated at the beginning of inspiration and that the beginning of expiration occurred at the end of the downstroke. Whether or not such phase locking confers any mechanical advantage to the process of lung ventilation remains an open question.

We note that Fig. 4.6 can be regarded as a stroboscopic plot: here the phase of respiration is observed at a fixed phase of the wings' motion (when the wings are fully elevated). The *phase stroboscope* is a very effective tool for detection of phase interrelation from measured data; we use it extensively in Chapter 6.

### 4.1.4   An example: transition between in-phase and anti-phase motion

With the following example we want to illustrate the above mentioned feature of mutual synchronization: for the same systems, the phase difference in the synchronous state may attain different values, depending on how the coupling is introduced.

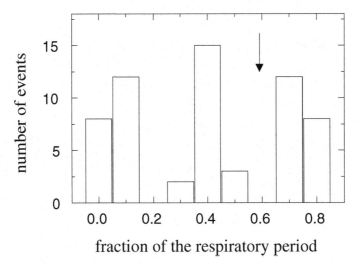

**Figure 4.6.** Histogram showing the position during the respiratory cycle at which the wings were fully elevated (called "events") during the flight of a barnacle goose. Twenty respiratory cycles were chosen, each of which occupied ten ciné frames. Each interval of the histogram therefore represents one frame. Frame zero is that in which the mouth was seen to open; the arrow marks the phase when the mouth closed. From [Butler and Woakes 1980] with the permission of Company of Biologists Ltd.

The original observation was made by J. A. S. Kelso; later this effect was studied by Haken, Kelso and co-workers (see [Haken *et al.* 1985; Haken 1988] for references and details). In their experiments, a subject was instructed to perform an anti-phase oscillatory movement of index fingers and gradually increase the frequency. It turned out that at higher frequency this movement becomes unstable and a rapid transition to the in-phase mode[3] is observed, see Fig. 4.7 and Fig. 4.8.

The explanation of this phenomenon exploits a hypothesis that rhythmic motion of a finger is controlled by the activity of a corresponding pacemaker in the central nervous system called a central pattern generator (CPG). Correspondingly, synchronous oscillatory motion of two fingers can be understood as locking of two CPGs. The natural assumption is that these generators are interacting; the fact that this interaction involves *coupling with a time delay* is important. Indeed, it is known that the brain receives information on the position and the velocity of limbs; this information is

---

[3] The motion is considered as in-phase when both flexor (extensor) muscles are activated simultaneously.

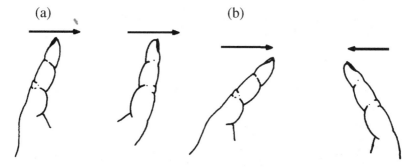

**Figure 4.7.** With an increase in speed, the anti-phase motion of index fingers (a) abruptly changes to the in-phase mode (b). From [Haken 1988], Fig. 11.1, Copyright Springer-Verlag.

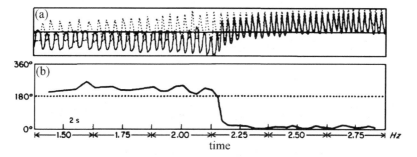

**Figure 4.8.** (a) The time course of the displacement of index fingers of left (solid curve) and right (dashed curve) hands. The subject starts in anti-phase mode and increases the frequency in response to the instructions from the experimentalist. (b) Phase difference between two oscillations. From [Haken 1988], Fig. 11.2, Copyright Springer-Verlag.

provided by special sensors (proprioceptors) located in muscles and tendons. Proprioceptive information is used by respective areas of the brain to control movements and is supplied with some delay. It is natural to suppose that such a signal from a finger influences both its "own" and the "foreign" generator. Suppose also that such a coupling leads to synchronization in, say, anti-phase. For a low frequency, the delay in the coupling loop is much smaller than the oscillatory period and can therefore be neglected. Otherwise, the delay is equivalent to the phase shift of the proprioceptive signal. If the time delay becomes comparable to half the period, it corresponds to a phase shift of $\pi$, which might be responsible for the transition observed in the experiments.

## 4.1.5 Concluding remarks and related effects

We have shown by means of simple argumentation, and we have illustrated it by a description of several experiments, that the effects of phase and frequency locking of two interacting oscillators are quite similar to the case of external forcing. Respectively, the synchronization of two systems can be characterized in the same way by means of frequency vs. detuning plots and families of Arnold tongues (synchronization regions). The onset of synchronization is again the transition from quasiperiodic motion with two incommensurate frequencies to motion with a single frequency; the time course of the phase difference near the border of the synchronization region is the same as for the case of a periodically forced system: intervals of almost constant phase difference are interrupted by relatively rapid phase slips (see Fig. 3.8).

There are also some distinctions between both types. First, for two interacting systems the frequency of the synchronous motion is not constant within an Arnold tongue (only the relation $n\Omega_1 = m\Omega_2$ holds). Indeed, two oscillators mutually adjust their frequencies until they achieve a "compromise"; the resulting frequency depends on the initial detuning. Second, the phase shift in the locked state depends on the type of coupling between the system, in particular, one can definitely distinguish between the in-phase and anti-phase regimes.

### Quenching

Mutual synchronization is not the only effect of interaction. If the coupling is relatively strong, it may cause the **quenching** of oscillation. (This effect is also called "oscillation death".) An example, an early observation by Lord Rayleigh who found that two organ-pipes mutually suppressed their vibration, was given in the Introduction. The reason for such a suppression is that interaction may introduce additional dissipation into the joint system. Imagine that two electronic generators, like those used by Appleton, are coupled via a resistor. The additional energy loss due to the current through this resistor may not be compensated by the supply of energy from the source, and as a result the oscillations die out.

### Multimode systems

Up to this point we have discussed self-oscillatory systems that generate oscillations with a single main frequency; such oscillations are represented by the motion of a point along the limit cycle. Generally, a complex self-sustained oscillator can generate two (or several) processes, called *modes*, which have (generally) incommensurate frequencies. Such a multimode generation can be expected if the dynamics of the oscillator are described by more than two variables, i.e., the evolution of dynamics is represented by the motion of a point in the $M$-dimensional phase space, where $M \geq 3$. For example, a triode generator like that studied by Appleton and van der Pol but with an additional oscillatory circuit is a four-dimensional system, and for certain parameter values it can generate two modes.

It is possible that (depending on the initial conditions) only one mode survives, due to their competition, but it is also possible that both coexist. In this case, one can speak of synchronization of these modes, although one cannot decompose the system into two independent generators. What one can do is to alter the coupling between the modes by varying a parameter of the system. With this variation one can achieve synchronization of the modes, often called *mode locking*. From the synchronization viewpoint, the modes can be considered to be oscillations of separate interacting generators.[4] This mode locking is frequently observed in lasers, see [Siegman 1986].

## 4.1.6  Relaxation oscillators. An example: true and latent pacemaker cells in the sino-atrial node

Two relaxation oscillators can interact in different ways. For example, each oscillator can alter the threshold of the other one; such a mechanism of synchronization was considered by Landa [1980] for two saw-tooth voltage generators. Here we briefly discuss *pulse coupled* integrate-and-fire oscillators, a model that is important for many biological applications.

We suppose that the oscillators are independent during the integration epochs, and that when one of them fires, it instantly increases the slowly growing variable of the other one (cf. the case of pulse forcing shown in Fig. 3.25). It is quite obvious that the fastest oscillator imposes its frequency on the slow partner (certainly, this takes place only if the detuning is not too large). It is also clear that the same mechanism results in synchronization in an *ensemble* of elements, where firing of each one forces all the others to fire as well. This is a particular case of *global coupling* that we discuss in Section 4.3; here we illustrate synchronization of pulse coupled oscillators by the following example.

---

[4]  In the phase space mode locking corresponds to the transition from motion on the torus to motion on the limit cycle that lies on that torus. From this topological viewpoint this is indistinguishable from the case of two coupled oscillators.

We have already mentioned that the sino-atrial node of the heart can be considered a relaxation oscillator that (normally) causes the heart to contract. The frequency of this oscillator is regulated by the autonomous nervous system; in cases of severe pathology, when its frequency drops too low, the sino-atrial node can be entrained by an artificial pacemaker (see Section 3.3).

On a microscopical level the sino-atrial node can be considered as an assembly of a large number of pacemaker cells. When one of these cells (true pacemaker) fires, the current it produces forces the other cells (latent pacemakers) to generate the action potential earlier than they would have done without interaction (Fig. 4.9). Thus, the rhythm generated by the sino-atrial node is formed by synchronous collective discharge of a large number of cells [Schmidt and Thews 1983]; this rhythm is determined by the cells that have the highest frequency [Gel'fand *et al.* 1963]. This makes the functioning of the primary cardiac pacemaker very robust: if, due to some reason, the fastest pacemaker cell fails to fire, the pulse generation will be triggered by some other cell.

### 4.1.7  Synchronization of noisy systems. An example: brain and muscle activity of a Parkinsonian patient

The effect of noise on mutual synchronization of two oscillators is similar to the case of entrainment by an external force (see Section 3.4). The typical features of interacting noisy oscillators are as follows.

■ There is no distinct border between synchronous and nonsynchronous states, the synchronization transition is smeared.

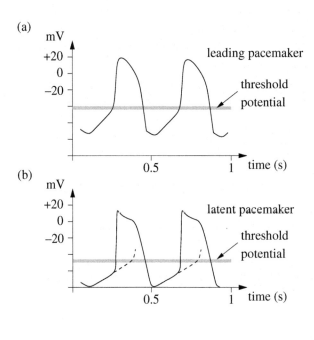

**Figure 4.9.** Synchronization of a true cell and a latent pacemaker cell of the sino-atrial node of the heart. The membrane potential of each cell slowly increases until it reaches the critical value; then, an action potential is generated, i.e., the cell fires. The threshold is first reached by the true pacemaker cell (a), and its discharge makes the latent pacemaker (b) generate the action potential before its threshold level is reached. Without interaction, the latent pacemaker would fire later (dashed curve). From [Dudel and Trautwein 1958], see also [Schmidt and Thews 1983].

■ In a synchronous state, epochs of almost constant but fluctuating phase difference are interrupted by phase slips when the phase difference comparatively rapidly increases or decreases by $2\pi$.

Let us illustrate these comments by an experimental example.

For this purpose we present an analysis of data that reflect the brain and muscle activity of a Parkinsonian patient: the recordings of the magnetic field outside the head noninvasively measured by means of whole-head magnetoencephalography (MEG) [Hämäläinen *et al.* 1993] and simultaneously registered electromyogram (EMG). Details of the experiment, preprocessing of the data and discussion of the results can be found in [Tass *et al.* 1998, 1999; Tass 1999]. The goal of this experiment was to study the electrophysiology of Parkinsonian resting tremor, which is an involuntary shaking with a frequency of around 3–8 Hz that predominantly affects the distal portion of the upper limb and normally decreases or vanishes during voluntary action [Elble and Koller 1990]. As yet, the dynamics of the resting tremor generation remain a matter of debate. It is known that several interacting oscillatory systems are involved [Volkmann *et al.* 1996], but the role of synchronization between these systems in the appearance of this pathology is not clear. In particular, an important open question is whether the cortical areas synchronize their activity during epochs of tremor, and whether the muscle activity becomes synchronized with the brain activity. An analysis of signals coming from two elements of this complex network (Fig. 4.10) is now discussed.

For the particular data, the frequencies of the EMG and MEG signals are $\approx6$ Hz and $\approx12$ Hz, respectively,[5] therefore we expect to find synchronization of order 1 : 2. Figure 4.11a shows the phase difference $\phi_{MEG} - 2\phi_{EMG}$ between the signal from the motor cortex and the EMG.[6] We note that the analyzed system is *nonstationary*, i.e., its parameters (e.g., frequencies of oscillators, strength of the coupling between them) vary with time. A qualitative change occurs at $\approx50$ s, when the tremor activity starts. This nonstationarity is reflected in the behavior of the phase difference: it grows during some epochs (nonsynchronous state), and fluctuates around a constant level at other times (synchronous state). We see that during the synchronous epochs the phase difference resembles a random walk motion, as we expect for noisy oscillators. Due to nonstationarity, the degree of synchronization varies with time. In the time interval shown in Fig. 4.11b the phase difference increases (except for the first 5 s), and the distribution of the cyclic relative phase is broad, which is typical for the absence of synchronization (cf. Fig. 3.38). Within the other time intervals in Fig. 4.11c and Fig. 4.11d the oscillators can be considered synchronized. In Fig. 4.11c we see two phase jumps, and in Fig. 4.11d they are more frequent, so that during the latter interval

---

[5] The frequencies were estimated by spectral analysis. We note also that the data were previously filtered by a bandpass filter in order to extract the relevant signals from the overall activity.

[6] The *instantaneous* phases are computed from MEG and EMG signals with the help of the Hilbert transform based technique. Technical details are extensively discussed in Chapter 6 and in Appendix A2.

the oscillators are "less synchronized" (remembering that for noisy systems there is no well-defined border of synchronization).

Rather rapid $2\pi$ phase jumps that are clearly seen in Fig. 4.11 (a, c and d) are what we call phase slips. Close investigation demonstrates that a slip is not an instant event but takes place within several oscillation periods (Fig. 4.12). Indeed, outside the marked region one EMG cycle corresponds exactly to two cycles of the MEG, whereas inside that region we observe 8 and 17 cycles, respectively.

## 4.1.8    Synchronization of rotators. An example: Josephson junctions

Here we briefly discuss a particular class of systems that are not self-sustained (and are, therefore, not in the mainstream of our presentation). These systems are **rotators**; an example is a disk on a shaft of an electric motor. The rotation frequency can be considered as an analog of the frequency of oscillations, and one can look whether two (or many) coupled rotators adjust their frequencies. (Note that there is no notion of amplitude for such systems.)

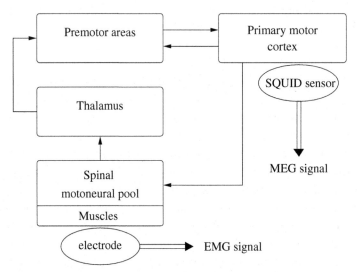

**Figure 4.10.** Generation of Parkinsonian resting tremor involves cerebral areas, spinal cord, peripheral nerves and muscles. As suggested by Volkmann *et al.* [1996] rhythmic thalamic activity drives premotor areas (premotor cortex and supplementary motor area) which drive the primary motor cortex. The latter drives the spinal motoneuron pool which gives rise to rhythmic muscle activity. The peripheral feedback reaches the motor cortex via the thalamus. The activity of the motor cortex is characterized by the magnetic field (MEG signal) registered outside the head by means of a superconductive interference device (SQUID), whereas the activity of the spinal motoneuronal pool is reflected in the electrical activity of the muscles measured by surface electrodes (EMG signal). By analyzing these signals we address the interaction between the corresponding oscillators.

Already at the end of the nineteenth century it was known that at certain conditions generators of electrical power, acting on a common load, synchronize. This effect has an extremely high practical importance for the normal functioning of electricity networks. Indeed, loss of synchronization by one of the generators leads to serious disruptions or even to catastrophe. Theoretically this was studied by Ollendorf and Peters [1925–1926].

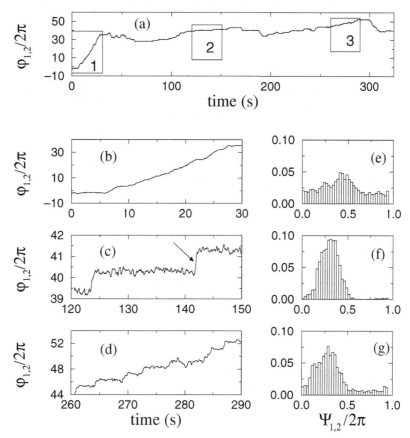

**Figure 4.11.** (a) The time course of the 1 : 2 phase difference $\varphi_{1,2} = \phi_{MEG} - 2\phi_{EMG}$ demonstrates the behavior that is typical of noisy interacting oscillators. Three 30 s intervals marked by boxes 1, 2 and 3 are enlarged in (b), (c) and (d), respectively. These intervals are essentially different due to the nonstationarity of the system (the tremor activity starts at $\approx 50$ s). Indeed, within the first 30 s the oscillators can be considered as nonsynchronized (b): the phase difference increases and the distribution of the cyclic relative phase $\Psi_{1,2} = \varphi_{1,2} \bmod 2\pi$ is almost uniform (e). The intervals shown in (c) and (d) and the corresponding unimodal distributions of $\Psi$ in (f) and (g) reveal synchronization of interacting systems, although the strength of synchronization is different. The phase slip indicated by an arrow in (c) is enlarged in Fig. 4.12. From [Rosenblum *et al.* 2000], Fig. 2, Copyright Springer-Verlag.

In the 1940s I. I. Blekhman and co-workers observed, in an experiment, synchronization of disbalanced rotors driven by motors: two rotors placed on a common vibrating base rotate with the same frequency. Indeed, as the rotors are disbalanced, they force the base to vibrate with the frequencies of their rotation, so that the systems "feel" each other. This effect of mutual synchronization of rotating electromechanical systems has found a number of applications in mechanical engineering (see [Blekhman 1971, 1981; Blekhman *et al.* 1995] for description, theory and original references).

Recently much attention has been paid to the superconductive electronic devices, Josephson junctions. It turns out that if such a junction, shunted by a capacitor, is fed by a constant external current, then its dynamics are described by the same equations as the dynamics of a mechanical pendulum driven by a constant torque (see Section 7.4). Therefore, the Josephson junction can be regarded as a kind of "electrical rotator". Experiments show that "rotations" can be synchronized by an external periodic current (Fig. 4.13), or by coupling of two junctions (Fig. 4.14). A review of these phenomena can be found in [Jain *et al.* 1984].

On a qualitative level, synchronization of rotators can be explained by demonstrating a certain similarity to oscillations. First, we note that the state of a rotator is unambiguously determined by two quantities, the angle and the rotation velocity. As the angle is a $2\pi$-periodic variable, the phase space of a rotator is not a plane, as in the case of simplest oscillators, but a cylindrical surface. Rotation caused by a constant torque corresponds to the motion of a point along a closed curve on that cylinder. This curve has the same property of neutral stability with respect to the perturbations along the trajectory as the limit cycle of a self-sustained oscillator.[7] Consequently, rotation can be entrained.

---

[7]  In other words, the closed trajectory on the cylinder has a zero Lyapunov exponent.

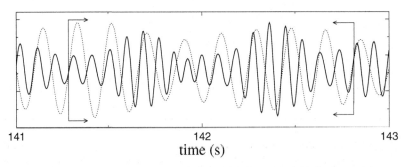

141                                                  142                                                  143
time (s)

**Figure 4.12.** Phase slip takes several periods. The slow signal (dashed curve) is EMG, the fast signal (bold curve) is MEG (both are in arbitrary units). Within the marked region there are 8 and 17 cycles of EMG and MEG, respectively. On a larger time scale, the corresponding increase in the phase difference appears as a jump (cf. Fig. 4.11).

## 4.1.9   Several oscillators

Synchronization can also be observed if more than two oscillators interact. We postpone discussion of the case of large ensembles to Sections 4.2 and 4.3. Here we just make a note on the synchronization of several systems.

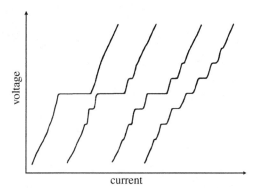

**Figure 4.13.** Voltage–current characteristics of a periodically driven Josephson junction. The leftmost curve corresponds to the autonomous junction, whereas the three other curves are obtained for increasing external force. (These curves are arbitrarily shifted along the current axis for better visualization.) The current and the voltage correspond to the torque and the rotation frequency of the rotator, respectively. For small current (torque) there are no rotations because of the dissipation, so that the voltage is zero (see the plateau in the first curve). Larger current causes rotation in different directions, depending on its sign; this is reflected by nonzero voltage. If an additional periodic external current is applied, additional plateaux are observed; they correspond to regions of frequency locking to the frequency of external current. The stronger the forcing, the larger the number of synchronization regions that are seen. These plateaux are known as Shapiro steps [Shapiro 1963]. From [Kulik and Yanson 1970].

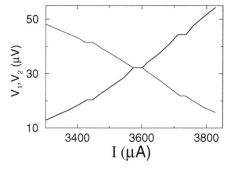

**Figure 4.14.** Voltage–current characteristics of two coupled Josephson junctions exhibit mutual synchronization. Regions of $1 : 1$ and $1 : 2$ locking where $nV_1 = mV_2$ are seen as the current is varied (the voltage is proportional to the rotator's frequency). Reprinted from *Physics Reports*, 1984, Vol. 109, A. K. Jain, Mutual phase-locking in Josephson junction arrays, pp. 309–426. Copyright 1984, with permission from Elsevier Science.

The features of synchronization depend on the number of oscillators as well as on the strength and the type of interaction between each pair. Thus, the problem is characterized by many parameters, and different synchronous states are possible, not only in-phase and anti-phase locking. Moreover, complicated forms of coupling may lead to a coexistence of several stable phase configurations. In this case the coupled system is multistable: which of these configurations is realized depends on the initial conditions (in the phase space this corresponds to existence of several attractors with different basins).

We do not approach this general problem (some cases are considered in [Landa 1980; Collins and Stewart 1993]) but just present an example to demonstrate that even the behavior of a few identical oscillators is rather complex. So, already for three oscillators arranged in a ring, three stable synchronous configurations are possible depending on the parameters of the coupling:[8] (i) all elements are synchronized in-phase; (ii) they are shifted in phase by $2\pi/3$ with respect to the neighbor; (iii) two oscillators have the same phase and the third element has a different one [Landa 1980].

As another illustration we consider four identical oscillators coupled in a ring (Fig. 4.15a). If the coupling is phase-attractive, then all generators would synchronize with the same phase (Fig. 4.15b). With phase-repulsive coupling the elements form two in-phase synchronized pairs (Fig. 4.15c), whereas the symmetric configuration with the phase difference $\pi/2$ between the neighbors is unstable.

The fact that several synchronized oscillators may exhibit different patterns of phase shifts was used in a hypothesis that this property is exploited by the central nervous system to implement different gaits [Collins and Stewart 1993; Strogatz and Stewart 1993]. According to this hypothesis, each leg is controlled by a corresponding oscillator (central pattern generator or CPG). In the case of bipedal gaits, in-phase and

[8] The interaction in this example corresponds to the case when the electronic generators are coupled both via resistors and capacitors and the coupling is generally not symmetric. Hence, it is characterized by several parameters.

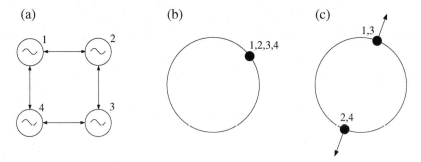

**Figure 4.15.** (a) Four identical oscillators arranged in a ring so that each system interacts with its two neighbors. (b) For a phase-attractive coupling, all oscillators are locked in-phase. (c) Phase-repulsive coupling results in the configuration when noninteracting elements (1 and 3, 2 and 4) have the same phases, and the interacting neighbors are in anti-phase.

anti-phase locking of these generators correspond to hopping and walking. Different gaits of quadrupeds (pace, trot, gallop, etc.) correspond to different synchronous states in a network of four systems.

## 4.2 Chains, lattices and oscillatory media

In this section we describe synchronization effects in large spatially ordered ensembles of oscillators. This means that the systems are arranged in a regular spatial structure. The simplest example is a chain, where each element interacts with its nearest neighbors; if the first and the last elements of the chain are also coupled, then we encounter a ring structure. Generally, both the spatial ordering and the interaction is more complicated, e.g., the oscillators can interact with several neighbors. An interesting particular case, when each element of a large ensemble interacts with all others, is considered separately in Section 4.3. Another special object of our discussion is an oscillatory medium that is continuous in space.

Suppose the oscillators have slightly different frequencies that are somehow distributed over the ensemble. What kind of collective behavior can be expected in such a population? Certainly, if the interaction is very weak, there will be no synchronization so that all the systems will oscillate with their own frequencies. We can also imagine that sufficiently strong coupling can synchronize the whole ensemble, provided the natural frequencies are not too different (Section 4.2.1). For an intermediate coupling or a broader distribution of natural frequencies of elements we can expect some partially synchronous states. Indeed, it may be that several oscillators synchronize and oscillate with a common frequency, whereas their neighbors have their own, different, frequencies (Section 4.2.2). There may appear several such groups, or **clusters** of synchronized elements. These considerations should also be valid for continuous media, where each point can be treated as an oscillator (Sections 4.2.3 and 4.2.4). We illustrate these expectations with a description of several experiments.

### 4.2.1 Synchronization in a lattice. An example: laser arrays

Synchronization of laser arrays is a basic tool used to create a source of high-intensity radiation. This can be achieved by coupling the lasers in a linear array so that they either interact with their nearest neighbors or with all other elements in the structure. Here we present the results of experiments with $CO_2$ wave guide lasers conducted by Glova *et al.* [1996], where five lasers were coupled with the help of a spatial filter that was placed between the array and an outer mirror. Within such a setup, each laser interacts with the other four, but the strength of this interaction varies depending on the distance between the lasers (Fig. 4.16). The results, presented in Fig. 4.17, clearly indicate synchronization. Indeed, if the lasers were not synchronized then the radiation intensity in the far-field zone, as the sum of noncoherent oscillations, would be spatially uniform. The nonuniform distribution in Fig. 4.17 appears because phase locking sets in; this is a typical interference image.

The same team performed an experiment with a multibeam $CO_2$ wave guide laser consisting of 61 glass tubes in a honeycomb arrangement [Antyukhov *et al.* 1986]; this is an example of a two-dimensional lattice (Fig. 4.18a). The lasers were coupled by means of an external coupling mirror. Again, a nonuniform intensity of the radiation far from the array indicates synchronization between the lasers (Fig. 4.18b, c and d).

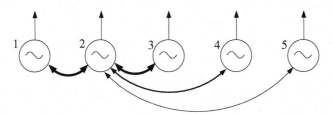

**Figure 4.16.** Schematic representation of a laser experiment by Glova *et al.* [1996]. The coupling strength in the array decreases with the distance between the lasers; this is reflected by the thickness of the arcs that denote coupling (shown only for laser 2).

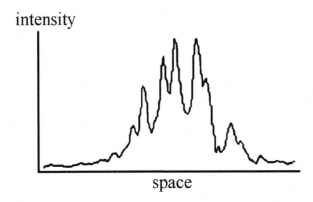

**Figure 4.17.** Radiation intensity in the far-field zone of a weakly coupled laser array. The interference occurs due to the phase locking of individual lasers. From [Glova *et al.* 1996].

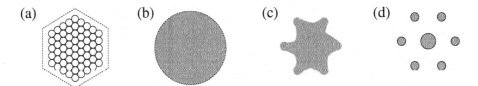

**Figure 4.18.** Synchronization in a lattice of 61 laser oscillators arranged in a honeycomb (a). For low coupling, the intensity in the focal spot of the array output is approximately uniform (b). Stronger coupling results in synchronization that manifests itself as a spatially ordered intensity distribution (d). The case of intermediate coupling is shown in (c). Schematically drawn after [Antyukhov *et al.* 1986].

## 4.2.2 Formation of clusters. An example: electrical activity of mammalian intestine

Intestine consists of layers of muscle fibers supporting propagation of traveling waves of electrical activity that run from the oral to aboral end. These waves trigger the waves of muscular contraction. Diamant and Bortoff [1969] experimentally investigated the distribution of frequencies of electrical activity along the intestine. The majority of experiments was performed on cats, with most of the basic observations being repeated in dogs and rhesus monkeys. Each frequency determination represents the average over a 5 min period.

From the physical viewpoint, if we consider electrical activity only, the intestine can be regarded as a one-dimensional continuous medium, where each point is oscillatory. Indeed, Diamant and Bortoff [1969] found that if a section of intestine is cut into pieces, each piece is capable of maintaining nearly sinusoidal oscillations of a constant frequency. There exists an approximately linear gradient of these frequencies, so that they decrease from the oral to aboral end. Measured *in situ* and plotted as a function of the coordinate along the intestine, the frequency of electrical activity typically exhibits plateaux (Fig. 4.19). This indicates the existence of clusters of synchronous

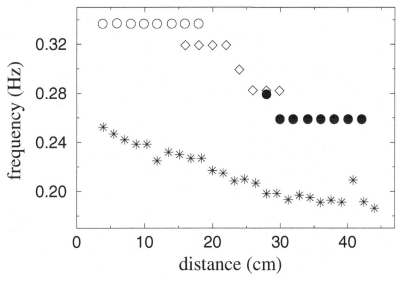

**Figure 4.19.** Synchronous clusters in a mammalian intestine. The frequency of slow electrical muscle activity plotted as a function of distance along the intestine typically shows a step-wise structure. (The distance is measured from the ligament of Treitz.) The ○, ◇, and ● symbols represent the three consecutive (at 30 min intervals) measurements of frequency along the intestine *in situ*; for each measurement the electrodes were re-positioned. The stars show the frequency of the consecutive segments of the same intestine *in vitro*. From [Diamant and Bortoff 1969].

activity.[9] Within each cluster, the phase shift between the oscillations increases with the spatial coordinate (in accordance with the gradient of frequencies in the pieces of intestine); neighboring clusters are separated by regions of modulated oscillations, or beats (Fig. 4.20). In the physiological literature such a time course is often termed a waxing and waning pattern. Note that the clusters primarily form at the oral end of the intestine, where the relative detuning $\Delta f / f$ is smaller.

### 4.2.3   Clusters and beats in a medium: extended discussion

The formation of clusters in an oscillatory medium like intestine is a result of the counter-play of two factors: inhomogeneity in the distribution of natural frequencies and coupling (which is often a consequence of the diffusion that tries to make the states of two points equal, and is therefore called diffusive). Let us consider what happens at the border between two clusters that have different frequencies. Here it is important to distinguish between discrete lattices and continuous media.

[9] Formation of clusters in a chain of loosely coupled van der Pol oscillators with application to modeling the synchronization on a mammalian intestine was numerically studied by Ermentrout and Kopell [1984].

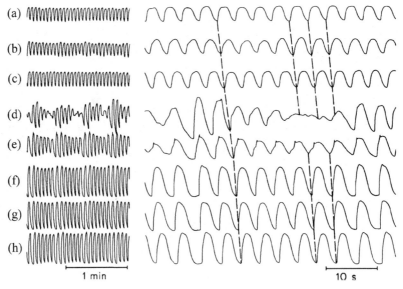

**Figure 4.20.** Electrical activity of the intestine of a cat, measured *in situ* by means of eight equidistantly spaced electrodes. The first three measurement points (a, b and c) belong to one cluster with the oscillation frequency 0.29 Hz. The three last electrodes (f, g and h) reflect the activity within another cluster, with a frequency of 0.26 Hz. In the zone between these two clusters an amplitude modulation of oscillation, or beats, is observed (d and e). Dashed lines show the lines of equal phase. From [Diamant and Bortoff 1969].

In a discrete lattice the border between two clusters is formed by two oscillators having different frequencies. This simply means that these oscillators are not locked: each rotates with its own frequency. Contrary to this, in a continuous medium, if the oscillations at two points have different frequencies, there must be a continuous connection between these two regimes. At first glance, one can just draw a continuous frequency profile between these two points. A closer inspection shows, however, that this is impossible. Indeed, different frequencies mean different velocities of the phase rotation. Hence, the phase difference between the points belonging to two clusters grows in time proportionally to the frequency difference. Therefore, the phase profile becomes steeper and steeper. On the other hand, a continuous steep phase profile means nothing else but a waveform pattern with a small wavelength in the medium. Growth of the phase difference between the clusters leads to shortening of the wavelength with time. It is clear that such a process cannot continue for long – and indeed, the medium finds a way out of this short-wavelength catastrophe. The growing phase gradient is reduced via *space–time defects*. These defects appear when the amplitude of the oscillations vanishes and thus helps to keep the phase gradients finite.

To demonstrate how such a space–time defect occurs, let us assume that the phase difference between the points 1 and 2 belonging to different clusters has reached a value $\approx 2\pi$. If there were no medium between 1 and 2, we would simply consider the states at these points to be nearly identical. But in the medium there is a continuous phase profile between these points. Representing both the amplitude and the phase in polar coordinates, we can depict the field between the points 1 and 2 as a circumference, see Fig. 4.21. Let us now look at the effect of coupling in the medium on this phase and amplitude profile. A typical coupling is diffusive (or at least has a diffusive component); it tends to reduce the differences between the states of the nearest neighbors, i.e., to reduce the difference between the points in Fig. 4.21. The only way to reduce these differences is to make the amplitude of the oscillations smaller. From Fig. 4.21 one can easily see that reduction of the amplitude indeed

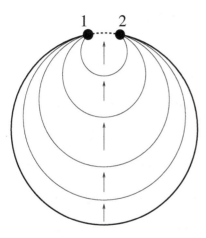

**Figure 4.21.** Illustration of a space–time defect. The initial phase and amplitude profile between points 1 and 2 is shown with a bold solid curve. With time the amplitude reduces and the profile evolves as depicted by the arrows. In the end state (bold dashed curve) the phase difference between points 1 and 2 is nearly zero.

transforms the field profile between 1 and 2 from a large circumference to nearly a point. In the end state the phases at points 1 and 2, that initially differed by $2\pi$, are nearly equal. Thus, one can say that the space–time defect "eats" the phase difference $2\pi$. As is clear from Fig. 4.21, this is only possible if at some spatial point at some moment of time the amplitude of oscillations vanishes. One can also understand this in the following way: if the amplitude is always and everywhere finite, then the phase is always and everywhere well-defined, and there is no way to remove large phase gradients. Contrary to this, in a state with a vanishing amplitude the phase is not defined, thus the only way to restructure the phase profile is to pass through such a state. We refer the reader to Fig. 4.20d for an example of space–time defects in a real system.

### 4.2.4    Periodically forced oscillatory medium. An example: forced Belousov–Zhabotinsky reaction

The Belousov–Zhabotinsky reaction is a well-known example of an oscillatory chemical process (see, e.g., [Kapral and Showalter 1995] and references therein). The time course of the reaction is followed by a periodic change of the color of the medium. Petrov *et al.* [1997] performed experiments with a light-sensitive form of the reaction, using periodic optical forcing. The reaction medium was a thin membrane sandwiched between two reservoirs of reagents. For the conditions of the experiment, the natural frequency (i.e., without forcing) was $f_0 = 0.028$ Hz. If the medium was exposed for some time to bright light that reset all the points to the same initial conditions and was removed afterwards, the reactor oscillated homogeneously for several cycles, so that the natural frequency could be measured. However, the presence of boundaries and small imperfections in the reactor always led to the destruction of homogeneous oscillation and the appearance of rotating spirals.

If the reaction was perturbed with pulses of spatially uniform light at the frequency $f_e$, several $n : m$ synchronous states were observed with variation of $f_e$. In this case, forcing destroys spiral waves and different spatial structures appear. So, for $1 : 1$ locking, the medium oscillates homogeneously with the frequency of the forcing. Within the $1 : 2$ synchronization region the system is bistable: depending on the initial conditions, either two clumps[10] of synchronized oscillations are observed, with a phase shift of $\pi$ between the clumps, or a labyrinthine structure occurs (Fig. 4.22). At $1 : 3$ locking, three homogeneously oscillating clumps, with phases differing by $2\pi/3$ were found. The structure of the Arnold tongues in a periodically forced Belousov–Zhabotinsky system was also investigated by Lin *et al.* [2000].

We note that the reason for clump formation in the forced Belousov–Zhabotinsky medium differs from that of cluster formation in the intestine. In the latter case the

---

[10]  We prefer not to use the term "clusters" here, because different clumps have the same frequency and differ only by their phases.

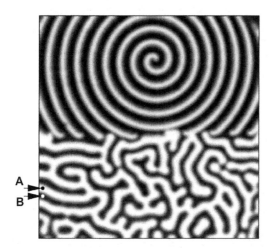

**Figure 4.22.** An example of a labyrinthine 1 : 2 locked pattern. The upper part of the reactor is kept in the dark; a spiral wave existing in this part is typical for an autonomous oscillatory medium. The lower part of the reactor is illuminated with light pulses at the frequency $f_e \approx 2f_0$. The time course of oscillations in point A and B is shown in Fig. 4.23. Reprinted with permission from *Nature* [Petrov *et al.* 1997]. Copyright 1997 Macmillan Magazines Limited.

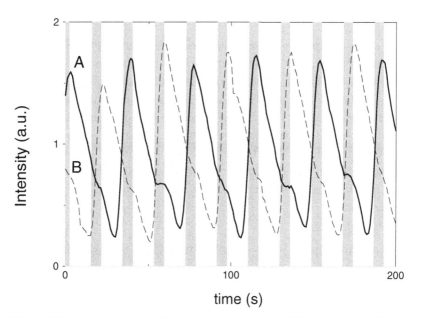

**Figure 4.23.** Time series of medium intensity in point A (solid curve) and B (dashed curve) in Fig. 4.22. The gray-shaded stripes indicate light perturbation. Thus, the white and dark stripes in the labyrinthine structure represent points that are synchronized with the forcing, with a $\pi$ phase shift between these clumps. Reprinted with permission from *Nature* [Petrov *et al.* 1997]. Copyright 1997 Macmillan Magazines Limited.

clustering occurs due to inhomogeneity of the medium (the points with close frequencies try to group). In the present example the mechanism is different: each point tends to synchronize with the external force as well as with the neighboring points. If there were no interaction within the medium (in this case we would come back to the case of many noninteracting oscillators entrained by a common force, Section 3.6.4), then, for $1 : m$ locking, the neighboring points would with equal probability have phase differences $2\pi/m \cdot i$, where $i = 1, \ldots, m - 1$.[11] Because of the interaction, the points tend to have the same phase as the neighbors, and the compromise is achieved via clumping. Obviously, the phase difference between the clumps is $2\pi/m \cdot i$.

### Synchronization of the motion of a spiral wave tip

Another synchronization effect in a continuous medium was studied both experimentally and theoretically by Steinbock *et al.* [1993]. In their setup the oscillatory chemical system (Belousov–Zhabotinsky reaction) was subject to periodically modulated illumination. An interesting regime was observed, when the spiral core followed one of a set of open or closed hypercycloidal trajectories, in phase with the applied illumination.

## 4.3    Globally coupled oscillators

Now we study synchronization phenomena in large ensembles of oscillators, where each element interacts with all others (Fig. 4.24a). This is usually denoted as *global*, or all-to-all coupling. As representative examples we have already mentioned synchronous flashing in a population of fireflies and synchronous applause in a large audience. Indeed, each firefly is influenced by the light field that is created by the whole population. Similarly, each applauding person hears the sound that is produced by all other people in the hall. In another example, singing snowy tree crickets (see [Walker 1969]) are influenced by the songs of their neighbors. Below we explain why such coupling results in synchronization in the ensemble and we illustrate it with further examples.

### 4.3.1    Kuramoto self-synchronization transition

All fireflies are not identical, in the same way that we can consider a community of human beings each having their own personality. Hence, to understand the phenomenon of collective synchrony, we consider an *ensemble of nonidentical oscillators*. We know that a pair of systems can synchronize (if the detuning is not too strong), and therefore we can expect that synchrony extends to a whole population, or at least to a large portion of it.

---

[11]  Indeed, the shift in time of the slower signal for a multiple of the period of the fastest one does not change the phase difference between these signals.

As an auxiliary step, we redraw Fig. 4.24a in the equivalent form that is shown in Fig. 4.24b. Here we just have denoted the driving inputs that come to an oscillator from all others by one input from the whole ensemble. Indeed, we can say that each oscillator is driven by a force that is proportional to the sum of outputs of all oscillators in the ensemble.[12]

The scheme presented in Fig. 4.24b strongly resembles Fig. 3.51. Indeed, in both cases all oscillators are driven by a common force and, as we know already, this force can entrain many oscillators if their frequencies are close. The problem is that in the case of global coupling this force, or the mean field, is not predetermined, but arises from interaction within the ensemble. This force determines whether the systems synchronize, but it itself depends on their oscillation – it is a typical example of self-organization. To explain qualitatively the appearance of this force (or to compute it, as is done in Section 12.1), we consider this problem self-consistently.

First, assume for the moment that the mean field is zero. Then all the elements in the population oscillate independently, and, as we know from Section 3.6.4, their contributions to the mean field nearly cancel each other. Even if the frequencies of these oscillations are identical, but their phases are independent, the average of the outputs of all elements of the ensemble is small if compared with the amplitude of a single oscillator.[13] Thus, the asynchronous, zero mean field state obeys the self-consistency condition.

Next, to demonstrate that synchronization in the population is also possible, we suppose that the mean field is nonvanishing. Then, naturally, it entrains at least some

---

[12] Let us denote these outputs by $x_k(t)$, where $k = 1, \ldots, N$ is the index of an oscillator, and $N$ is the number of elements in the ensemble; $x$ can be variation of light intensity or of the acoustic field around some average value, or, generally, any other oscillating quantity. Then the force that drives each oscillator is proportional to $\sum_k x_k(t)$. It is conventional to write this proportionality as $\varepsilon N^{-1} \sum_k x_k(t)$, so that it includes the normalization by the number of oscillators $N$. The term $N^{-1} \sum_k x_k(t)$ is just an arithmetic mean of all oscillations, and therefore this type of coupling is often called **mean field** coupling.

[13] According to the law of large numbers, it tends to zero when the number of interacting oscillators tends to infinity; the fluctuations of the mean field are of the order $N^{-1/2}$.

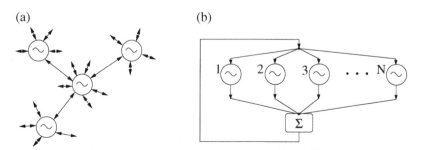

**Figure 4.24.** (a) Each oscillator in a large population interacts with all others. Such an interaction is denoted all-to-all, or global coupling. (b) An equivalent representation of globally coupled oscillators: each element of the ensemble is driven by the mean field that is formed by all elements.

part of the population, the outputs of these entrained elements sum up coherently (cf. Section 3.6.4), and the mean field is indeed nonzero, as assumed. Which of these two states – synchronous or asynchronous – is realized, or, in other words, which one is stable, depends on the strength of interaction between each pair and on how different the elements are. The interplay between these two factors, the coupling strength and the distribution of the natural frequencies, also determines how many oscillators are synchronized, and, hence, how strong the mean field is.

We discuss now how the synchronization transition occurs, taking the applause in an audience as an example. Initially, each person claps with an individual frequency, and the sound they all produce is noisy.[14] As long as this sound is weak, and contains no characteristic frequency, it does not essentially affect the ensemble. Each oscillator has its own frequency $\omega_k$, each person applauds and each firefly flashes with its individual rate, but there always exists some value of it that is preferred by the majority. Definitely, some elements behave in a very individualistic manner, but the main part of the population tends to be "like the neighbor". So, the frequencies $\omega_k$ are distributed over some range, and this distribution has a maximum around the most probable frequency. Therefore, there are always at least two oscillators that have very close frequencies and, hence, easily synchronize. As a result, the contribution to the mean field at the frequency of these synchronous oscillations increases. This increased component of the driving force naturally entrains other elements that have close frequencies, this leads to the growth of the synchronized cluster and to a further increase of the component of the mean field at a certain frequency. This process develops (quickly for relaxation oscillators, relatively slow for quasilinear ones), and eventually almost all elements join the majority and oscillate in synchrony, and their common output – the mean field – is not noisy any more, but rhythmic.

The physical mechanism we described is known as the Kuramoto self-synchronization transition [Kuramoto 1975]. The scenarios of this transition may be more complicated, e.g., if the distribution of the individual frequencies $\omega_k$ has several maxima. Then several synchronous clusters can be formed; they can eventually merge or coexist. Clustering can also happen if, say, the strength of interaction of an element of the population with its nearest (in space) neighbors is larger than with those that are far away.

The scenario of the Kuramoto transition does not depend on the origin of the oscillators (biological, electronic, etc.) or on the origin of interaction. In the above presented examples the coupling occurred via an optical or acoustic field. Global coupling of electronic systems can be implemented via a common load (Fig. 4.25); in this case the voltage applied to individual systems depends on the sum of the

---

[14] Naturally, the common (mean) acoustic field is nonzero, because each individual oscillation is always positive; the intensity of the sound cannot be negative, it oscillates between zero and some maximal value. Correspondingly, the sum of these oscillations contains some rather large constant component, and it is the deviation from this constant that we consider as the oscillation of the mean field and that is small. Therefore, the applause is perceived as some noise of almost constant intensity.

currents of all elements. (An example with a sequential connection of the Josephson junctions is presented in Section 12.3.) Chemical oscillators can be coupled via a common medium, where concentration of a reagent depends on the reaction in each oscillator and, on the other hand, influences these reactions. One example of such a coupling is given by Vanag *et al.* [2000] who studied the dynamics of the photosensitive Belousov–Zhabotinsky reaction in a thin layer subject to illumination that was a function of the average concentration of one of the reagents;[15] for a sufficiently strong interaction, depending on the initial conditions, synchronous states with different structure of clusters were observed.

Sometimes the exact mechanism of the interaction between the elements of an ensemble is not quite clear. This is illustrated in the following example.

### 4.3.2   An example: synchronization of menstrual cycles

We illustrate this theoretical consideration with the results reported by McClintock [1971] who performed a detailed study of the menstrual cycles of 135 females aged 17–22 years, all residents of a dormitory in a women's college, and showed an increase in synchrony in the onset of cycles during the academic year (September–April). This confirmed earlier indirect observations that social groupings influence some aspects of the menstrual cycle, and agreed with the results of the mice experiments (see citations in [McClintock 1971]) demonstrating that such grouping influences the balance of the endocrine system. In particular, it was demonstrated that the dispersion of the times of the menstruation onset within different groups of "close friends" within the collective under investigation decreased during the academic year; the term "close friends" was adopted for persons that listed each other as those whom they saw most often and with whom they spent most of the time. The study indicated the existence in humans of some interpersonal physiological process which affects the menstrual cycle. Whether the mechanism underlying this phenomenon is pheromonal[16] or is mediated by awareness of cycles among friends or some other process, remains an open question.

---

[15] In this system global coupling coexists with local diffusive coupling.

[16] Pheromones are aromatic substances produced by animals and humans that influence sexual behavior and related functions of the organism.

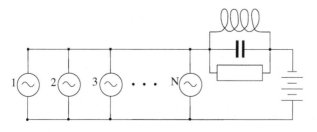

**Figure 4.25.** An ensemble of electronic systems globally coupled via a common load. The mean field here is proportional to the current through the load.

### 4.3.3    An example: synchronization of glycolytic oscillations in a population of yeast cells

We have already mentioned that chemical oscillators can be coupled via a common medium. Indeed, suppose that some reagent is produced at a reaction and that the reaction rate depends on the concentration of that reagent. If the medium is constantly stirred, then the concentration is spatially homogeneous and is determined by all oscillators (Fig. 4.26). Hence, it can be considered as the mean field. We now describe an example of such a system.

Under certain conditions sustained glycolitic oscillations can be observed in a suspension of yeast cells in a stirred cuvette (see [Richard *et al.* 1996] and references therein). The oscillations can be followed by measuring the fluorescence of one of the metabolites, namely NADH.[17]

Richard *et al.* [1996] considered two alternative hypotheses on the origin of macroscopic NADH oscillation. First, one can assume that it arises due to the summation of simultaneously induced oscillations of individual cells. Indeed, the glycolytic oscillation is induced by adding glucose to the starved cell culture. As the cells are not too different and begin to oscillate at the same instant of time, one can expect that at least for some time the cells remain approximately in phase. The alternative hypothesis is synchronization of chemical oscillators globally coupled via the common medium. Richard *et al.* [1996] confirmed the second alternative by performing the following experiment. They initiated glycolytic oscillations in two populations of cells, so that the phase shift between them was about $\pi$, and then mixed these two populations together. If the cells were oscillating independently, the oscillations would cancel each other. If the cells are coupled, one expects synchronization to occur in the mixture of two (previously synchronous) populations. This latter effect was indeed observed in experiments: immediately after mixing there was no oscillation of NADH, but it

---

[17] HADH (nikotinamid-adenin-dinucleotid) has the property to absorb the light of a certain wavelength; therefore, concentration of NADH can be followed spectroscopically.

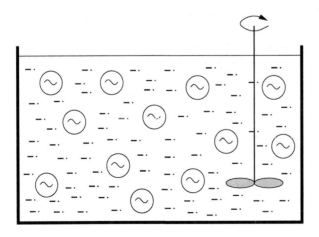

**Figure 4.26.** Chemical oscillators in a stirred tank are globally coupled via a common medium.

re-appeared after approximately 3 min (the characteristic oscillation period is about 40 s). Next, Richard *et al.* [1996] demonstrated that the extracellular free acetaldehyde concentration oscillates at the frequency of intracellular glycolytic oscillations. They concluded that this chemical plays the role of the communicator between the cells, or what we call the mean field. This conclusion is confirmed by two facts. First, the yeast cells respond to acetaldehyde pulses. Added during oscillations, acetaldehyde induces a phase shift that depends on the concentration of the addition (i.e., strength of the pulse) and its phase (cf. the discussion of phase resetting by pulses in Section 3.2). Second, the acetaldehyde is secreted by oscillating cells.

### 4.3.4 Experimental study of rhythmic hand clapping

The formation of rhythmic applause has been recently studied experimentally by Néda *et al.* [2000], who recorded several opera performances in Romania and Hungary. They measured the global noise intensity using a microphone placed on the ceiling of the opera hall as well as the sounds of individual clapping by means of a microphone hidden in the vicinity of a spectator. Néda *et al.* [2000] emphasize that the onset of synchronous applause is preceded by an approximate doubling of the period of clapping. A plausible explanation is that at lower speed (i) the individuals are capable of maintaining a rather stable rhythm of their own (this means that each "oscillator" becomes less noisy due to a reduction in frequency fluctuation) and (ii) the dispersion of frequencies decreases. Both factors, reduction of the noise and of the detuning promote the synchronization transition. Interestingly, the reduction of the frequency aimed at inducing synchrony is a voluntary act of the individuals. As Néda *et al.* [2000] mentioned, such a collective behavior is typical for the culturally more homogeneous Eastern European communities and happens only sporadically in Western European and North American audiences.

## 4.4 Diverse examples

In this section we illustrate the effect of mutual synchronization with further examples.

### 4.4.1 Running and breathing in mammals

Bramble and Carrier [1983] systematically investigated synchronization between breathing and locomotion in running mammals. Phase locking of limb and respiratory frequency has been recorded during treadmill running in jackrabbits and during locomotion on solid ground in dogs, horses and humans. It was found that at high speed, running quadrupedal species normally synchronize the locomotor and respiratory cycles at a constant ratio 1 : 1 (strides per breath) in both the trot and gallop (Fig. 4.27). Human runners employ several phase locked patterns (4 : 1, 3 : 1, 2 : 1,

1 : 1, 5 : 2 and 3 : 2), although a 2 : 1 regime appears to be favored (Fig. 4.28). It
remains unclear whether synchronization between breathing and locomotor activities

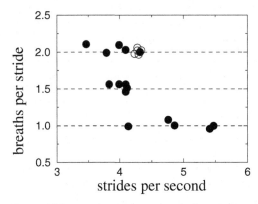

**Figure 4.27.** Relation between breathing and stride frequency in a young hare *Lepus
californicus* running on a treadmill. Each data point is the mean of three consecutive
strides. Filled circles are data obtained when the animal was immature; open circles
were obtained when it was sub-adult. At lower speeds, there were two complete
breathing cycles per locomotor cycle. At higher speeds, the hare abruptly switched to
1 : 1 synchronous regime, thereby halving its breathing frequency. When forced to
run near the transition speed of approximately four strides per second, the animal
repeatedly alternated two patterns, thus exhibiting 3 : 2 locking. Plotted using data
from [Bramble and Carrier 1983].

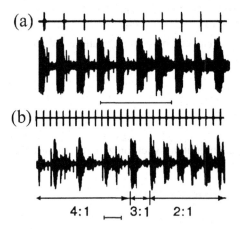

**Figure 4.28.** Oscilloscope record of gait and breathing in free running mammals.
The upper trace is the footfall of the right forelimb (horse) or leg (human). The lower
trace is breathing as recorded from a microphone in a face mask. (a) Horse at gallop.
(b) Human during shift from 4 : 1 to 2 : 1 pattern. Horizontal bars denote the time
scale (2 s). Reprinted with permission from Bramble and Carrier, *Science*, Vol. 219,
1983, pp. 251–256. Copyright 1983 American Association for the Advancement of
Science.

enhances performance, or whether it is just a physiologically irrelevant consequence of a general property of coupled oscillators.

Synchronization of heartbeat, respiration and locomotion in humans walking or running on a treadmill was studied by Niizeki *et al.* [1993].

### 4.4.2 Synchronization of two salt-water oscillators

A salt-water oscillator consists of a cylindrical plastic cup with NaCl aqueous solution placed in an outer vessel containing pure water. The cup has a small orifice in the bottom. If one measures the electric potential in the cup, it turns out that it oscillates due to oscillatory in- and outflow of the pure water and salt-water, respectively. If two such cups are placed in a common outer vessel, anti-phase synchronization of oscillations is observed if the relation between the surfaces of the cup and the outer vessel is large enough [Nakata *et al.* 1998].

### 4.4.3 Entrainment of tubular pressure oscillations in nephrons

Synchronization effects associated with renal blood flow control were reviewed by Yip and Holstein-Rathlou [1996]. Normal functioning of a kidney requires maintenance of a constant renal blood flow. This is provided by a special feedback mechanism (tubuloglomerular feedback) that operates at the level of a single nephron; this regulation often results in excitation of oscillations. These oscillations are autonomous and independent of cardiac and respiratory rhythms. Experiments demonstrate that the neighboring nephrons that belong to the same cortical radial artery commonly synchronize their oscillations. It was hypothesized that the interaction of nephrons was due to the spread of the signal from the feedback system along the preglomerular vasculature.

The entrainment phenomenon probably also underlies the resonance seen in renal blood flow when the arterial pressure is made to oscillate at a frequency close to the natural frequencies of the individual nephrons. In this situation the kidney does not autoregulate. Instead, strong oscillations in renal blood flow and in the tubular pressures synchronous with external forcing can be found. Most probably, this is due to an induced synchrony of the majority of nephrons within the kidney [Yip and Holstein-Rathlou 1996].

### 4.4.4 Populations of cells

Brodsky [1997] reviewed many experiments on rhythms in populations of cells. In particular, ultradian (nearly 1-hour) rhythms of protein synthesis were observed in cell cultures. It was concluded that these rhythms are due to the synchronization of the oscillations of many cells interacting via a common medium (i.e., globally cou-pled); the particular mechanisms of interaction are also discussed by Brodsky [1997]. Mitjushin *et al.* [1967] investigated variation of the nucleus size in a population of cells of Ehrlich ascites carcinoma. They reported that in that population there are groups of cells pulsating synchronously.

Vasiliev *et al.* [1966] found that the cell complexes (groups of several contacting cells) in mouse ascites hepatoma are characterized either by complete synchrony in progression through the mitotic cycle, or by considerable asynchrony. The authors em-phasize "that the synchrony is so precise that it is natural to suggest that it is supported by some interactions of the contacting cells which counteract random fluctuations of the cycle". They conclude that mitotic synchrony may play an important role in the development of tissue anaplasia (change in the cell morphology) characteristic in malignant tumors. A possible mechanism for cell cycle synchronization is discussed in [Polezhaev and Volkov 1981].

Milan *et al.* [1996] found that in metamorphosing wing disks in *Drosophila*, pro-gression through the cell cycle takes place in clusters of cells synchronized in the same cell cycle stage. (It is emphasized that these clusters are nonclonally derived, i.e., the cells in a cluster are progeny of different cells.)

Synchronization in cultures of spontaneously beating ventricular cells was studied by Soen *et al.* [1999]. It was found that network behavior often consists of periodic beating that is synchronized throughout the field of view. However, sufficiently long recordings reveal complex rhythm disorders.

### 4.4.5   Synchronization of predator–prey cycles

A well-known phenomenon in ecology is oscillation in the abundances of species. One of the most studied examples is the Canadian hare–lynx cycle (see Fig. 1.13 and its discussion, as well as [Elton and Nicholson 1942; Blasius *et al.* 1999] and references therein). A striking fact is that the abundances in different regions of Canada perfectly synchronize in phase, although the amplitudes are irregular and remain quite different (Fig. 4.29). Blasius *et al.* [1999] assumed that the irregularity of the amplitudes is due to chaotic dynamics in the predator–prey system, and the interaction between the populations in adjacent regions occurs because of the migration of animals. They ex-plained the phenomenon as synchronization in a lattice of coupled chaotic oscillators.

### 4.4.6   Synchronization in neuronal systems

Investigation of synchronization in large ensembles of neurons becomes an important problem in neuroscience; here we do not provide a comprehensive review of the field, but just mention the main effects and give some references.

Synchronization phenomena are related to several central issues of neuroscience (see, e.g., [Singer and Gray 1995; Singer 1999]). For instance, synchronization seems to be a central mechanism for neuronal information processing within a brain area as well as for communication between different brain areas. Results of animal experiments indicate that synchronization of neuronal activity in the visual cortex appears to be responsible for the binding of different but related visual features so that a visual pattern can be recognized as a whole (see [Gray *et al.* 1989; Singer and Gray 1995] and references therein). Further evidence is that synchronization of oscillatory activity in the sensorimotor cortex may serve for the integration and coordination of information underlying motor control [MacKay 1997].

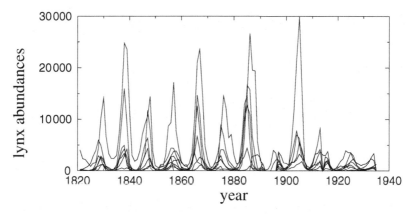

**Figure 4.29.** Time series of lynx abundances from nine regions in Canada. Plotted using data from [Elton and Nicholson 1942].

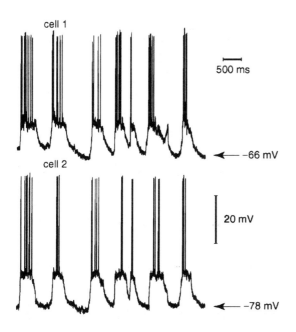

**Figure 4.30.** Synchronization of slow fluctuations of the membrane potentials of striatal spiny neurons. Note that the spiking (generation of the short high-amplitude pulses seen on the top of the slow oscillation) is not synchronous. Reprinted with permission from *Nature* [Stern *et al.* 1998]. Copyright 1998 Macmillan Magazines Limited.

Synchronization is the mechanism that maintains vital rhythms like that of respiration. Koshiya and Smith [1999] have shown that this rhythm is generated by a network of synaptically coupled pacemaker neurons in the lower brainstem. (Being decoupled by means of the pharmacological block of the synaptic transmission, the neurons continue to burst rhythmically but asynchronously.) On the other hand, synchronization is responsible for the generation of pathological tremors [Freund 1983; Elble and Koller 1990] and plays an important role in several neurological diseases like epilepsy [Engel and Pedley 1975].

Firing of many neurons, if they are synchronized, gives rise to measurable fluctuations of the electroencephalography (EEG) signal. Spectral analysis of EEG shows that neurons can oscillate synchronously in various frequency bands (from less than 2 to greater than 60 Hz) [Singer 1999]. Simultaneous spiking in a neuronal population is a typical response to different stimuli: visual [Gray *et al.* 1989], odorous [Stopfer *et al.* 1997] or tactile [Steinmetz *et al.* 2000].

Another kind of synchronization of neurons was reported by Stern *et al.* [1998] who recorded a membrane potential of striatal spiny neurons in anesthetized animals *in vivo*. The potential fluctuates between two subthreshold states and during the "up" state a neuron fires (Fig. 4.30). It can be seen that slow fluctuations are synchronous, whereas the spiking is asynchronous, with the instants of spiking determined by the noise-like fluctuations within the "up" state. (Similar patterns were observed by Elson *et al.* [1998] in their experiments with two coupled neurons from the lobster stomatogastric ganglion.)

Stern *et al.* [1998] noted that slow subthreshold "up–down" fluctuations in cortical cells are correlated with a slow ($\approx 1$ Hz) rhythm of the EEG during sleep, and had been shown to be dependent on the level of anesthesia. As the level of anesthesia becomes shallower, the slow fluctuations in different cortical neurons become less synchronized, and the slow EEG rhythm becomes progressively more shallow until it is lost, although the fluctuations in individual neurons are still present (see [Stern *et al.* 1998] and references therein). We note that in our terms this means that the anesthesia decouples individual oscillators and therefore the mean field (slow EEG rhythm) decreases.

# Chapter 5

# Synchronization of chaotic systems

In this chapter we describe synchronization effects in chaotic systems. We start with a brief description of chaotic oscillations in dissipative dynamical systems, emphasizing the properties that are important for the onset of synchronization. Next, we describe different types of synchronization: phase, complete, generalized, etc. In studies of these phenomena computers are widely used, therefore in our illustrations we often use the results of computer simulations, but also show some real experimental data. Whenever possible, we try to underline a similarity to synchronization of periodic oscillations.

## 5.1    Chaotic oscillators

One of the most important achievements of nonlinear dynamics within the last few decades was the discovery of complex, **chaotic** motion in rather simple oscillators. Now this phenomenon is well-studied and is a subject of undergraduate and high-school courses; nevertheless some introductory presentation is pertinent. The term "chaotic" means that the long-term behavior of a dynamical system cannot be predicted even if there were no natural fluctuations of the system's parameters or influence of a noisy environment. Irregularity and unpredictability result from the internal deterministic dynamics of the system, however contradictory this may sound. If we describe the oscillation of dissipative, self-sustained chaotic systems in the phase space, then we find that it does not correspond to such simple geometrical objects like a limit cycle any more, but rather to complex structures that are called **strange attractors** (in contrast to limit cycles that are simple attractors).

### 5.1.1   An exemplar: the Lorenz model

In 1963 the meteorologist Ed Lorenz published his famous work, where a strange attractor was found in numerical experiments in the context of studies of turbulent convection. Fortunately, there is a simple physical realization of the *Lorenz model*: convection in a vertical loop [Gorman *et al.* 1984, 1986], see Fig. 5.1. The fluid is heated from below, and for strong enough heating convection sets in. Just beyond the onset, the motion is steady, with a constant velocity $V$. Clearly, due to symmetry, motions in both the clockwise and counter-clockwise directions are possible. If the heating from below increases, the steady rotation becomes unstable and reversions of the flow are observed. Moreover, these reversions are irregular and do not repeat themselves: the motion is not periodic, but chaotic.

To describe chaos theoretically, one can model the system with ordinary differential equations. Thus far we have always considered systems that can be represented on the phase plane, i.e., with two independent variables. This is the minimal dimension of the phase space for a limit cycle oscillator, but it is not enough for chaotic motion. Because the trajectories cannot intersect in a phase space (this would contradict determinism – only one trajectory can evolve from a given point in a phase space), it is not possible to encounter something more complex than a limit cycle on a phase plane. A chaotic model must have at least three dimensions, i.e., the state of the oscillator must be described by at least three coordinates. The Lorenz model has exactly three variables $x, y, z$ having the following physical meaning: $x$ is proportional to the horizontal temperature difference $T_3 - T_1$; $y$ is proportional to the flow velocity $V$; $z$ is proportional to the vertical temperature difference $T_4 - T_2$. Following the title of this Part, we will not write the equations here (see Part II, Eqs. (10.4)), but just present a piece of their numerical solution in Fig. 5.2. The time dependence of all variables demonstrates erratic oscillations with switchings between the convection motions in the clockwise (negative $y$) and counter-clockwise (positive $y$) directions.

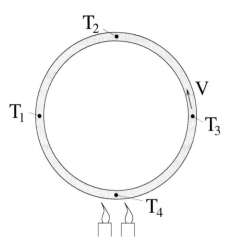

**Figure 5.1.** Convection of a viscous fluid in a thin loop heated from below is described well by the Lorenz system.

We have taken the Lorenz model as a representative example of autonomous chaotic oscillators. There are many other systems (e.g., electronic circuits, lasers, chemical reactions) demonstrating chaos that can be modeled by simple systems of differential equations; one can find their descriptions in numerous books on chaos, see Section 1.4 for citations. Moreover, observations of some natural irregular processes allow us to assume that there are chaotic dynamics behind them, see examples in [Kantz and Schreiber 1997].

Here, like in the case of periodic oscillations, it is important to distinguish between self-sustained and forced systems. Self-sustained chaotic oscillators are described by *autonomous* equations, therefore all instants of time are equivalent. One can say that here the time symmetry (in the sense of independence of the dynamics to time shifts) is continuous. There are many examples of chaotic motion in periodically driven nonlinear systems, described by nonautonomous equations. Here the force breaks the continuous time symmetry, which becomes discrete (only instants of time shifted by the period of the force are equivalent); see also the corresponding discussion for periodic oscillations in Section 2.3.2. Another popular class of chaotic models – mappings – are in this respect equivalent to periodically driven systems. Here the time symmetry is obviously discrete, because time itself is discrete. For many properties of chaos, and in particular for the effect of complete synchronization (Section 5.3), the

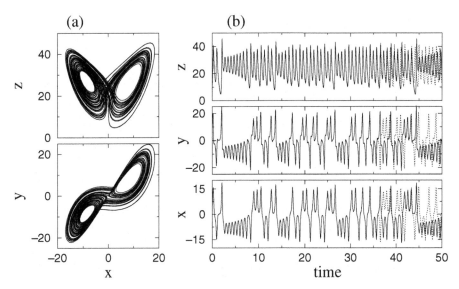

**Figure 5.2.** The dynamics of the Lorenz system (see Eqs. (10.4)). (a) Projections of the phase portrait on planes $(x, y)$ and $(x, z)$. Note the symmetry $x \rightarrow -x$, $y \rightarrow -y$. (b) The time series of the variables $x$, $y$, $z$ (solid curves). In this figure we also demonstrate the sensitivity to small perturbations: at time $t = 25$ a perturbation $10^{-4}$ is added to the $x$ variable. The perturbed evolution, shown with the dashed curve, very soon diverges from the original one and demonstrates a different pattern of oscillations.

difference between self-sustained and forced systems is irrelevant. However, for phase synchronization of chaos (Section 5.2) it is crucial.

## 5.1.2  Sensitive dependence on initial conditions

Irregularity of a chaotic system does not mean complete nonpredictability. Indeed, the processes in Fig. 5.2 appear to be rather predictable on a small (much less than one characteristic period) time scale; this follows from the regularity of Fig. 5.2a within one oscillating pattern. This completely agrees with the deterministic nature of the process: if one knows the state $x, y, z$ at time $t = 0$ precisely (i.e., with infinite accuracy), the dynamics for all the time $t > 0$ are determined uniquely. These dynamics are defined by ordinary differential equations of the motion; practically, one calculates them using some numerical integration scheme. The dynamics of chaotic systems have, however, one essential feature: they are *sensitive to small perturbations of initial conditions*. This means that if we take two close but different points in the phase space and follow their evolution, then we see that the two phase trajectories starting from these points eventually diverge (Fig. 5.2b). In other words, even if we know the state of a chaotic oscillator with a very high, but finite precision, we can predict its future only for a finite, depending on the precision, time interval, and cannot predict its state for longer times.

The sensitivity pertains to any point of the trajectory. This means that all motions on the strange attractor are unstable. Quantitatively, the instability is measured with the largest **Lyapunov exponent**. The inverse of this exponent is the characteristic time of instability; roughly speaking the perturbation doubles within such time interval. In Fig. 5.3 we illustrate why the sensitivity leads to irregularity. First, we restrict our attention to nontransient (recurrent) states, i.e., to those that nearly repeat themselves. Suppose that such a state 1, after some evolution, nearly repeats itself, coming to a neighboring position 2. This neighboring position can be considered as a perturbation of 1. Due to the instability, the evolution of state 2 does not follow the evolution starting from 1, but diverges from it.[1] Thus, all similarities in the states of the system are temporary, and no pattern can regularly repeat itself. The instability also implies a mixing on the chaotic attractors: if a set of close initial points is chosen, after some time (inversely proportional to the largest Lyapunov exponent) the points will be spread over the whole attractor.

The stability characteristics of trajectories can be followed in more detail. Indeed, one can make small perturbations of a state in the phase space in all possible directions. The number of independent components in this linear evolution is exactly the number of independent variables of the dynamics, and for each such component one can define the growth (or decay) rate of instability (stability); these rates are called Lyapunov

---

[1]  In the stable case the trajectories starting from 1 and 2 come closer to each other; the necessary consequence is the existence of a stable limit cycle in the vicinity.

exponents. In the Lorenz model we have three variables, therefore it has three Lyapunov exponents.[2] For chaotic three-dimensional systems one exponent is positive (corresponding to the sensitivity described above), one is negative (this corresponds to the property of the attractor to catch nearby states), and one is exactly zero, which corresponds to shifts along the trajectory – clearly such perturbations neither grow nor decay.[3] We illustrate these local stability properties of a chaotic state in Fig. 5.4. We remind the reader that periodic oscillators have one zero Lyapunov exponent, and all others are negative (cf. Fig. 2.6). The latter correspond to the convergence of the trajectories towards the attractor (the limit cycle), while the former corresponds to a shift of the point along the limit cycle, which is equivalent to a shift of the phase of the oscillator. This suggests the introduction of phase for chaotic oscillators as well, as a variable corresponding to the zero Lyapunov exponent, or, in other words, to the coordinate along the trajectory. We will show now that many effects are possible where the phase dynamics are important, e.g., the frequency entrainment of chaotic oscillators. We call these effects "phase synchronization" to distinguish them from other types of synchronization of chaotic oscillators, to be described in Section 5.3.

## 5.2    Phase synchronization of chaotic oscillators

In this section we first demonstrate that at least some chaotic oscillators can be characterized with a time-dependent phase and frequency. Next we argue that synchronization of these oscillators in the sense of frequency locking is possible. We discuss in detail entrainment by external forcing and illustrate it by an experimental example.

---

[2]    An $M$-dimensional dynamical system has $M$ Lyapunov exponents.

[3]    The property to have at least one zero Lyapunov exponent holds for autonomous systems, where any shifts of time are possible. In the case of periodically forced systems or mappings only discrete shifts of time can be made (multiples of the period or natural numbers, respectively); there are no neutral small perturbations along the trajectory and generally all Lyapunov exponents are nonzero.

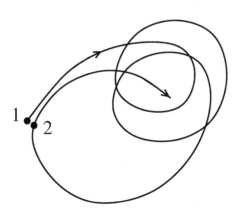

**Figure 5.3.** An illustration of why the instability of trajectories leads to irregularity: nearly repeated states 1 and 2 eventually diverge, making all repetitions temporary.

### 5.2.1   Phase and average frequency of a chaotic oscillator

The main observation behind the definition of phase and frequency is that chaotic oscillations of many systems look like irregularly modulated nearly periodic ones. For example, if we take for the Lorenz model the coordinates $z$ and $u = \sqrt{x^2 + y^2}$ (in fact this corresponds to a special two-dimensional projection of the phase portrait), then the trajectory on the plane $(z, u)$ looks like a smeared limit cycle (Fig. 5.5a). The time dependencies of $z$ and $u$ resemble periodic oscillations, with a varying "amplitude" and "period". Let us focus on the latter characteristics of the oscillations. As the process is irregular, we cannot define its period in the way we have done for periodic oscillations.[4] Instead, we can determine a time between two similar events in the time series, e.g., between two maxima of the variable $z$. In terms of the theory of dynamical systems, this can be interpreted as constructing a Poincaré map according to the condition that the variable $z$ has a maximum, and looking on the times between the successive piercings of the Poincaré surface (Fig. 5.5b). These **return times** are not equal: they depend on the values of the variables in the Poincaré surface. The latter are chaotic, therefore the times are irregular as well. Interpreting the return times as *"instantaneous" periods* of the oscillations, we can define the mean (averaged) period of the process $z(t)$. The simplest way to do this is to take a long interval of time $\tau$ and to count the number $N(\tau)$ of maxima of $z$ within this interval (or to count any other events chosen for the construction of the Poincaré map); the ratio $\tau/N(\tau)$ gives the *mean period*. Correspondingly, the mean angular frequency of the oscillations can be

---

[4] Moreover, from the power spectrum of a chaotic process one can conclude that the motion contains many frequencies.

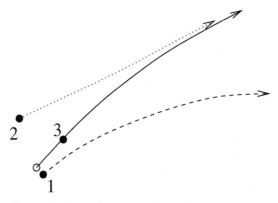

**Figure 5.4.** Diverging, converging and neutral perturbations of a chaotic system. The state of a three-dimensional chaotic system (open circle) is perturbed in one of three ways corresponding to three directions in the phase space (filled circles). The unperturbed trajectory is shown with the solid curve. The perturbation 1 grows: the corresponding trajectory (the dashed curve) moves away from the original one. The perturbation 2 decays: its trajectory (the dotted line) converges to the unperturbed one. The perturbation 3 lies on the same trajectory and neither grows nor decays.

estimated as $\langle \omega \rangle = 2\pi N(\tau)/\tau$. The main idea of phase synchronization of chaotic oscillators is that this frequency can be entrained by a periodic external force, or by coupling to another chaotic oscillator. To discuss it in detail, it is beneficial to define the phase of a chaotic oscillator.

Acting in the same spirit as we did while introducing the phase for periodic oscillators, we say that each cycle in Fig. 5.5 corresponds to a phase gain of $2\pi$. Using the Poincaré map, we can say that each piercing of the Poincaré surface of section corresponds to some particular phase (naturally, we will assume it to be 0). During the rotation between the successive piercings the phase grows by $2\pi$. Because the return times are irregular, the instantaneous frequency (defined as the inverse return time) is a fluctuating quantity. In other words, the phase rotates not uniformly, like in the case of periodic oscillators, but it experiences irregular decelerations and accelerations. As a result, *the phase diffuses like in the case of a noise-driven periodic oscillator* (see Section 3.4). The total phase dynamics can be represented as a combination of two processes: rotations with the mean frequency and diffusion (random walk), with the rate of the diffusion being proportional to the variation of the return times. We illustrate these dynamics in Fig. 5.6, to be compared with the dynamics of a noisy oscillator shown in Fig. 3.35.

We emphasize that the diffusion of the phase is weaker than the divergence of neighboring trajectories due to intrinsic instability of chaos. In a nonbiased diffusion process the deviation from the initial position is roughly proportional to the square root of the time; the same law describes the distance between two neighboring points. Contrary to this, the instability is exponentially fast. Moreover, if the difference

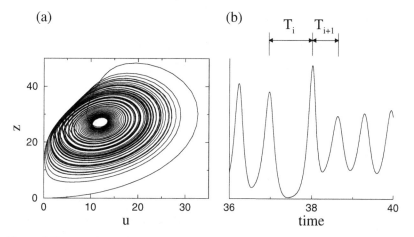

**Figure 5.5.** (a) In the variables $z$, $u$ the dynamics of the Lorenz model look like rotations around a center (at $u \approx 12$, $z \approx 27$), with an irregular amplitude and an irregular return time. (b) The return times $T_i$ are shown in the plot of $z$ vs. time as the distance between the maxima; $T_i$ can be considered as instantaneous periods of chaotic oscillation.

between the return times is small, the diffusion constant can be small as well, in this case the chaotic oscillations appear in a phase plane projection as relatively regular rotations with a chaotic modulation of the amplitude. Such oscillations are often called coherent; they have a sharp frequency peak in the power spectrum (an example is the Rössler model (see Sections 1.3 and 10.1)). Note also that calculation of the phase is a nonlinear transformation, or some kind of "nonlinear filtering". Indeed, in calculating the phase we neglect all variations of the amplitude that usually contribute to a broad-band part of the power spectrum of a chaotic system. The phase diffusion then characterizes the width of a spectral peak at the mean frequency.

Another way to look on the phase dynamics of a chaotic system is to consider an ensemble (a cloud) of initial conditions, and follow it in the phase space. As chaotic systems are characterized by mixing, an initially localized cloud will be eventually spread over the chaotic attractor. This spreading includes both a fast expansion due to chaotic instability, and a diffusion corresponding to the phase (Fig. 5.7).

### 5.2.2   Entrainment by a periodic force. An example: forced chaotic plasma discharge

Suppose now that a chaotic oscillator is subject to external periodic forcing. In the case of the Lorenz model one can, e.g., make the heating periodic in time; such force periodically drives the variable $z$. If the period of the forcing is close to the mean return time, then the motions advanced in the phase are slowed down, while the lagged ones are accelerated. As a result, the phase is locked by the external force, as is illustrated in Fig. 5.8. The synchronization can be also characterized as the frequency entrainment: the mean frequency of chaotic oscillations coincides (or nearly coincides) with the frequency of the external force.

Phase synchronization of chaotic oscillators as described above occurs for moderate amplitudes of the forcing. On the one hand, the force should be strong enough to

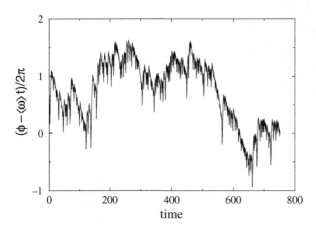

**Figure 5.6.** The deviation of the phase of the Lorenz system from uniform rotation demonstrates a typical pattern of a diffusion process (random walk).

suppress the phase diffusion and to entrain the frequency. On the other hand, the force should be small enough not to destroy chaos. If the forcing is very strong, typically a stable periodic motion appears instead of chaos (see Section 5.3.4).

As a next remark, we would like to describe phase synchronization of chaotic oscillators in the general context of phase locking. In Fig. 5.9 we compare periodic, noisy and chaotic oscillators. While phase locking of periodic oscillators is perfect and can be achieved by arbitrary small forcing, locking of noisy or chaotic oscillators requires suppression of the phase diffusion, thus there is usually a synchronization

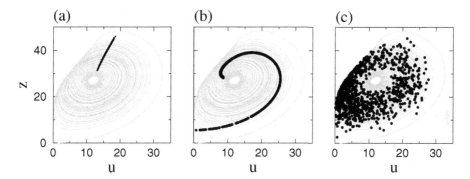

**Figure 5.7.** Phase diffusion in the Lorenz model. The attractor is shown in gray. (a) A set of initial conditions (black dots) is chosen on the Poincaré surface of section (variable $z$ is maximal), all the phases are zero. The evolution of this set is followed in (b) and (c). (b) After time $t = 0.75$ (one mean return time) some points lag while others advance, the phases are distributed in a finite range less than $2\pi$. (c) After a further time interval $t = 3.75$ the trajectories become distributed along the attractor, which means that the phases are nearly uniformly distributed from 0 to $2\pi$.

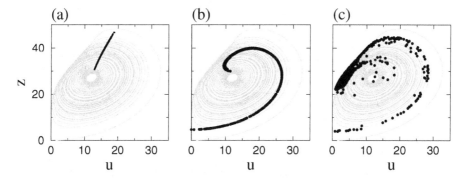

**Figure 5.8.** The same as Fig. 5.7, but in the periodically forced (with dimensionless frequency 8.3) Lorenz model. (a) Initial conditions are the same as in Fig. 5.7a. (b) After one period of forcing the phases are slightly more concentrated around the mean value; in fact, a difference to Fig. 5.7b can hardly be seen. Nevertheless, the effect accumulates and after a further 50 forcing periods this concentration is clearly seen in (c). A large number of points rotate synchronously with forcing, although some are not entrained.

threshold with respect to the forcing amplitude. Note also, that the entrained oscillators remain noisy or chaotic, respectively: the force tends to bring some order (a perfect rhythm) into the motion, but does not make it fully regular. In general, we can say that the phase synchronization of chaotic systems is very similar to that of noisy ones; this allows a rather unrestricted interpretation of observations of phase locking in irregular processes (see Chapter 6): to detect synchronization in an experiment, one does not have to decide whether the process is noisy or chaotic.

Phase synchronization of chaos has been observed in several experiments. Pikovsky [1985] indirectly demonstrated this effect by comparing the power spectra of a free and a forced electronic oscillator. He found that the forcing made the spectral peak more narrow (remember that the width of the peak characterizes phase diffusion, and as we already know, forcing suppresses diffusion). Frequency entrainment of a chaotic electronic oscillator was demonstrated by Parlitz *et al.* [1996], see also [Rulkov 1996]; laser experiments have been conducted by Tang *et al.* [1998a,c].

For illustration we have chosen experiments by Rosa Jr. *et al.* [2000] and Ticos *et al.* [2000], who studied phase synchronization of a chaotic gas discharge by periodic forcing. The discharge was caused by applying an 800 V dc voltage to a helium tube, and forcing was implemented by connecting an ac source in series with the dc one. The amplitude of the ac voltage was 0.4 V. The stroboscopic pictures for the unforced and forced discharge (Fig. 5.10) provide evidence of synchronization; here $I$ is the

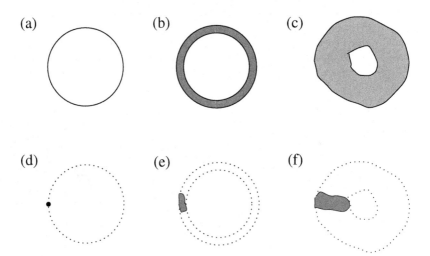

**Figure 5.9.** A sketch of common features of phase synchronization in periodic (a), (d), noisy (b), (e), and chaotic (c), (f) oscillators. When a periodic oscillator (a) is entrained by a periodic force, the stroboscopically (with the period of the force) observed phase has a definite value (d). This perfect picture is partially spoiled by a bounded noise, but nevertheless entrainment is possible: the fluctuations of the phase are bounded (e). A chaotic oscillator (c) resembles the noisy one (b); here the diffusion of the phase can be also suppressed, yielding a state with a chaotic amplitude but bounded phase (f).

intensity of light emitted by the tube, and the phase portrait of the system is plotted in the delayed coordinates $(I(t), I(t + \tau))$.[5] Systematic variation of the amplitude and frequency of the external pacing allows determination of the synchronization region. The shape of this region, presented in Fig. 5 in [Ticos *et al.* 2000], is similar to that shown in Fig. 3.39: a very weak force cannot overcome the phase diffusion, and even for vanishing detuning synchronization is not possible.

We conclude this section by mentioning that mutual phase synchronization of chaotic oscillators is also possible. If two chaotic systems have different parameters, then their mean frequencies are different as well. Coupling between the systems introduces some interaction between the phases, so that they can be mutually entrained. As in the case of periodic oscillators, small coupling affects the phases only. As a result, two synchronized oscillators have the same mean frequencies, but each system preserves its own chaos in the amplitude. Generally, for chaotic systems, one can observe all the effects that we have described for periodic noisy oscillators, e.g., the formation of clusters in an oscillatory lattice, or the Kuramoto synchronization transition in an ensemble.

## 5.3    Complete synchronization of chaotic oscillators

Strong mutual coupling of chaotic oscillators leads to their **complete synchronization**. Contrary to phase synchronization, it can be observed in any chaotic systems, not necessarily autonomous, in particular in periodically driven oscillators or in discrete-time systems (maps). In fact, this phenomenon is not close to the classical synchronization of periodic oscillations, as here we do not have adjustment of rhythms (and,

---

[5]  Time here is expressed in the units of the sampling rate. For the technique of attractor reconstruction from a scalar time series see, e.g., [Abarbanel 1996; Kantz and Schreiber 1997].

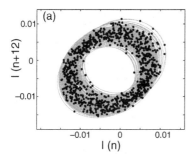 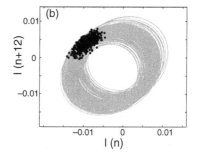

**Figure 5.10.** Stroboscopic observation of the unforced (a) and forced (b) chaotic gas discharge. The attractor is shown in gray, and the states of the system observed with the period of the force are shown by circles. Concentration of the points in (b) indicates phase synchronization (cf. Figs. 5.8 and 5.9). From [Rosa Jr. *et al.* 2000].

in particular, we cannot put it in the framework illustrated by Fig. 5.9). Instead, complete synchronization means suppression of differences in coupled *identical* systems. Therefore, this effect cannot be described as entrainment or locking; it is closer to the onset of symmetry.[6]

## 5.3.1  Complete synchronization of identical systems. An example: synchronization of two lasers

To describe this phenomenon, let us take two *identical* chaotic systems (e.g., two Lorenz systems), and couple them in such a way that the coupling tends to make the corresponding variables equal. We use indices $1, 2$ to describe the two systems; attractive coupling tends to make the differences $|x_1 - x_2|$, $|y_1 - y_2|$ and $|z_1 - z_2|$ smaller. Moreover, we demand that the coupling force is proportional to the differences of the states of the two oscillators (i.e., to $x_1 - x_2$, $y_1 - y_2$, $z_1 - z_2$), and vanishes for coinciding states $x_1 = x_2$, etc. Thus, in the case of complete coincidence of the variables, each system does not feel the other one, and performs chaotic motion as if it were uncoupled. Because the systems are identical, the coincidence of states is preserved with time. This is exactly the situation that is called **complete synchronization**: the states of two systems coincide and vary chaotically in time.

Clearly, a completely synchronized state can occur for any value of coupling strength, but only for sufficiently strong coupling can we expect it to be stable. Indeed, let us slightly perturb the full identity of the states, i.e., set $x_1 \neq x_2$, etc. What will happen to a small difference $x_2 - x_1$? If there is no coupling, the answer can be deduced from the instability property of chaos: because the two states $x_1, y_1, z_1$ and $x_2, y_2, z_2$ can be considered as two states in one system (due to identity), they will diverge exponentially in time, the growth rate being determined by the largest Lyapunov exponent. With small coupling, the divergence will be slower, due to "attraction" between the two states. If the coupling is strong enough, the attraction wins and a small difference will decay, eventually leading to a completely synchronized state.

We see that complete synchronization is a threshold phenomenon: it occurs only when the coupling exceeds some critical level that is proportional to the Lyapunov exponent of the individual system. Below the threshold, the states of two chaotic systems are different but close to one another. Above the threshold, they are identical and chaotic in time. We demonstrate this transition with two coupled Lorenz systems in Figs. 5.11 and 5.12.

Roy and Thornburg [1994] observed synchronization of chaotic intensity fluctuations of two Nd:YAG lasers with modulated pump beams. The coupling was implemented by overlapping the intracavity laser fields and varied during the experiment.

---

[6]  Maybe another word instead of "synchronization" would better serve for underlining this difference; we will follow the nowadays accepted terminology, using the adjective "complete" to avoid ambiguity.

For strong coupling, the intensities became identical, although they continued to vary in time chaotically (Fig. 5.13).

## 5.3.2    Synchronization of nonidentical systems

The consideration above is, strictly speaking, not valid for coupled *nonidentical* systems. Clearly, in this case the states cannot coincide exactly, but they can be rather close to each other. In particular, it may be that for large enough coupling there is a functional relation $\mathbf{x}_2 = \mathbf{F}(\mathbf{x}_1)$ between the states of two systems. This means, that knowing the functions $\mathbf{F}$ one can uniquely determine the state of the second system, if the first system is known. This regime is called **generalized synchronization** [Rulkov

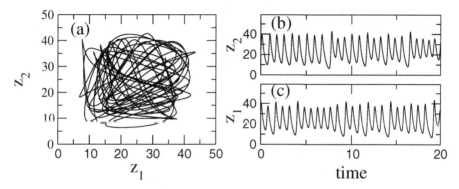

**Figure 5.11.** Weakly coupled Lorenz models (below the synchronization threshold). The simplest way to see the difference between the two systems is to plot the variables of one vs. the variables of the other one (a Lissajous-type plot), as in (a). The difference between two chaotic states can be also seen from the time series (b) and (c).

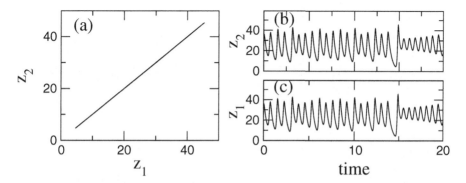

**Figure 5.12.** Complete synchronization in coupled Lorenz models. (a) The states of the systems are identical, as can be easily seen on the plane $z_1$ vs. $z_2$: the trajectory lies on the diagonal $z_1 = z_2$. The time series of two systems are chaotic in time, but completely coinciding ((b) and (c)).

*et al.* 1995]. Complete synchronization is a particular case of the generalized one, when the functions **F** are simple identity functions. Typically, generalized synchronization is observed for *unidirectional coupling*, when the first (driving) system forces the second (driven) one, but there is no back-action. Such a situation is often called master–slave coupling. The onset of generalized synchronization can be interpreted as the suppression of the dynamics of the driven system by the driving one, so that the "slave" passively follows its "master".

### 5.3.3    Complete synchronization in a general context. An example: synchronization and clustering of globally coupled electrochemical oscillators

Several generalizations of the phenomenon described above are of particular interest. One is based on the observation that the synchronization transition can be considered as the onset of a symmetric regime in a symmetry-possessing system. Indeed, demanding that identical systems interact in such a way, that the coupling force vanishes for coinciding states, we in fact demand certain symmetry between the two systems.

(a)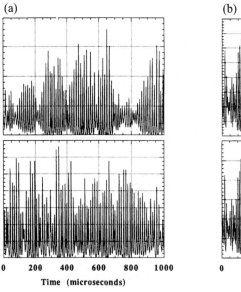          (b)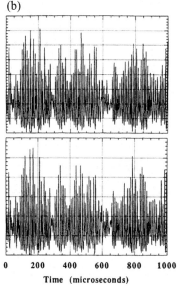

**Figure 5.13.** Complete synchronization in coupled lasers. (a) Intensities (arbitrary units) of uncoupled lasers fluctuate chaotically, and the time course is different, although both lasers experience the same pump modulation. (b) Under strong coupling both lasers remain chaotic, but now the oscillations are nearly identical, i.e., complete synchronization sets in. From Roy and Thornburg, *Physical Review Letters*, Vol. 72, 1994, pp. 2009–2012. Copyright 1994 by the American Physical Society.

One can thus consider a more general situation, where in a large chaotic system there is some internal symmetry, e.g., the equations of motion do not change if some variables are exchanged (say, in a four-dimensional system with variables $x$, $y$, $z$, $v$ the exchange $z \leftrightarrow v$ leaves the system invariant). Then the regime, where these variables are exactly equal (in our example $z = v$), is a solution, and possibly a chaotic one. If the corresponding small perturbations $z - v$ decay in time, the symmetric solution will be stable and one can say that the subsystems $z$ and $v$ are synchronized [Pecora and Carroll 1990]. Such synchronization inside chaos is sometimes called *master–slave synchronization*; we will mainly use the term *replica synchronization* because typically one constructs an equation for $v$ as a replica of the equation for $z$. We emphasize that here we do not have two distinct systems that can operate either separately or being coupled. Instead, some variables inside one large system can coincide, bringing partial order in chaos.

A general symmetry allowing complete synchronization can occur for a large number of interacting chaotic systems. Wang *et al.* [2000a] experimentally studied synchronization of 64 electrochemical oscillators. Global coupling was achieved by connecting all the electrodes via a common load; this is equivalent to the scheme shown in Fig. 4.25. The experimentalists took care to prepare the elements of the ensemble in the same manner. The phase portrait of an individual oscillator for zero coupling is shown in Fig. 5.14a. Being observed at some moment of time, the uncoupled elements have different states (Fig. 5.14b). These states scatter in the phase space approximately in the same way as a trajectory of one system. If a sufficiently strong coupling is introduced and a snapshot of all states is taken, the phase points describing the motion of the elements form a small cloud, nearly a point (Fig. 5.14c), which indicates complete synchronization in the ensemble. For some intermediate values of coupling, two or three such clouds were observed, i.e., several synchronous clusters were formed in the ensemble.

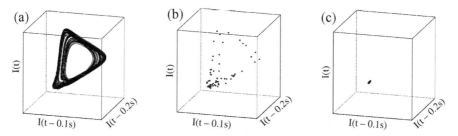

**Figure 5.14.** Complete synchronization of globally coupled electrochemical oscillators. (a) Attractor of an individual oscillator reconstructed from the current measurements of a single electrode. (b) Snapshot of position in state space constructed from the currents of all electrodes in the absence of global coupling. (c) Snapshot in the case of sufficiently strong coupling indicates complete synchronization of 64 oscillators. From [Wang *et al.* 2000a].

### 5.3.4   Chaos-destroying synchronization

A particular form of synchronization is observed if a comparatively strong periodic force acts on a chaotic system. This force can suppress chaos and make the system oscillate periodically with the period of the force. Such a regime can be interpreted as chaos-destroying synchronization.

We illustrate this with the results of Bezruchko and co-workers [1979, 1980, 1981], who experimentally studied the effect of a periodic external force on an ultrahigh frequency electronic generator, the electron beam–backscattered electromagnetic wave system. For certain values of parameters, this generator, being autonomous, exhibits chaotic oscillations. The regions where the force makes the system periodic are shown in Fig. 5.15. This plot resembles the picture of synchronization regions (Arnold tongues) discussed extensively in Section 3.2. Note that contrary to the case of periodically forced limit cycle oscillators, the regions do not touch the frequency axis, i.e., synchronization sets in only if the amplitude of the forcing exceeds some threshold value. We remind the reader that we encountered a similar situation while considering synchronization of a noisy oscillator (see Fig. 3.39).

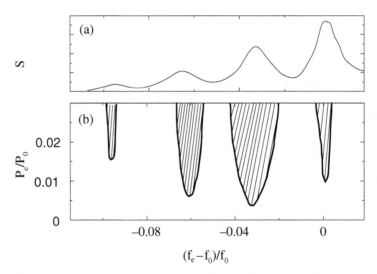

**Figure 5.15.** Chaos-destroying synchronization of an electron beam–backscattered electromagnetic wave system by a periodic force. (a) Power spectrum of the autonomous chaotic system shows several broad maxima. If a periodic force with a frequency close to that of one of the maxima acts on the system, it can suppress chaos. As a result, periodic oscillations with the frequency of the external force are observed within the shaded regions (b). Note that the force should be sufficiently strong. From [Bezruchko 1980; Bezruchko *et al.* 1981].

# Chapter 6

## Detecting synchronization in experiments

In this chapter we dwell on techniques of experimental studies of synchronization and give some practical hints for experimentalists. Previously, presenting different features of this phenomenon, we illustrated the theory with the results of a number of experiments and observations. In those examples the presence (or absence) of synchronization was quite obvious, but this is not always the case. Actually, detection of synchronization of irregular oscillators is not an easy task. A simple visual inspection of signals, as was done by Huygens in his experiments with clocks, is not always sufficient, and special techniques of data analysis are required. Indeed, the mere estimation of phase and frequency from a complex time series, especially from a nonstationary one, is a complicated problem, and we begin with its discussion. Next, we proceed in two directions: first, we summarize how to determine the synchronization properties of oscillator(s) experimentally; second, we use the idea of synchronization to analyze the interdependence between two (or more) scalar signals. Some technical details of data processing are given in Appendix A2.

## 6.1 Estimating phases and frequencies from data

Synchronization arises as the appearance of a relationship between phases and frequencies of interacting oscillators. For periodic oscillators these relations (phase and frequency locking) are rather simple (see Eqs. (3.3) and (3.2)); for noisy and chaotic systems the definition of synchronization is not so trivial. Anyway, in order to analyze synchronization in an experiment, we have to estimate phases and frequencies from the data we measure. To be not too abstract, we consider a human electrocardiogram (ECG) and a respiratory signal (air flow measured at the nose of the subject) as examples (Fig. 6.1).

### 6.1.1    Phase of a spike train. An example: electrocardiogram

An essential feature of the ECG is that every (normal) cardiocycle contains a well-pronounced sharp peak that can be with high precision localized in time; it is traditionally denoted the R-peak (Fig. 6.1a). The series of R-peaks can be considered as a sequence of point events taking place at times $t_k$, $k = 1, 2, \ldots$. The phase of such a process can be easily obtained. Indeed, the time interval between two R-peaks corresponds to one complete cardiocycle;[1] therefore the phase increase during this time interval is exactly $2\pi$. Hence, we can assign to the times $t_k$ the values of the phase $\phi(t_k) = 2\pi k$, and for an arbitrary instant of time $t_k < t < t_{k+1}$ determine the phase as a linear interpolation between these values

$$\phi(t) = 2\pi k + 2\pi \frac{t - t_k}{t_{k+1} - t_k}. \tag{6.1}$$

This method can be effectively applied to any process that contains distinct marker events and can therefore be reduced to a spike train. Determination of the phase via marker events in a time series can be regarded as an analogy to the technique of the

---

[1]  From the physiologist's viewpoint the cardiocycle starts with a P-wave that reflects the beginning of excitation in the atria. This does not contradict our procedure: we understand a cycle as the interval between two nearly identical states of the system.

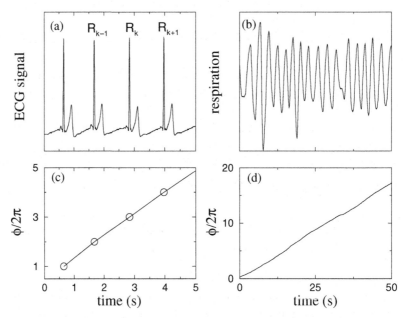

**Figure 6.1.** Short segments of (a) an electrocardiogram (ECG) with R-peaks marked and (b) of a respiratory signal; both signals are shown in arbitrary units. (c) Phase of the ECG computed according to Eq. (6.1) is a piece-wise linear function of time; the instants when the R-peaks occur are shown by circles. (d) Phase of respiration obtained via the Hilbert transform (Eq. (6.2)).

Poincaré map (see Section 5.2), although we do not need to assume that the system under study is noise-free, i.e., a dynamical one.

## 6.1.2 Phase of a narrow-band signal. An example: respiration

Now we consider the much smoother respiratory signal (Fig. 6.1b); it resembles a sine wave with slowly varying frequency and amplitude. The phase of such a narrow-band signal can be obtained by means of the **analytic signal concept** originally introduced by Gabor [1946]. To implement it, one has to construct from the scalar signal $s(t)$ a complex process

$$\zeta(t) = s(t) + is_{\mathrm{H}}(t) = A(t)e^{i\phi(t)}, \tag{6.2}$$

where $s_{\mathrm{H}}(t)$ is the Hilbert transform of $s(t)$. The **instantaneous** phase $\phi(t)$ and amplitude $A(t)$ of the signal are thus uniquely determined from (6.2). Note that although formally this can be done for an arbitrary $s(t)$, $A(t)$ and $\phi(t)$ have clear physical meaning only if $s(t)$ is a narrow-band signal (see the discussion of properties and practical implementation of the Hilbert transform and analytic signal in Appendix A2).

## 6.1.3 Several practical remarks

An important practical question is: which method of phase estimation should be chosen for the analysis of particular experimental data? Theoretically, the two techniques (Hilbert transform and Poincaré map) are both suitable,[2] but in an experimental situation, where we have to estimate the phases from short, noisy and nonstationary records, the numerical problems become a decisive factor. From our experience, if the signal has very well-defined marker events, like the ECG, the Poincaré map technique is the best choice. It could also be applied to an "oscillatory" signal, like the respiratory signal: here it is also possible to define the "events" (e.g., as the times of zero crossing) and to compute the phase according to Eq. (6.1). However, we do not recommend this procedure: the drawback is that the determination of an event from a slowly varying signal is strongly influenced by noise and trends in the signal. Besides, we only have one event per characteristic period, and if the record is short, then the statistics are poor. In such a case the technique based on the Hilbert transform is much more effective because it provides the phase for every point of the time series. Hence, we have many points per characteristic period and can therefore smooth out the influence of noise and obtain sufficient statistics for the determination of phase relationships. Finally, we mention that the technique based on the Hilbert transform is sensitive to

---

[2]  Although they give phases that vary microscopically, i.e., on time scale less than one (quasi)period, the average frequencies obtained from these phases coincide, and these are exactly the frequencies that are primarily important for a description of synchronization.

low-frequency trends (see Appendix A2); we also recommend the use of a filter if the relevant signal is mixed with signals in other frequency bands.

Another important point is that even if we can unambiguously compute the phase of a signal, we cannot avoid uncertainty in the estimation of the phase of an oscillator.[3] The estimate depends on the observable used; however, "good" observables provide equivalent phases (i.e., the average frequencies defined from these observables coincide). In an experiment we are rarely free in the choice of an observable. Therefore, one should always be very careful in formal application of the presented methods and in the interpretation of the results.

The frequency of a signal can be determined either by computation of the number of cycles per time unit (if the signal can be reduced to a spike train and the phase is computed according to Eq. (6.1)), or as a slope of the phase growth (if the analytic signal concept is used); see Appendix A2 for details.

## 6.2  Data analysis in "active" and "passive" experiments

Here we discuss two typical experimental tasks and related problems of data analysis. The first task is to find out whether a given oscillator can be entrained by a certain external force, or whether two oscillators can mutually synchronize for a given type of interaction. The second experimental problem is to analyze the signals coming from two oscillators in order to conclude whether these oscillators are coupled or independent.

### 6.2.1  "Active" experiment

We understand synchronization as the appearance of certain relations between the phases and frequencies of interacting objects. These relations must hold for a certain range of detuning and strength of coupling (this range denotes the synchronization region). Therefore, in order to establish the onset of synchronization in a particular experiment, we must either have access to some parameters of the oscillators (or an oscillator and external force) that govern the frequency detuning, or be able to vary the coupling. We have to look what happens to the frequencies and/or phase difference under variation of one of these parameters, i.e., we have to perform an "active" experiment and to control the system. The classical work of Appleton [1922] (see Section 4.1) is a good example of such an experiment.

A full description of the synchronization properties of systems under study requires the determination of the Arnold tongue(s). Nevertheless, if we can vary only one parameter and observe a transition, i.e., the adjustment of frequencies with a decrease of the detuning for a fixed coupling, or with an increase of coupling for a fixed

---

[3] Although one can compute several phases from different observables of the same oscillator, there exists only one phase of that system corresponding to its zero Lyapunov exponent.

detuning, we can be sure that these systems are synchronized. The straightforward way to characterize the synchronization transition is to compute the frequencies of coupled oscillators $\Omega_1$ and $\Omega_2$ and to plot their difference vs. the varied parameter as Appleton did in his experiment (see Fig. 4.4).

In contemporary experiments the signals are often stored in a computer and afterwards processed off-line; in this case the frequencies can be computed according to the recommendations given in the previous section and Appendix A2. We note also that if the signals are nearly periodic, then synchronization can be detected in a very simple way by observing the classical Lissajous figures, i.e., by plotting one signal against the other (cf. Figs. 3.9 and 3.23). When the frequencies of two oscillators are locked, the resulting plot represents a closed curve.

## 6.2.2   "Passive" experiment

Now we discuss experiments where one cannot change the parameters of the systems and/or coupling, but just observe the signals under free-running conditions; this situation is frequently encountered in investigations of biological or geophysical systems. A representative example is the registration of the human ECG and of the respiratory signal considered in the previous section; these bivariate data reflect the interaction of the cardiovascular and the respiratory systems. The general problem is, what kind of information can be obtained from a passive experiment? In particular, the natural question appears: can one detect synchronization by analyzing bivariate data?

Generally, the answer to the above question is negative. As synchronization is not a state, but a process of adjustment of phases and frequencies, its presence or absence cannot be established from a single observation. Besides, for noisy systems (and real data are inevitably noisy!), the synchronization transition is smeared and there is no distinct difference between synchronous and asynchronous states. Thus, without additional assumptions, we cannot detect synchronization from the data, even if we reveal some relations between the phases and frequencies (that should be estimated from the signals as described in the previous section). Nevertheless, a synchronization analysis of bivariate data may provide useful information on the interrelation of the systems that generate the signals.

We note that the determination of *interdependence between two (or more) signals* is a typical problem in time series analysis. Traditionally this is done by means of linear cross-correlation (cross-spectrum) techniques [Rabiner and Gold 1975] or nonlinear statistical measures like mutual information or maximal correlation [Rényi 1970; Pompe 1993; Voss and Kurths 1997].

As illustrated in Fig. 6.2, the problem we address here is different: we try to reveal the *interaction between the oscillators* that generate bivariate data. We emphasize that our synchronization approach to data analysis explicitly uses the assumption that the data are generated by several (at least two) interacting self-sustained systems. If this assumption cannot be checked, then the techniques we describe below should be interpreted only as tools that reveal the relation between the signals and we cannot conclude about the underlying systems.

We remember also that interdependence of phases can result from a related phenomenon such as stochastic resonance (see Section 3.6.3). Another reason is, e.g., modulation of the propagation velocity in a channel transmitting the signal (see Section 3.3.5). With these remarks we emphasize that the interpretation of passive experiments must be made very carefully; further details can be found in Section 6.4 below.

### Coincidence of frequencies vs. frequency locking

In the case when an experimentalist indeed deals with data coming from two self-sustained oscillators and finds that their frequencies are close, the question is whether this is due to a hypothesized interaction or due to pure coincidence. Up to now, there

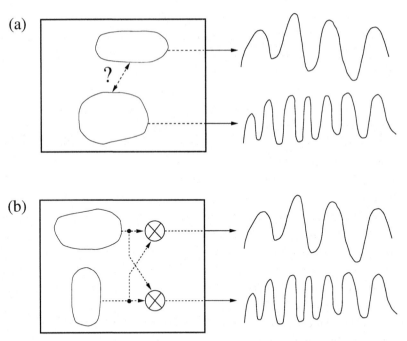

**Figure 6.2.** Illustration of the synchronization approach to the analysis of bivariate data. The goal of this analysis is to reveal the presence of a weak interaction between two subsystems from the signals at their outputs only. The assumption made is that the data are generated by two oscillators having their own rhythms (a). An alternative hypothesis is a mixture of signals generated by two uncoupled systems (b). These hypotheses cannot be distinguished by means of traditional techniques. In the former case, the interaction between the systems can be revealed by analysis of phases estimated from the data. Note that interdependence of phases can result not only from coupling of self-sustained oscillators, capable of synchronization, but also from modulation (cf. Fig. 3.31) or stochastic resonance (cf. Fig. 3.48).

has been no way to solve this problem; one can only get an indirect indication in favor of the hypothesis.

First, we note that the estimation of the frequency ratio does not help here. Indeed, if two frequencies are equal (within the precision of their determination), we cannot check whether this happens occasionally, or it is a manifestation of interaction. On the contrary, if we find that, e.g., $\Omega_1/\Omega_2 = 1.05$, it does not exclude the occurrence of $1 : 1$ synchronization, because frequent phase slips at the border of the locking region can make the frequencies different. (We remind the reader that in noisy systems the region of exact equality of frequencies can shrink to zero.) Therefore, usually the estimates of the frequencies $\Omega_{1,2}$ can only be used to estimate the possible order $n : m$ of synchronization. More information can be obtained from the analysis of phases.

There are typical experimental situations, which might be helpful in the analysis. We discuss two such aspects.

(i) **Noise** complicates the picture of synchronization, but in the particular case of the "passive" experiment it helps to distinguish the occasional coincidence of frequencies from true locking. Indeed, noise causes phase diffusion, and in the case of occasional coincidence of frequencies, the phase difference is not constant but performs a random-walk-like motion.[4] The distribution of $(\phi_1 - \phi_2) \bmod 2\pi$ is in this case nearly uniform, whereas the presence of interaction manifests itself in a peak in this distribution (see Section 3.4). This can be interpreted as the existence of a preferred value of the phase difference (taken modulo $2\pi$), i.e., as (statistically understood) phase locking. Thus, we have to analyze the phase difference; as we show below the stroboscopic technique is also very effective here.

(ii) **Nonstationarity** of the data due to slow variation of the parameters of experimental systems essentially complicates the analysis, but it also may give an additional evidence supporting the hypothesis of interacting oscillators. Indeed, if we observe that instantaneous frequencies of two signals change but their relation remains (approximately) constant then it is very unlikely that it happens by pure chance; in this case it is quite reasonable to conclude that we observe frequency locking. Another indication may be the variation of the frequency ratio from, say, $\approx 5/2$ to $\approx 3$: this is also unlikely to occur occasionally, it rather resembles the transition between adjacent Arnold tongues.

---

[4] Generally, if $n\Omega_1 \approx m\Omega_2$ we should speak of the generalized phase difference $n\phi_1 - m\phi_2$; here for simplicity of presentation we take $n = m = 1$.

## 6.3    Analyzing relations between the phases

In this section we elaborate on the techniques of phase relationship analysis and illustrate them by several examples. These techniques are based on the idea of synchronization and, therefore, we use the corresponding vocabulary, although generally speaking we can reveal only the presence of some interaction.

### 6.3.1    Straightforward analysis of the phase difference. An example: posture control in humans

The simplest approach to looking for synchronization is to plot the phase difference vs. time and look for horizontal plateaux in this presentation. Generally, we have to plot the *generalized phase difference*

$$\varphi_{n,m} = n\phi_1 - m\phi_2. \tag{6.3}$$

This straightforward method has turned out to be quite effective in the analysis of model systems and some experimental data sets.

To illustrate this, we describe the results of experiments on posture control in humans [Rosenblum *et al.* 1998]. During these tests a subject was asked to stay quietly on a special rigid force plate equipped with four tensoelectric transducers. The output of the setup provides current coordinates $(x, y)$ of the center of pressure under the feet of the standing subject. These bivariate data are called stabilograms; they are known to contain rich information on the state of the central nervous system [Gurfinkel *et al.* 1965; Cernacek 1980; Furman 1994; Lipp and Longridge 1994]. Every subject was asked to perform three tests of quiet upright standing (for three minutes) with (i) eyes opened and stationary visual surrounding (EO); (ii) eyes closed (EC); (iii) eyes opened and additional video-feedback (AF). 132 bivariate records obtained from three groups of subjects (17 healthy persons, 11 subjects with an organic pathology and 17 subjects with a psychogenic pathology) were analyzed by means of cross-spectra and generalized mutual information. It is important that interrelation between the body sway in anterior–posterior and lateral directions was found in pathological cases only. Another observation was that stabilograms could be qualitatively rated into two groups: noisy and oscillatory patterns. The latter appears considerably less frequently – only a few per cent of the records can be identified as oscillatory – and only in the case of pathology.

The appearance of oscillatory regimes in stabilograms suggests the excitation of self-sustained oscillations in the control system responsible for the maintenance of the constant upright posture; this system is known to contain several nonlinear feedback loops with a time delay. However, the independence of the body sway in two perpendicular directions for all healthy subjects and many cases of pathology suggests that two separate subsystems are involved in the regulation of the upright stance. A plausible hypothesis is that when self-sustained oscillations are excited in both these subsystems, synchronization may take place. To test whether the interdependence of two components of a stabilogram may be due to a relation between their phases, we performed an analysis of the phase difference.

Here we present the results for one trial (female subject, 39 years old, functional ataxia). We can see that in the EO and EC tests the stabilograms are clearly oscillatory (Figs. 6.3 and 6.4). The difference between these two records is that with eyes opened the oscillations in two directions are not synchronous during approximately the first 110 s, but exhibit strong interrelation between phases during the last 50 s. In the EC test, the phases of oscillations nearly coincide all the time. The behavior is essentially different in the AF test; stabilograms represent noisy patterns, and no phase relation is observed. We emphasize here that traditional techniques are not efficient in detecting the cross-dependence of these signals because of the nonstationarity and insufficient length of the time series.

### Remarks on the method

An important advantage of straightforward phase analysis is that by means of $\varphi_{n,m}(t)$ plots one can trace transitions between qualitatively different regimes that are due to nonstationarity of parameters of interacting systems and/or coupling (Fig. 6.3); this is possible even for very short records. Indeed, two different regimes that can

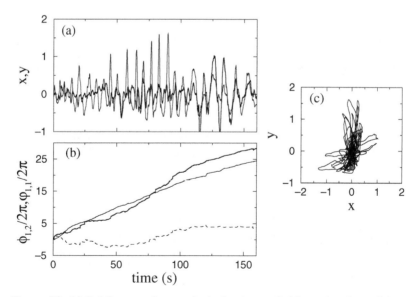

**Figure 6.3.** (a) Stabilogram of a neurological patient. $x$ (bold curve) and $y$ (solid curve) represent body sway while the subject was in a quiet stance with open eyes in anterior–posterior and lateral directions, respectively. The phases of these signals, and the phase difference are shown in (b) by bold, solid and dashed curves, respectively. The transition to a regime where the phase difference fluctuates around a constant is clearly seen at $\approx$110 s. A typical plot of $y$ vs. $x$ shows no structure that could indicate the interrelation between the signals (c). From [Pikovsky *et al.* 2000].

be distinguished in Fig. 6.3 contain only about ten characteristic periods, i.e., these epochs are too short for reliable application of conventional methods of time series analysis.

A disadvantage of the method is that synchronous regimes of order different from $n : m$, e.g., synchronization of order $n : (m+1)$, appear in this presentation as nonsynchronous epochs. Besides, there are no regular methods to determine the integers $n$ and $m$, so that they are usually found by trial and error. Correspondingly, in order to reveal all the regimes, one has to analyze a number of plots. In practice, the possible values of $n$ and $m$ can be estimated from the power spectra of the signals, or by computation of the frequencies along the lines given in the previous section, and are often restricted due to some additional knowledge on the system under study.

Another drawback of this technique is that if noise is relatively strong, this method becomes ineffective and may even be misleading. Indeed, frequent phase slips mask the presence of plateaux (cf. Fig. 4.11) and synchronization can be revealed only by a statistical approach, e.g., by analysis of the distribution of the cyclic relative phase to be defined below.

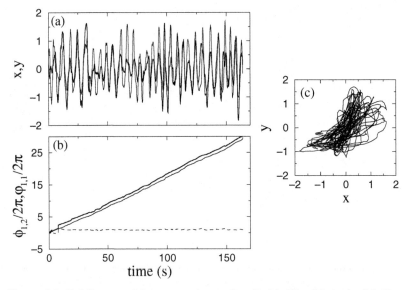

**Figure 6.4.** Stabilogram of the same patient as described in Fig. 6.3 obtained during the test with the eyes closed. All the notations are the same as in Fig. 6.3. From the phase difference one can see that the phases of the body sway in two directions are tightly interrelated within the whole test, although the amplitudes are irregular and essentially different. From [Pikovsky *et al.* 2000].

## 6.3.2  High level of noise

Strong noise causes frequent phase slips, and as a result the plot of the phase difference $\varphi_{n,m}(t)$ appears as short horizontal epochs intermingled with rapid up- and downward jumps. An example of such a behavior was given in Fig. 4.11. In this case the computation of the distribution of the cyclic relative phase $\Psi_{n,m} = \varphi_{n,m} \bmod 2\pi$ is quite helpful. Indeed, the modulo operation makes the states before and after phase jump equivalent, and interaction between the oscillators is reflected as a maximum in this distribution.

Visual inspection of Fig. 4.11 shows that the data are nonstationary: the degree of interaction varies with time. (Certainly we cannot find out why this is so; it may be that the strength of the coupling slowly varies with time, or frequencies of interacting systems, or both.) In this case it is appropriate to perform a *running window analysis*, sequentially computing the distributions in the time window $[t - \tau/2, t + \tau/2]$, where $\tau$ is the window length, for different $t$. One can characterize the time course of interaction by introducing a number that quantifies the deviation of the distribution from a uniform one. Computation of such indices is considered in [Tass *et al.* 1998; Rosenblum *et al.* 2001].

As a final remark we repeat that for systems with strong (or unbounded) noise it is impossible to say unambiguously whether the given state is synchronous or not. In this respect quantification of the strength of interaction from the data makes sense only for a relative comparison of different states of the same system. Particularly, this proved to be useful in the following two cases.

- Computation of a synchronization index as a function of time can reveal the time course of interaction. In the case of analysis of the brain and muscle activity data (see [Tass *et al.* 1998; Tass 1999] and Section 4.1.7) it reflected the intensity of the tremor activity; this fact allowed us to conclude that synchronization processes in the brain are responsible for the origination of tremor.

- In the case of multichannel data, i.e., when many oscillators interact, computation of a synchronization index for different pairs of signals helps in the determination the relative degree of interrelation between different signals. So, in the above mentioned example this approach allows one to establish which parts of the brain are involved in the generation of pathological tremor activity.

## 6.3.3  Stroboscopic technique

Here we illustrate the application of the stroboscopic technique that was initially introduced in Section 3.2 in the context of periodically kicked oscillators. With this method, the phase of the driven oscillator is observed with the period of external force, $\phi_k = \phi(t_0 + k \cdot T)$, where $k = 1, 2, \ldots$ and $t_0$ is the (arbitrary) time of the first observation. If the oscillator is entrained, the distribution of the $\phi_k$ is a $\delta$-function if the oscillator is periodic; it is narrow if the oscillator is noisy or chaotic. A nonsynchronous state implies that the stroboscopically observed phase attains an arbitrary value, and its distribution is therefore broad.

A simple generalization makes this technique a very effective tool for time series analysis. To this end, we consider two coupled oscillators and observe the phase of one oscillator not periodically in time, but periodically with respect to the phase of the other oscillator. One can say that the second oscillator plays the role of a time marker, "highlighting" $\phi_1$ every time when $\phi_2$ gains $2\pi$. In other words, we pick up $\phi_{1k}$ at the moments when $\phi_2(t) = \phi_0 + 2\pi \cdot k$. We refer to this tool as the *phase stroboscope*. Obviously, if the second oscillator is periodic, the phase and the time stroboscopes are equivalent. Certainly, it does not matter which oscillator is taken as the reference one (second in our notation); the choice solely depends on the convenience of the phase determination.

In the rest of this section we explain and illustrate how the stroboscopic technique can be used for the detection of interaction (provided that we know that the signals originate from interacting self-sustained oscillators) in the case when the frequencies of the signals obey $n\Omega_1 \approx m\Omega_2$, or, generally, for the detection of complex relations between the phases of two signals.

### 6.3.4   Phase stroboscope in the case $n\Omega_1 \approx m\Omega_2$. An example: cardiorespiratory interaction

Suppose first that we deal with two $n : 1$ synchronized oscillators that generate signals like those shown in Fig. 6.1 and let $n$ spikes[5] of the fastest signal occur within one cycle of the slow one, i.e., there is $n : 1$ locking. Then, we expect to find the spikes at $n$ different values of the phase of the slow signal. A similar picture can be observed if there is no synchronization, but one process is modulated by the other one. Therefore, in a particular experiment, we can use this idea to reveal a complex interaction, but we cannot distinguish between synchronization and modulation.

It is natural to observe the phase of the slow signal $\phi_1$ at the times of spiking. Thus, we plot the stroboscopically observed cyclic phase $\psi(t_k) = \phi_1(t_k)$ mod $2\pi$ vs. time and call such a plot a *synchrogram* (Fig. 6.5). The presence of interaction is reflected by the occurrence of $n$ stripes in this presentation.

The final step is to extend the stroboscopic technique to the general case of $n : m$ locking. Suppose again that we observe one oscillator, whenever the phase of the second one is a multiple of $2\pi$. Then, if interaction is present, we expect to make $n$ observations within $m$ cycles of the first oscillator. To construct a synchrogram we

---

[5] If there were no spikes, we can define the events as, say, zero crossing in one direction. In other words, we have to define the instants when the phase of the fastest oscillator attains some fixed value.

have to somehow distinguish the phases within $m$ adjacent cycles. For this purpose we make use of the fact that phase can be defined either on a circle, i.e., from 0 to $2\pi$, or on a real line. We have often intermingled these two definitions, and the range of the phase variation was clear from the context. Now we perform the following trick: we take the unwrapped (i.e., infinitely growing) phase and wrap it on the circle $[0, 2\pi m]$. In this way we consider $m$ cycles as one cycle, and then proceed as before, plotting $\psi_m(t_k) = \phi_1(t_k) \bmod 2\pi m$ vs. time (Fig. 6.5); the index $m$ indicates how the phase was wrapped. Note that only the value of $m$ should be chosen by trial and error, and different epochs, say with approximate frequency ratios $n : m$ and $(n + 1) : m$, can be seen within one synchrogram.

Here we present phase analysis of cardiorespiratory interaction in humans. It has been well-known for at least 150 years [Ludwig 1847], that cardiovascular and respiratory systems do not act independently; their interrelation is rather complex and still remains a subject of physiological research (see, e.g., [Koepchen 1991; Saul 1991] and references therein). As a result of this interaction, in healthy subjects the heart rate normally increases during inspiration and decreases during expiration, i.e., the heart rate is modulated by a respiratory-related rhythm. This frequency modulation of the heart rhythm (see Fig. 6.10) has been known for at least a century and is commonly referred to as "respiratory sinus arrhythmia" (RSA) (see, e.g., [Schmidt and Thews 1983] and references therein). Modulation, strong nonstationarity of the data, and a high level of noise make the interaction hardly detectable. The synchrogram technique

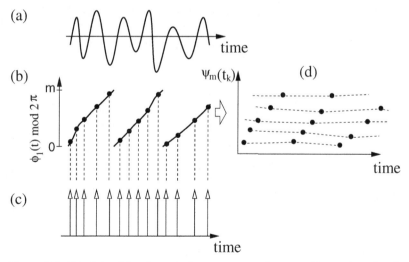

**Figure 6.5.** Principle of the phase stroboscope, or synchrogram. Here a slow signal (a) is observed in accordance with the phase of a fast signal (c). Measured at these instants, the phase $\phi_1$ of the slow signal wrapped modulo $2\pi m$, (i.e., $m$ adjacent cycles are taken as a one longer cycle) is plotted in (d); here $m = 2$. In this presentation $n : m$ phase synchronization shows up as $n$ nearly horizontal lines in (d); a similar picture appears in the case of modulation.

turned out to be a most useful tool in such an analysis [Schäfer *et al.* 1998, 1999; Rosenblum *et al.* 2001].

The bivariate data, namely the electrocardiogram (ECG) and respiratory signal, were introduced in Fig. 6.1. The signals we analyze were recorded from a healthy newborn baby [Mrowka *et al.* 2000]. Synchrograms reveal short alternating epochs of interaction with the approximate frequency relations 2 : 1 and 5 : 2 (Fig. 6.6).

### 6.3.5   Phase relations in the case of strong modulation. An example: spiking of electroreceptors of a paddlefish

Neiman *et al.* [1999a, 2000] studied spiking from electroreceptors of a paddlefish, *Polydon spathula*, in the presence of a periodic electric field. The field imitated the signals coming from zooplankton, the natural prey of the fish (see [Wilkens *et al.* 1997; Russell *et al.* 1999]). This is a typical active experiment (the parameters of the forcing can be varied) and therefore should be analyzed by means of the frequency–detuning plots. For this purpose, one just has to count the number of spikes within a fixed time interval and estimate the frequencies. Certainly, no signal analysis techniques are required here, and we use these data solely as a model example for illustration of the efficiency of the stroboscopic approach for detection of interrelation between the phases of two signals.

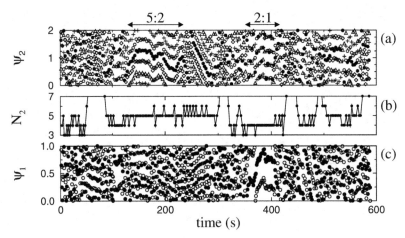

**Figure 6.6.** An example of alternation of interaction with approximate frequency relations 2 : 1 and 5 : 2 between heart rate and respiration of a healthy baby. (a) Two adjacent respiratory cycles are taken as one cycle. Therefore, epochs of 2 : 1 and 5 : 2 relation between phases appear as four- and five-line patterns. (b) Plot of the number of heartbeats within two respiratory cycles $N_2$ also indicates epochs of interdependence. (c) If the respiratory phase is wrapped modulo $2\pi$ then only the 2 : 1 relation is seen. The data points are shown by different symbols in cyclic order (five and two symbols in (a) and (c), respectively) for better visual impression. Plotted using data from [Mrowka *et al.* 2000].

The data were recorded by D. F. Russell, A. B. Neiman and F. Moss from an electrosensitive neuron. An essential feature of the data is modulation of spiking by the external electric field (Fig. 6.7c).[6] First, we estimate that $\Omega/\omega_e \approx 11$ and compute the 1 : 11 phase difference (Fig. 6.7a, b). It is convenient to determine it when the spikes occur; we obtain $\phi_{1,11}(t_k) = 11 \cdot \omega_e t_k - 2\pi k$, where $t_k$ is the time of the $k$th spike, and $\omega_e = 2\pi \cdot 5$; subscript e refers to the external field. The distribution of the cyclic relative phase $\Psi_{11,1} = \phi_{1,11} \bmod 2\pi$ is shown in Fig. 6.7d. Next, we use the stroboscopic technique; the results are shown in Fig. 6.8.

The presented example shows that in the case of modulation the stroboscopic technique is much more effective for detection of the relation between the phases of two signals than a simple analysis of the phase difference. Note that fluctuation of $\phi_{11,1}(t)$ is very strong (its amplitude is $\approx 2\pi$). As a result, the distribution of the cyclic relative phase is not unimodal and does not indicate interaction. On the contrary, 11 stripes in the synchrogram clearly show some complex relation; modulation just shows up as uneven spacing of these stripes.

[6] The external force is shown schematically as we do not know its exact shape and initial phase, only the frequency. Note that in the absence of an external field the interspike intervals do not vary much.

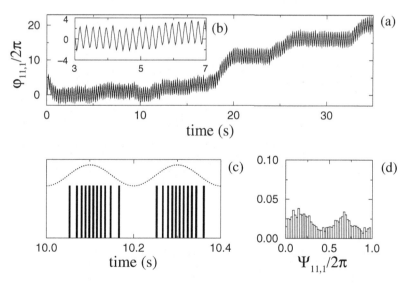

**Figure 6.7.** Phase relation between a neuron spiking and external electric field. (a, b) Plot of the phase difference supports the hypothesis of complex interrelation between the data. The raw signals are plotted in (c), two periods of the external force are shown; within this time interval there are two groups of 11 spikes. Note that due to strong modulation the fluctuation of the phase difference (a, b) is very strong, and the respective distribution (d) of the cyclic relative phase in not unimodal. From these plots it is not clear whether the signals are indeed interrelated; the synchrogram technique (Fig. 6.8) is much more effective here. Data courtesy of D. F. Russell, A. B. Neiman and F. Moss.

In summary, we cannot propose a unique recipe for the choice of analysis technique; this choice depends on the particular data set. In the case of complex noisy signals one should not rely on one method only. We recommend a combination of the simplest methods, e.g. counting the number of spikes within a cycle of external force with computation of the distribution of the relative phase and stroboscopic technique.

## 6.4    Concluding remarks and bibliographic notes

### 6.4.1    Several remarks on "passive" experiments

We conclude our discussion of experimental techniques with several notes. We emphasize again that the inverse problem – an attempt to reveal the interaction between oscillators without access to their parameters, only on the basis of data analysis – is ambiguous. We warn against blind application of synchronization analysis of bivariate data and careless interpretation of results. Four remarks are in order:

(i) **Synchronization analysis is based on an assumption.** We should not forget that synchronization analysis of bivariate data is based on the assumption that there are two self-sustained oscillators that either interact or oscillate

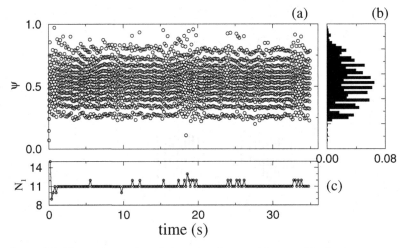

**Figure 6.8.** The stroboscopic technique (synchrogram) is very effective in an analysis of interrelations when the spiking frequency of the electrosensitive receptor of a paddlefish is much higher than that of an external electric field. (a) The 11-stripe structure of the synchrogram makes interrelation (with the exception of short epochs) obvious; this is confirmed by the corresponding distribution in (b). (c) Plot of the number of spikes per cycle of external force also confirms an approximate 11 : 1 relation between frequencies. Data courtesy of D. F. Russell, A. B. Neiman and F. Moss.

independently. Otherwise, this analysis makes no sense; this is illustrated by Fig. 6.9.

(ii) **Not all observables are appropriate.** As a counter-example we show two signals that do represent two interacting self-sustained oscillators, namely the cardiovascular and respiratory systems, but one of the signals is not suitable for phase determination. The data, the interbeat (RR) intervals and respiration, are shown in Fig. 6.10. One can say that both signals vary synchronously, but we emphasize that the RR time series reflects variation of the period of the heartbeat, not the heartbeat itself. Contraction of the heart with a constant rate would provide constant RR intervals. Thus, estimation of the phase from this time series by means of the Hilbert transform, although formally possible, provides an angle variable that is not related to the correct phase.

(iii) **We detect interaction, not synchronization.** Strictly speaking, the fact that for two coupled self-sustained systems the distribution of the cyclic relative phase $(\phi_1 - \phi_2)$ mod $2\pi$ has a maximum indicates only the presence of some interaction, but not synchronization. We remind the reader that for periodic systems, the growth of the phase difference outside the Arnold tongue, but in its vicinity, is not uniform (see Fig. 3.8). For noisy systems the border of synchronization is smeared, and the question of whether the state is synchronous is ambiguous. Moreover, maxima in the distribution of the cyclic

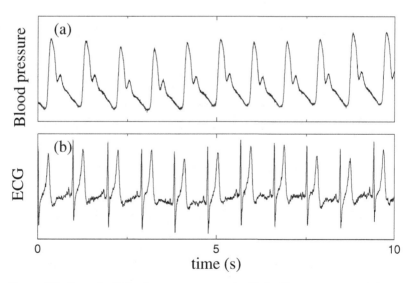

**Figure 6.9.** Records of blood pressure measured with the Finapress system at the index finger (a) and an electrocardiogram (b). One pulse in the blood pressure corresponds to one contraction of the heart, but it cannot be otherwise. The data represent two variables coming from one self-sustained system and one passive system. This is an example of the case when synchronization analysis is useless. Data courtesy of R. Mrowka.

relative phase can appear in the course of modulation of signals. Thus, it is not correct to speak of the detection of synchronization, one should always bear in mind that data analysis in a passive experiment can indicate only the presence of an interaction between the observed systems.

(iv) **Analysis of phase interrelation vs. other techniques.** It is worth noting that analysis of phase relations and cross-correlation (cross-spectral) techniques capture different aspects of interaction between systems: some examples show that two correlated (coherent) signals can originate from nonsynchronized oscillators [Tass *et al.* 1998]. This problem has not yet been studied systematically.

## 6.4.2  Quantification and significance of phase relation analysis

A natural problem in analyzing the relation between phases $\phi_{1,2}$ is quantification of this interrelation. Several measures have been recently proposed. Palus [1997] computed the mutual information between two phases. Tass *et al.* [1998] characterized the deviation of the distribution of the relative phase from a uniform one by means of the Shannon entropy; another measure is the intensity of the first Fourier mode of that distribution [Rosenblum *et al.* 2001]. A different approach is based on computation of the

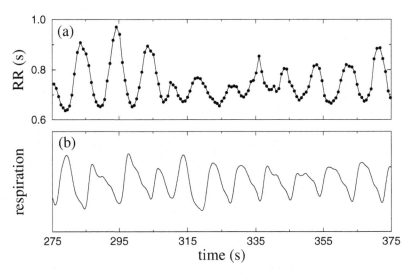

**Figure 6.10.** The cardiac interbeat intervals (a) oscillate in accordance with respiration (b) (in arbitrary units). This is an example of pronounced respiratory sinus arrhythmia in a young athlete. From the physical standpoint this oscillation is an effect of modulation that can be, or can be not accompanied by synchronization. From Schäfer *et al.*, *Physical Review* E, Vol. 60, 1999, pp. 857–870. Copyright 1999 by the American Physical Society.

conditional probability for $\phi_1$ to attain a certain value provided $\phi_2$ is constant; this is equivalent to quantification of the distribution of the stroboscopically observed phase [Tass *et al.* 1998; Rosenblum *et al.* 2001]. Toledo *et al.* [1999] quantified the flatness of stripes in the synchrogram. Neiman *et al.* [1999b] and Anishchenko *et al.* [2000] used the coefficient of phase diffusion as a measure of synchronization strength; computation of this coefficient requires a very long time series which complicates application of this measure to real experiments. Besides, this is a relative measure (one should compare the diffusion of uncoupled and coupled oscillators), so that it probably can be more efficiently used in active experiments, along with computation of frequency-detuning curves.

Estimation of the significance of phase analysis remains unanswered. Several attempts to address this problem exploit surrogate data tests. For the construction of surrogates, Seidel and Herzel [1998] used randomization of Fourier phases, which seems to be too weak a test. Tass *et al.* [1998] estimated the significance level of synchronization indices using as surrogates white or instrumental noise (empty-room measurements) filtered in the same way as the signal. In a study of cardiorespiratory interaction, Toledo *et al.* [1999] performed surrogate data tests taking the heart rate and respiration from different subjects, or inverting the interbeat series in time. The difficulties encountered in the construction of appropriate surrogates are discussed in [Schäfer *et al.* 1999; Rosenblum *et al.* 2001]. Note that surrogate data tests require a quantitative measure of interrelation.

### 6.4.3   Some related references

A particular kind of active experiment was described by [Glass and Mackey 1988]. They studied phase resetting of an oscillator by a single pulse (see Section 3.2.3). By applying the stimulus in different phases, one can experimentally obtain the dependence $\phi_{new} = \mathcal{F}(\phi_{old})$, i.e., the circle map. Numerical iteration of this map can predict the synchronization properties of a periodically kicked oscillator.

In our presentation of the synchronization approach to analysis of bivariate data we follow our previous works [Rosenblum *et al.* 1997a; Rosenblum and Kurths 1998; Rosenblum *et al.* 1998; Schäfer *et al.* 1998, 1999; Tass *et al.* 1998; Rosenblum *et al.* 2001]. This ansatz implies only estimation of some interdependence between the phases, whereas irregular amplitudes may remain uncorrelated. The irregularity of amplitudes can mask phase locking so that traditional techniques treating not the phases but the signals themselves may be less sensitive in the detection of the systems' interrelation. Mormann *et al.* [2000] exploited the concept of phase synchronization in an analysis of encephalograms recorded from patients with temporal lobe epilepsy. They observed characteristic spatial and temporal shifts in synchronization that appear to be strongly related to pathological activity.

Rodriguez *et al.* [1999] used the Gabor wavelet to estimate the phase of the signal; this method can be used in the case of 1 : 1 locking (if the same wavelet function is used for processing both signals). This technique seems to be similar to the analytical signal approach.

A graphical presentation similar to synchrograms was introduced in the context of cardiorespiratory interaction by Stutte and Hildebrandt [1966], Pessenhofer and Kenner [1975] and Kenner *et al.* [1976]. Instead of the phase, along the $y$-axis they plotted the time interval since the previous inspiration. No wrapping was used, so that this graphical tool allowed a search for $1 : m$ locking only. Such a simple variant of the synchrogram was also used by Hoyer *et al.* [1997], Schiek *et al.* [1998] and Seidel and Herzel [1998]. Bračič and Stefanovska [2000] used the synchrogram technique to analyze cardiorespiratory interaction in healthy relaxed subjects (nonathletes). Toledo *et al.* [1998] showed that the cardiorespiratory interaction can also be observed in heart transplant subjects. These subjects have no direct neural regulation of the heart rate by the autonomic nervous system, therefore in this case some other mechanisms are responsible for this phenomenon.

We also mention that Schiff *et al.* [1996] used the notion of dynamical interdependence introduced by Pecora *et al.* [1997a] and applied the mutual prediction technique to verify the assumption that measured bivariate data originate from two synchronized systems, where synchronization was understood to be the existence of a functional relationship between the states of two systems (generalized synchronization); see also [Arnhold *et al.* 1999]. As the state of phase synchronization occurs for lower values of coupling than the state of generalized synchronization, we expect phase synchronization analysis to be more sensitive.

It is interesting to note that an idea similar to the phase stroboscope was implemented 40 years ago in a device called a cardio-synchronizer, which allowed the operator to take X-ray images of a heart at an arbitrarily chosen phase of the cardiocycle [Tcetlin 1969]. However, this is certainly not synchronization in our sense, because the cardiovascular system was observed in accordance with its own phase.

**Part II**

Phase locking and frequency entrainment

# Chapter 7

## Synchronization of periodic oscillators by periodic external action

In this chapter we describe synchronization of periodic oscillators by a periodic external force. The main effect here is complete locking of the oscillation phase to that of the force, so that the observed oscillation frequency coincides exactly with the frequency of the forcing.

We start our consideration with the case of small forcing. In Section 7.1 we use a perturbation technique based on the phase dynamics approximation. This approach leads to a simple phase equation that can be treated analytically. This equation is, however, nonuniversal, as its form depends on the particular features of the oscillator. Another analytic approach is presented in Section 7.2; here we assume not only that the force is small, but also that the periodic oscillations are weakly nonlinear. This enables us to use a method of averaging and to obtain universal equations depending on a few parameters. Historically, this is the first analytical approach to synchronization going back to the works of Appleton [1922], van der Pol [1927] and Andronov and Vitt [1930a,b]. The averaged equations can be analyzed in full detail, but their applicability is limited: in fact, quantitative predictions are possible only for small-amplitude self-sustained oscillations near the Hopf bifurcation point of their appearance.

Generally, when the forcing is not small and/or the oscillations are strongly non-linear, we have to rely on the qualitative theory of dynamical systems. The tools used here are the annulus and the circle maps described in Section 7.3. This approach gives a general description up to the transition to chaos, it allows one to find limits of the analytical methods and provides a framework for numerical investigations of particular systems.

In Section 7.4 we discuss the synchronization of rotators. These systems are de-scribed by a phase-like variable, and the synchronization properties are similar to those of self-sustained oscillators. Finally, we describe a technical system (phase locked loop) that can be interpreted as a particular example of a driven oscillator.

## 7.1    Phase dynamics

In this section we consider the effect of a weak periodic external force on periodic self-sustained oscillators. The main idea here is that a small force influences only the phase, not the amplitude, so that we can describe the dynamics with a phase equation. In its derivation we follow the method developed by Malkin [1956] and Kuramoto [1984]. Although the method is quite general, the resulting phase equation is very simple and easy to investigate. This will allow us to describe many important properties of synchronization analytically.

### 7.1.1    A limit cycle and the phase of oscillations

Consider a general $M$-dimensional ($M \geq 2$) dissipative autonomous[1] system of ordinary differential equations

$$\frac{d\mathbf{x}}{dt} = \mathbf{f}(\mathbf{x}), \qquad \mathbf{x} = (x_1, \ldots, x_M), \tag{7.1}$$

and suppose that this system has a stable periodic (with period $T_0$) solution $\mathbf{x}_0(t) = \mathbf{x}_0(t + T_0)$. In the *phase space* (the space of all variables $\mathbf{x}$) this solution is an isolated closed attractive trajectory, called the *limit cycle* (Fig. 7.1). The point in the phase

---

[1] Formally, a driven system can be written as an autonomous one, if one introduces an additional equation for a variable that is equivalent to the time. Physically, such a manipulation does not make the system really autonomous, because the "time variable" cannot be influenced.

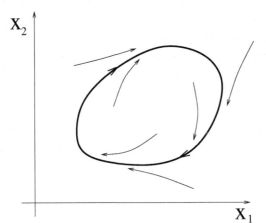

**X$_2$**

**X$_1$**

**Figure 7.1.** A stable limit cycle (bold curve), here shown for a two-dimensional dynamical system. Its form can be very different from a circular one; in a high-dimensional phase space it can be even knotted. The neighboring trajectories are attracted to the cycle.

space moving along the cycle represents the *self-sustained oscillations*.[2] The most popular classical example of a self-oscillating system is the van der Pol equation [van der Pol 1920, 1927]

$$\ddot{x} - 2\mu\dot{x}(1 - \beta x^2) + \omega_0^2 x = 0. \tag{7.2}$$

For small $\mu$ this oscillator is quasilinear, while for large $\mu$ it is of relaxation type.

Our first aim is to describe this motion in terms of the phase. We introduce the phase $\phi$ as a coordinate along the limit cycle, such that it grows monotonically in the direction of the motion and gains $2\pi$ during each rotation. Moreover, we demand that the phase grows uniformly in time so that it obeys the equation

$$\frac{d\phi}{dt} = \omega_0, \tag{7.3}$$

where $\omega_0 = 2\pi/T_0$ is the frequency of self-sustained oscillations. In the following, when the period of oscillations is influenced by forcing and/or coupling, we will need to refer to the frequency of this isolated oscillator; thus we call $\omega_0$ the *natural frequency*. Note that such a uniformly rotating phase always exists, and can be obtained from any nonuniformly rotating $2\pi$-periodic angle variable $\theta$ on the cycle through the transformation

$$\psi = \omega_0 \int_0^\theta \left[\frac{d\theta}{dt}\right]^{-1} d\theta. \tag{7.4}$$

The system variables $\mathbf{x}$ on the cycle are $2\pi$-periodic functions of the phase.

From Eq. (7.3) follows a very important property of the phase: it is a neutrally stable variable. Indeed, a perturbation in the phase remains constant: it neither grows nor decays in time. In terms of trajectory stability this means that a stable limit cycle has one zero Lyapunov exponent corresponding to perturbations along the cycle (the other exponents correspond to transverse perturbations and are negative). This reflects the property of autonomous dynamical systems to be invariant under time shifts: if $\mathbf{x}(t)$ is a time-dependent solution, then the same function of time with a shifted argument $\mathbf{x}(t + \Delta t)$ is a solution, too. On the limit cycle, the time shift $\Delta t$ is equivalent to the phase shift $\Delta\phi = \omega_0\Delta t$. In mathematical language, one can say that the phase is stable but not asymptotically stable.

## 7.1.2  Small perturbations and isochrones

We now consider the effect of a small external periodic force on the self-sustained oscillations. We describe the forced system by the equations

$$\frac{d\mathbf{x}}{dt} = \mathbf{f}(\mathbf{x}) + \varepsilon\mathbf{p}(\mathbf{x}, t), \tag{7.5}$$

---

[2] This should be contrasted to periodic motions in conservative integrable systems that are typically neither isolated nor attractive. Some relations between the frequencies in such systems can be observed (e.g., between the periods of the planets in the solar system), but we do not consider these relations as synchronization, but rather as resonances.

where the force $\varepsilon \mathbf{p}(\mathbf{x}, t) = \varepsilon \mathbf{p}(\mathbf{x}, t + T)$ has a period $T$, which is in general different from $T_0$. The force is proportional to a small parameter $\varepsilon$, and below we consider only first-order effects in $\varepsilon$.

The external force drives the trajectory away from the limit cycle, but because it is small and the cycle is stable, the trajectory only slightly deviates from the original one $\mathbf{x}_0(t)$, i.e., it lies in the near vicinity of the stable limit cycle.[3] Thus, perturbations in the directions transverse to the cycle are small.[4] Contrary to this, the phase perturbation can be large: the force can easily drive the phase point along the cycle. This qualitative picture suggests a description of the perturbed dynamics with the phase variable only, resolving the perturbations transverse to the limit cycle with the help of a perturbation technique. To this end we need to introduce the phase of the autonomous system (Eq. (7.1)) not only on the limit cycle, as we have done above, but also in its near vicinity. A natural and convenient definition has been suggested by Winfree [1980] and Guckenheimer [1975], see also [Kuramoto 1984].

The key idea is to define the phase variable in such a way that it rotates uniformly according to Eq. (7.3) not only on the cycle, but in its neighborhood as well. To accomplish this, we need to define the so-called *isochrones* [Winfree 1967; Guckenheimer 1975]. The construction of these curves in the vicinity of the limit cycle is illustrated in Fig. 7.2. Let us observe the dynamical system (7.1) stroboscopically, with the time interval being exactly the period of the limit cycle $T_0$. Thus we get from (7.1) a mapping

$$\mathbf{x}(t) \rightarrow \mathbf{x}(t + T_0) \equiv \Phi(\mathbf{x}).$$

This mapping has all points on the limit cycle as fixed points, and all points in the vicinity of the cycle are attracted to it. Let us choose a point $\mathbf{x}^*$ on the cycle and consider all the points in the vicinity that are attracted to $\mathbf{x}^*$ under the action of $\Phi$. They form a $(M - 1)$-dimensional hypersurface $I$, called an *isochrone*, crossing the limit cycle at $\mathbf{x}^*$. An isochrone hypersurface can be drawn through every point on the cycle. Thus, we can parameterize the hypersurfaces according to the phase as $I(\phi)$ (Fig. 7.2). We now extend the definition of the phase to the vicinity of the limit cycle, demanding that the phase is constant on each isochrone $I(\phi)$. In this way we define the phase in a neighborhood of the limit cycle, at least in that neighborhood where the isochrones exist.

It is clear why the hypersurfaces $I(\phi)$ are called isochrones: the flow of the dynamical system (7.1) transforms these hypersurfaces into each other. From this construction it follows immediately that the phase obeys Eq. (7.3), because isochrones rotate with the same velocity as a point on the cycle. Moreover, the evolution during the cycle period $T_0$ keeps these hypersurfaces invariant. They have thus a remarkable property: if

---

[3] For relaxation oscillators, which have extremely stable limit cycles, deviations from the limit cycle are small even for moderate force. These oscillators can be well-described with the phase variable only, we will discuss them in Section 7.3.

[4] One may call the variable transverse to the limit cycle amplitude, this definition is, however, ambiguous in a higher-dimensional phase space.

we choose such a surface of section for the Poincaré map, this map will have the same return time for all points on the surface. Note also that the isochrones are well-defined for a stable cycle as well as for a fully unstable one (it is unstable in all transverse directions and thus becomes stable in the inverse time, so that the isochrones can be defined using the inverse time), but are not defined for saddle cycles which have both stable and unstable manifolds.

### 7.1.3    An example: complex amplitude equation

We now consider a particular example of a system with a limit cycle and define its phase and the isochrones. We write the system in the complex form as a first-order differential equation for the complex variable $A$. As we will see in Section 7.2, this equation describes weakly nonlinear self-sustained oscillations and $A$ is a complex amplitude, cf. Eq. (7.41). In different contexts this equation is called the Landau–Stuart equation, or a "lambda–omega model":

$$\frac{dA}{dt} = (1 + i\eta)A - (1 + i\alpha)|A|^2 A. \qquad (7.6)$$

Rewriting this equation in polar coordinates, $A = Re^{i\theta}$, we obtain the second-order system

$$\frac{dR}{dt} = R(1 - R^2), \qquad (7.7)$$

$$\frac{d\theta}{dt} = \eta - \alpha R^2, \qquad (7.8)$$

which is readily solvable. The limit cycle here is the unit circle $R = 1$ and the solution

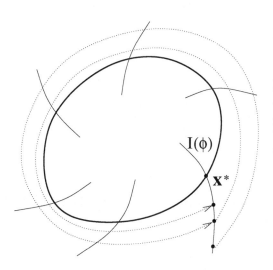

**Figure 7.2.** Isochrones $I(\phi)$ in the vicinity of the stable limit cycle. The isochrones are invariant under the stroboscopic (with the period of the cycle $T_0$) mapping, which is formed by trajectories shown with dotted arrowed lines.

with arbitrary initial conditions $R_0 = R(0)$, $\theta_0 = \theta(0)$ reads

$$R(t) = \left[1 + \frac{1 - R_0^2}{R_0^2}e^{-2t}\right]^{-1/2},$$

$$\theta(t) = \theta_0 + (\eta - \alpha)t - \frac{\alpha}{2}\ln(R_0^2 + (1 - R_0^2)e^{-2t}). \tag{7.9}$$

On the limit cycle the angle variable $\theta$ rotates with a constant velocity $\omega_0 = \eta - \alpha$ and, hence, can be taken as the phase $\phi$. However, if the initial amplitude deviates from unity, an additional phase shift occurs due to the term proportional to $\alpha$ in (7.8). It is easy to see from (7.9) that the additional phase shift is equal to $-\alpha \ln R_0$. Thus, the phase can be defined on the whole plane $(R, \theta)$ as

$$\phi(R, \theta) = \theta - \alpha \ln R. \tag{7.10}$$

One can check that this phase indeed rotates uniformly:

$$\frac{d\phi}{dt} = \frac{d\theta}{dt} - \alpha\frac{\dot{R}}{R} = \eta - \alpha.$$

The isochrones are the lines of constant phase $\phi$; on the $(R, \theta)$ plane these are logarithmic spirals

$$\theta - \alpha \ln R = \text{constant}.$$

For the case $\alpha = 0$ instead of spirals we have straight lines $\theta = \phi$. This example presents a good opportunity to discuss the property of isochronity of oscillations. Physically, one often says that isochronous oscillations are those with an amplitude-independent frequency, and for nonisochronous oscillations the frequency depends on the amplitude (the variable $R$ in our example). This definition is, however, ambiguous, because outside of the limit cycle one can define the phase and, correspondingly, the frequency in different ways. If we take the above definition based on isochrones, then the frequency is constant and all oscillators appear to be isochronous. From the other side, the frequency, defined as the rotation velocity of the angle variable $\theta$ in the example above, is given by Eq. (7.8), which is amplitude-dependent. We prefer the latter approach, and call the oscillator (7.6) isochronous if $\alpha = 0$, and non-isochronous otherwise. In terms of isochrones, one can call the oscillator isochronous if the isochrones are orthogonal to the limit cycle, and nonisochronous otherwise. Note that this definition is still ambiguous as it is noninvariant under transformations of coordinates.

## 7.1.4    The equation for the phase dynamics

Having defined the phase in some neighborhood of the limit cycle, we can write Eq. (7.3) in this vicinity as

$$\frac{d\phi(\mathbf{x})}{dt} = \omega_0. \tag{7.11}$$

As the phase is a smooth function of the coordinates $\mathbf{x}$, we can represent its time derivative as

$$\frac{d\phi(\mathbf{x})}{dt} = \sum_k \frac{\partial\phi}{\partial x_k} \frac{dx_k}{dt}, \tag{7.12}$$

which gives, together with (7.1), the relation

$$\sum_k \frac{\partial\phi}{\partial x_k} f_k(\mathbf{x}) = \omega_0.$$

We consider now the perturbed system (7.5). Using the "unperturbed" definition of the phase and substituting Eq. (7.5) in (7.12) we obtain

$$\frac{d\phi(\mathbf{x})}{dt} = \sum_k \frac{\partial\phi}{\partial x_k}(f_k(\mathbf{x}) + \varepsilon p_k(\mathbf{x}, t)) = \omega_0 + \varepsilon \sum_k \frac{\partial\phi}{\partial x_k} p_k(\mathbf{x}, t). \tag{7.13}$$

The second term on the right-hand side (r.h.s.) is small (proportional to $\varepsilon$), and the deviations of $\mathbf{x}$ from the limit cycle $\mathbf{x}_0$ are small too. Thus, in the first approximation we can neglect these deviations and calculate the r.h.s. on the limit cycle:

$$\frac{d\phi(\mathbf{x})}{dt} = \omega_0 + \varepsilon \sum_k \frac{\partial\phi(\mathbf{x}_0)}{\partial x_k} p_k(\mathbf{x}_0, t). \tag{7.14}$$

Because the points on the limit cycle are in one-to-one correspondence with the phase $\phi$, we get a closed equation for the phase:

$$\frac{d\phi}{dt} = \omega_0 + \varepsilon Q(\phi, t), \tag{7.15}$$

where

$$Q(\phi, t) = \sum_k \frac{\partial\phi(\mathbf{x}_0(\phi))}{\partial x_k} p_k(\mathbf{x}_0(\phi), t).$$

Note that $Q$ is a $2\pi$-periodic function of $\phi$ and a $T$-periodic function of $t$.

## 7.1.5   An example: forced complex amplitude equations

As an example, we periodically force the nonlinear oscillator described by Eq. (7.6), which we rewrite as a system of real equations:

$$\frac{dx}{dt} = x - \eta y - (x^2 + y^2)(x - \alpha y) + \varepsilon \cos \omega t,$$

$$\frac{dy}{dt} = y + \eta x - (x^2 + y^2)(y + \alpha x).$$

Rewriting the expression for the phase (7.10) as

$$\phi = \tan^{-1} \frac{y}{x} - \frac{\alpha}{2} \ln(x^2 + y^2),$$

we obtain a particular case of the phase equation (7.15):

$$\frac{d\phi}{dt} = \omega_0 + \varepsilon \frac{\partial\phi}{\partial x} \cos \omega t = \eta - \alpha - \varepsilon(\alpha \cos \phi + \sin \phi) \cos \omega t,$$

yielding, with $\tan \phi_0 = 1/\alpha$,

$$\frac{d\phi}{dt} = \eta - \alpha - \varepsilon\sqrt{1+\alpha^2}\cos(\phi - \phi_0)\cos\omega t \qquad (7.16)$$

Equation (7.15) is the basic equation describing the dynamics of the phase of a self-sustained periodic oscillator in the presence of a small periodic external force. There are different ways to investigate it; its study without further simplification is presented in Section 7.3. However, before that we shall exploit the smallness of the parameter $\varepsilon$ and make further approximations.

## 7.1.6    Slow phase dynamics

In the "zero-order" approximation, when we neglect the effect of the external force (i.e., $\varepsilon = 0$), Eq. (7.15) has the solution

$$\phi = \omega_0 t + \phi_0. \qquad (7.17)$$

Let us substitute this solution in the function $Q$. As this function is $2\pi$-periodic in $\phi$ and $T$-periodic in $t$, we can represent it as a double Fourier series and write

$$Q(\phi, t) = \sum_{l,k} a_{l,k} e^{ik\phi + il\omega t}, \qquad (7.18)$$

where $\omega = 2\pi/T$ is the frequency of the external force. Substitution of (7.17) in (7.18) gives

$$Q(\phi, t) = \sum_{l,k} a_{l,k} e^{ik\phi_0} e^{i(k\omega_0 + l\omega)t}. \qquad (7.19)$$

We see that the function $Q$ contains fast oscillating (compared to the time scale $1/\varepsilon$) terms, as well as slowly varying terms. The latter are those satisfying the *resonance condition*

$$k\omega_0 + l\omega \approx 0.$$

Being substituted in (7.15), the fast oscillating terms lead to phase deviations of order $O(\varepsilon)$, while the resonant terms in the sum (7.19) can lead to large (although slow, due to the small parameter $\varepsilon$) variations of the phase and are mostly important for the dynamics. Thus, in order to keep the essential dynamics only, we have to average the forcing (7.19) leaving only the resonant terms. Which terms are resonant depends on the relation between the frequency of the external force $\omega$ and the natural frequency $\omega_0$. The simplest case is when these frequencies are nearly equal, $\omega \approx \omega_0$. Then only the terms with $k = -l$ are resonant. Summation of these terms gives a new, averaged, forcing

$$\sum_{l=-k} a_{l,k} e^{ik\phi + il\omega t} = \sum_k a_{-k,k} e^{ik(\phi - \omega t)} = q(\phi - \omega t). \qquad (7.20)$$

The averaged forcing $q$ is a $2\pi$-periodic function of its argument and contains all the resonant terms. Substituting it in (7.15) we get

$$\frac{d\phi}{dt} = \omega_0 + \varepsilon q(\phi - \omega t). \tag{7.21}$$

We now define the new variable, namely the difference between the phase of oscillations and the phase of the external force:

$$\psi = \phi - \omega t. \tag{7.22}$$

One can look on $\psi$ as the slow phase in the rotating reference frame. We introduce further the frequency detuning according to

$$\nu = \omega - \omega_0 \tag{7.23}$$

to obtain finally

$$\frac{d\psi}{dt} = -\nu + \varepsilon q(\psi). \tag{7.24}$$

Before proceeding with an analysis of this equation, we demonstrate that it also describes the situation in a general case, when the resonance condition between the frequency of the external force $\omega$ and the internal frequency $\omega_0$ has the form

$$\omega \approx \frac{m}{n}\omega_0, \tag{7.25}$$

where $m$ and $n$ are integers without a common divisor. It is easy to see that in this case the resonant terms in (7.19) contain expressions like $e^{i(jm\omega_0 - jn\omega)t}$. Thus instead of (7.20) we obtain

$$\sum_{l=-nj,k=mj} a_{l,k} e^{i(k\phi + l\omega t)} = \sum_j a_{-nj,mj} e^{ij(m\phi - n\omega t)} = \hat{q}(m\phi - n\omega t), \tag{7.26}$$

with a $2\pi$-periodic function $\hat{q}(\cdot)$. The equation for the phase now takes the form

$$\frac{d\phi}{dt} = \omega_0 + \varepsilon \hat{q}(m\phi - n\omega t). \tag{7.27}$$

Introducing the phase difference as

$$\psi = m\phi - n\omega t,$$

yields

$$\frac{d\psi}{dt} = -\nu + \varepsilon m \hat{q}(\psi) \tag{7.28}$$

with the detuning $\nu = n\omega - m\omega_0$. This equation has the same form as (7.24). The simplest $2\pi$-periodic function is $\sin(\cdot)$, so that the simplest form of the averaged phase equation is

$$\frac{d\psi}{dt} = -\nu + \varepsilon \sin \psi. \tag{7.29}$$

This is often called the Adler equation after Adler [1946].

### 7.1.7 Slow phase dynamics: phase locking and synchronization region

We now discuss the solutions of the basic equation (7.24) which is a nonlinear ordinary first-order differential equation. There are two ways to introduce its phase space: the phase $\psi$ can be considered either as varying from $-\infty$ to $\infty$, or, using the $2\pi$-periodicity of the function $q$, one can take the circle $0 \leq \psi < 2\pi$ as the phase space. As these two representations are equivalent, we will often intermingle them. Equation (7.24) depends on two parameters, $\varepsilon$ and $\nu$. According to our initial Eqs. (7.5), $\varepsilon$ can be interpreted as the amplitude of the external force. The parameter $\nu$, according to (7.23), is the frequency detuning, or mismatch, i.e., the difference between the natural frequency and that of the forcing. In the derivation of (7.24) it was assumed that the detuning is small, in fact of the order $\varepsilon$. Note also that particular properties of the limit cycle in the autonomous system (7.1) and particular properties of the external force are combined in the function $q(\psi)$.

According to Eq. (7.24), there are two cases in the dynamics of the phase $\psi$, as depicted in Fig. 7.3. The function $q(\psi)$ is a $2\pi$-periodic function of $\psi$ and thus has in the interval $[0, 2\pi)$ a maximum $q_{max}$ and a minimum $q_{min}$; typically these extrema are nondegenerate. So, if the frequency detuning $\nu$ lies in the interval

$$\varepsilon q_{min} < \nu < \varepsilon q_{max}, \tag{7.30}$$

then there is at least one pair of fixed points of Eq. (7.24), i.e., a pair of stationary solutions for $\psi$. It is easy to see that one of these fixed points is (asymptotically) stable and the other one unstable; generally there can be several pairs of stable and unstable fixed points if the function $q$ has more than two extrema. Therefore, if (7.30) is fulfilled, the system evolves to one of the stable fixed points and stays there, so that the rotating phase $\psi = \psi_s$ is a constant. For the phase $\phi$ this means a constant rotation with the frequency of the external force:

$$\phi = \omega t + \psi_s, \tag{7.31}$$

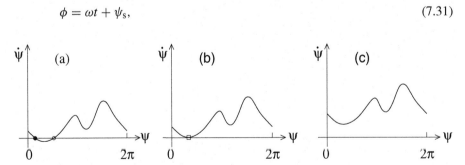

**Figure 7.3.** The right-hand side of Eq. (7.24) inside (a), at the border (b), and outside the synchronization region (c). The stable and unstable fixed points are shown with the filled and open circles. In panel (b) the synchronization transition is shown, here the stable and the unstable points collide and form a semistable point (box).

and this is the *regime of synchronization*. It exists inside the domain (7.30) on the parameter plane $(v, \varepsilon)$, called *synchronization region* (Fig. 7.4a). One often says that the phase of the self-sustained oscillator $\phi$ is locked by the phase of the external force $\omega t$, and this regime is called *phase locking*. Another often used term is *frequency entrainment*, meaning that the frequency of oscillations coincides with that of the external force.

Another situation is observed if the frequency detuning lies outside the range (7.30). Then the time derivative of the phase $\psi$ is permanently positive (or negative) and the oscillation frequency differs from the frequency of the external force $\omega$. The solution of (7.24) can be formally written as

$$\int^{\psi} \frac{d\psi}{\varepsilon q(\psi) - v} = t,$$

which defines the slow phase as a function of time $\psi = \psi(t)$. This function has a period $T_\psi$ defined by the equation

$$T_\psi = \left| \int_0^{2\pi} \frac{d\psi}{\varepsilon q(\psi) - v} \right|. \tag{7.32}$$

The phase $\phi$ rotates now nonuniformly,

$$\phi = \omega t + \psi(t), \tag{7.33}$$

and the state of the system $\mathbf{x}(\phi)$ is in general quasiperiodic (with two incommensurate periods).[5]

An important characteristic of the dynamics outside the synchronization region is the mean velocity of phase rotation, we call it the *observed frequency*. As the phase $\psi$ gains $\pm 2\pi$ during time $T_\psi$, the mean frequency of the rotations of the slow phase $\psi$ (often called the *beat frequency*) is

$$\Omega_\psi = 2\pi \left( \int_0^{2\pi} \frac{d\psi}{\varepsilon q(\psi) - v} \right)^{-1}. \tag{7.34}$$

Correspondingly, the observed frequency $\Omega$ of the original phase $\phi$ is

$$\langle \dot\phi \rangle = \Omega = \omega + \Omega_\psi.$$

(The brackets $\langle \rangle$ here denote time averaging.) The beat frequency is the difference between the observed frequency of the oscillator and that of the external force.

One can easily see that the beat frequency $\Omega_\psi$ depends monotonically on the frequency detuning $v$. Moreover, in the vicinity of the synchronization transition we can estimate this dependence analytically. As the parameter $v$ changes, the synchronization

---

[5] We can write any state variable $x_i$ as a $2\pi$-periodic function of variables $\theta_1 = \omega t$ and $\theta_2 = \Omega_\psi t$, which in the case of incommensurate frequencies $\omega$ and $\Omega_\psi = 2\pi/T_\psi$ gives a quasiperiodic function of time.

transition happens at $\nu = \varepsilon q_{max,min}$, where the stable and the unstable fixed points collide and disappear through a saddle-node bifurcation, see Fig. 7.3. Consider, for definiteness, the transition at $\nu_{max} = \varepsilon q_{max}$. If $\nu - \nu_{max}$ is small, the expression $|\varepsilon q(\psi) - \nu|$ is very small in the vicinity of the point $\psi_{max}$, so that this vicinity dominates in the integral (7.34). Expanding the function $q(\psi)$ in a Taylor series at $\psi_{max}$ and setting the integration limits to infinity yields a square-root behavior of the beat frequency (7.34)

$$|\Omega_\psi| \approx 2\pi \left| \int_{-\infty}^{\infty} \frac{d\psi}{\frac{\varepsilon}{2} q''(\psi_{max})\psi^2 - (\nu - \nu_{max})} \right|^{-1}$$
$$= \sqrt{\varepsilon |q''(\psi_{max})|} \cdot (\nu - \nu_{max}) \sim \sqrt{(\nu - \nu_{max})}. \tag{7.35}$$

We show a typical dependence of the beat frequency on the frequency detuning $\nu$ in Fig. 7.4b.

It is worth noting that in the vicinity of the transition point the dynamics of the phase $\psi$ are highly nonuniform in time (Fig. 7.5). Indeed, near the bifurcation point the trajectory spends a long time (proportional to $(\nu - \nu_{max})^{-1/2}$) in the neighborhood of the point $\psi_{max}$, where the r.h.s. of (7.24) is nearly zero. These large epochs of nearly constant phase $\psi \approx \psi_{max}$ intermingle regularly with relatively short time intervals where the phase $\psi$ increases (or decreases) by $2\pi$; these events are called *phase slips*. The phase rotation can thus be represented as a periodic (with period $T_\psi$ (7.32)) sequence of phase slips. Between the slips the oscillator is almost synchronized by the external force and its phase is nearly locked to the external phase. During the slip the phase of the oscillator makes one more (or one less) rotation with respect to the external force. Note that in our approximation (slow dynamics of the rotating phase $\psi$) the duration of the slip is much longer than the oscillation period, although near the transition point it is much less than the time interval between the slips. A transition to synchronization appears then as an increase of the time intervals between these slips according to (7.35), unless these intervals diverge at the bifurcation point.

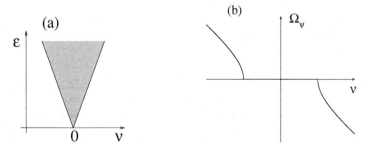

**Figure 7.4.** (a) The synchronization region on the plane of parameters $\nu, \varepsilon$. According to (7.30), the borders of the synchronization region are straight lines. (b) The dependence of the observed frequency on frequency detuning for a fixed amplitude of the forcing.

### 7.1.8    Summary of the phase dynamics

In this section we have shown in full detail how to describe the dynamics of periodic self-sustained oscillations under a small periodic external force. In the first-order approximation in $\varepsilon$ (the amplitude of the force), the force significantly influences the phase of the oscillations but has only a small effect on the amplitude. On the plane of forcing parameters $\nu$, $\varepsilon$ (frequency detuning–amplitude) there exists a synchronization region (7.30) (see Fig. 7.4a). This region is bounded by two straight lines, the slopes of which are determined by the extrema of the function $q$ (7.20). Inside this synchronization region the slow phase $\psi$ has a definite stable value (or one of the possible stable values), and the phase of the oscillator $\phi$ rotates exactly with the frequency of the external force, the process $\mathbf{x}(t)$ being periodic with the period of the external force. Outside this synchronization region the phase $\phi$ rotates with a frequency different from that of the external force, and the process $\mathbf{x}(t)$ is generally quasiperiodic. There, one basic frequency is that of the external force, the other one, the so-called beat frequency, is given by (7.34). At the synchronization threshold this beat frequency grows as a square root (see (7.35)); the time evolution here looks like a periodic sequence of synchronous epochs with $2\pi$ phase slips in between.

The depiction above was obtained under the condition of a small forcing amplitude $\varepsilon$. Below we briefly discuss what changes if one considers moderate and large amplitudes of the force; the detailed discussion is given in Section 7.3.

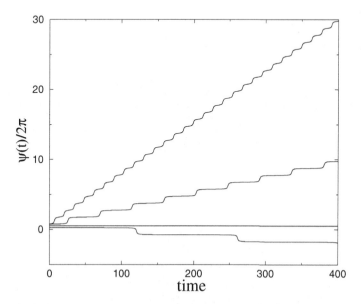

**Figure 7.5.** The dynamics of the phase described by Eq. (7.24) for $q(\psi) = \sin\psi$, $\varepsilon = 1$ (here the parameter $\varepsilon$ is rescaled to unity from a small value by virtue of rescaling the time variable) and different values of the frequency detuning: from bottom to top $\nu = 1.001, 0, -1.01, -1.1$.

## Moderate amplitudes of the force

Here the dynamics remain qualitatively the same: inside the synchronization region there is a periodic motion with the period of the external force; outside this region one observes quasiperiodic behavior. Quantitatively, the behavior of the beat frequency at the synchronization transition – the square-root dependence (7.35) – does not change, as it is determined by the type of bifurcation (and the bifurcation remains of the saddle-node type for moderate amplitudes as well). However, two other main features are slightly modified as follows.

(i)  The boundaries of the synchronization region are no longer straight lines for moderate $\varepsilon$, but, generally, curved lines. If the resonant terms are absent in the Fourier series (7.18), they can appear in the higher approximations, i.e., as terms proportional to $\varepsilon^2$, $\varepsilon^3$, etc. In these cases the width of the synchronization region is extremely small as $\varepsilon \to 0$.

(ii)  In the synchronous regime the difference between the phase of the oscillations $\phi$ and that of the external force is not constant any more, as follows from (7.31), but becomes a periodic (with the period of the external force) function of time. This is due to nonresonant terms in the expansion (7.18).

## Large forcing amplitudes

Here even qualitative changes in the described picture can occur. Transition to synchronization may happen through other bifurcations. Moreover, more complex regimes, including transition to chaos, may be observed, as is discussed in Section 7.3.

It is important to mention that, although here we have considered the case of small forcing, synchronization cannot be described in the framework of *linear response theory*. Indeed, if a physicist were to look at Eq. (7.5), the first idea would be to apply a perturbation method and to get the response as a series in $\varepsilon$. We already know the answer, and can now say that this approach will work only outside the synchronization region, where the quasiperiodic motion can be roughly represented as a combination of unperturbed oscillations with frequency $\omega_0$ and of the forced oscillations with frequency $\omega$. Inside the synchronization region the process has only one frequency $\omega$, thus formally the response at the forcing frequency is $O(1)$ and cannot be obtained as a series in $\varepsilon$. The failure of a linear or weakly nonlinear response theory is due to the singular nature of the unperturbed free oscillations, where the phase is neutral and exhibits large, $O(1)$, deviations for arbitrary small forcing. Another manifestation of this singularity is a rather unusual (for a physical problem) form of the dependence of the observed frequency on the forcing one: it has a perfectly horizontal plateau with sharp ends, Fig. 7.4b. In many cases, when such plateaux appear in physics, they can be explained in terms of synchronization (cf. discussion of Shapiro steps in voltage–current characteristics of Josephson junctions in Section 4.1.8).

## 7.2     Weakly nonlinear oscillator

In the previous section we used the smallness of the forcing to obtain a closed equation for the phase dynamics. In general, a periodic external force influences both the phase and the amplitude, and the latter effect cannot be neglected. In many situations, especially when both the amplitude of oscillations and the forcing are large, the only way to investigate the dynamics is to simulate them on a computer. Here we consider one important case when the properties of synchronization can be analyzed to a large extent analytically: we assume that both the force and the amplitude of self-sustained oscillations are small. The smallness of the amplitude we understand as the closeness of the self-sustained oscillations to linear ones; the dimensionless small parameter is the ratio between the oscillation period and the characteristic time scales of amplitude variations. The existence of this small parameter allows us to describe the problem with the so-called averaged (over the period of oscillations) amplitude equations, and these equations are universal. Thus, the analysis of the amplitude equations gives a description valid for the whole class of weakly nonlinear oscillating systems.

### 7.2.1     The amplitude equation

There is an abundant literature devoted to the description of weakly nonlinear systems (e.g., [Bogoliubov and Mitropolsky 1961; Minorsky 1962; Hayashi 1964; Nayfeh and Mook 1979; Glendinning 1994]). Without going into technical details, we shall outline the derivation of the corresponding amplitude equations. The method is applicable to weakly nonlinear oscillators, i.e., to systems that can be represented as a linear oscillator with frequency $\omega_0$ subject to small perturbations (the terms on the r.h.s.):

$$\ddot{x} + \omega_0^2 x = f(x, \dot{x}) + \varepsilon p(t). \tag{7.36}$$

This is not the most general form of the perturbation, as we have assumed it to consist of two parts: the nonlinear function $f(x, \dot{x})$ determines the properties of autonomous self-sustained oscillations, while the external force is described by a periodic function of time $p(t) = p(t + T)$ having the frequency $\omega = 2\pi/T$. We have chosen the forcing term to be explicitly proportional to the small parameter $\varepsilon$. The term $f$ should be also small; we shall write a condition for this later.

As Eq. (7.36) is close to that of a linear oscillator, we can expect that its solution has a nearly sinusoidal (harmonic) form with still unknown amplitude, frequency and phase. All these quantities should be determined eventually, but at this stage we can choose any representation of the solution, and the first important step is to choose the most suitable one. As we expect that the frequency of the oscillations will be (at least for some values of parameters) entrained by the frequency of the external force $\omega$, we will look for a solution in the complex form

$$x(t) = \tfrac{1}{2}(A(t)e^{i\omega t} + c.c.), \tag{7.37}$$

i.e., in the form of harmonic oscillations with the "basic" frequency $\omega$ and a time-dependent complex amplitude $A(t)$. Note that we make no restriction on $x(t)$ here, as the "observed" frequency may well deviate from $\omega$ if the amplitude $A$ rotates on the complex plane.

It is convenient to represent Eq. (7.36) as that of a linear oscillator with the frequency $\omega$, obtaining thus an additional term on the r.h.s.:

$$\ddot{x} + \omega^2 x = (\omega^2 - \omega_0^2)x + f(x, \dot{x}) + \varepsilon p(t). \tag{7.38}$$

This new term should be also small, i.e., our treatment is valid if the frequency detuning $\omega - \omega_0$ is small.

Rewriting Eq. (7.38) as a system

$$\dot{x} = y,$$
$$\dot{y} = -\omega^2 x + (\omega^2 - \omega_0^2)x + f(x, y) + \varepsilon p(t),$$

and introducing the following relation[6] between $y$ and $A$

$$y = \tfrac{1}{2}(i\omega A(t)e^{i\omega t} + c.c.), \tag{7.39}$$

we get, resolving (7.37) and (7.39), the equation for the complex amplitude

$$\dot{A} = \frac{e^{-i\omega t}}{i\omega} \cdot [(\omega^2 - \omega_0^2)x + f(x, y) + \varepsilon p(t)]. \tag{7.40}$$

### Averaging the amplitude equation

Until now, this transformation is exact, but the new equation is not easier to solve than the original Eq. (7.36). We now use the small parameters to obtain an analytically solvable approximate equation for the evolution of $A$. There are many ways to perform the analysis on a rigorous mathematical basis (different variants are known as the asymptotic method [Bogoliubov and Mitropolsky 1961], method of averaging [Nayfeh and Mook 1979], multiscale expansion [Kahn 1990]), but here we only outline the idea. As the r.h.s. of (7.40) is small, the variations of $A$ can be either slow (if they are large) or small (if they are fast, e.g., with the frequency $\omega$). We restrict ourselves to large and slow variations, i.e., we neglect all the fast terms on the r.h.s. of (7.40). Neglecting the terms containing the fast oscillations ($e^{\pm i\omega t}$, $e^{\pm i2\omega t}$, etc.) can also be considered as averaging over the period of the oscillations $T = 2\pi/\omega$; thus this method is often called the method of averaging. Performing averaging is straight-forward: one substitutes $x$ and $y$ expressed via $A$ in (7.40), and neglects all oscillating terms. For any particular choice of the functions $f$ and $p$ this can be done explicitly,

---

[6] As the new variable $A$ is complex, we need two relations to have a one-to-one correspondence between $(x, y)$ and $A$.

but we want to argue that the result is universal for a large class of systems. First note, that averaging of the term

$$\frac{e^{-i\omega t}\varepsilon p(t)}{i\omega}$$

means that we take the first Fourier harmonic of the periodic function $p(t)$, in general this harmonic is nonzero and this term gives a complex constant $-i\varepsilon E$.

Next, let us consider the contribution of the function $f$:

$$\frac{e^{-i\omega t}}{i\omega}f(x, y).$$

Suppose that $f$ is a polynomial in $x$, $y$, so it will be a polynomial in $Ae^{i\omega t}$, $A^*e^{-i\omega t}$ too. From all powers of the type $(Ae^{i\omega t})^n (A^*e^{-i\omega t})^m$ after multiplication with $e^{-i\omega t}$ and averaging, only the terms with $m = n - 1$ do not vanish. Thus, the only possible result of averaging should have the form $g(|A|^2) \cdot A$ with an arbitrary function $g$. For small amplitudes of oscillation only the linear ($\propto A$) and the first nonlinear ($\propto |A|^2 A$) terms are important.

Finally, the averaging of the first term on the r.h.s. of (7.40) gives a term linear in $A$. Summarizing, we get the amplitude equation in the form

$$\dot{A} = -i\frac{\omega^2 - \omega_0^2}{2\omega}A + \mu A - (\gamma + i\kappa)|A|^2 A - i\varepsilon E. \tag{7.41}$$

Here we assume the parameter $\mu$ to be real, as the imaginary part can be absorbed by the first term on the r.h.s. For example, for the van der Pol equation (7.2) one obtains Eq. (7.41) with $\kappa = 0$ and $\gamma = \mu\beta/4$.

The new parameters have a clear physical meaning. The parameters $\mu$ and $\gamma$ describe the linear and nonlinear growth/decay of oscillations. For self-sustained oscillations to be stable, we need growth for small amplitudes and decay for large amplitudes, which corresponds to $\mu > 0$, $\gamma > 0$. The parameter $\kappa$ describes the nonlinear dependence of the oscillation frequency on the amplitude; it can be both positive and negative, and vanishes in the isochronous case (cf. discussion in Section 7.1.3).

Now we return to the validity conditions of the method of averaging. For Eq. (7.41) they imply that all the terms on the r.h.s. should be small. This is the case if the frequency detuning $|\omega - \omega_0|$ and the linear growth rate $\mu$ are small compared to the frequency $\omega_0$. This ensures that the nonlinear term on the r.h.s. is small as well, because the amplitude of unforced oscillations is $|A_0|^2 = \mu/\gamma$, so that the nonlinear term is of the same order of magnitude as the linear term $\mu A$. The smallness of the parameter $\mu$ means that the instability of the fixed point $A = 0$ is weak. Usually this is the case near the bifurcation point at which a limit cycle appears (Hopf bifurcation). Thus, the amplitude equation (7.41) is a universal equation (normal form) for a periodically driven system near the Hopf bifurcation point.

By choosing appropriate scales for the amplitude $A$ and for time, we can reduce the number of parameters in Eq. (7.41). Which parameters have to be chosen for scaling, depends on the physical formulation of the problem. As we focus on the properties of synchronization with the parameters of the external force to be varied, it is reasonable to use the parameters of the self-sustained oscillations to normalize the equation. Introducing the new amplitude and the new time according to the relations

$$A = \sqrt{\frac{\mu}{\gamma}}a, \qquad t = \mu^{-1}\tau, \tag{7.42}$$

we obtain

$$\dot{a} = -i\nu a + a - |a|^2 a - i\alpha|a|^2 a - ie, \tag{7.43}$$

where

$$\nu = \frac{\omega^2 - \omega_0^2}{2\omega\mu} \approx (\omega - \omega_0)/\mu,$$
$$\alpha = \kappa/\gamma, \qquad e = \varepsilon E \gamma^{1/2} \mu^{-3/2}. \tag{7.44}$$

Note the nontrivial dependence on the parameters $\varepsilon$ (the amplitude of the external force) and $\mu$ (the squared amplitude of the unforced oscillations): they appear in the "effective" force amplitude $e$ in the combination $\varepsilon\mu^{-3/2}$. We discuss below the regimes of weak ($e \lesssim 1$) and strong ($e \gtrsim 1$) forcing; in the terms of the original parameters this corresponds to $\varepsilon \lesssim \mu^{3/2}$ and $\varepsilon \gtrsim \mu^{3/2}$.

Before proceeding with the investigation of Eq. (7.43), we first briefly discuss the case of vanishing force $e = 0$. Then the problem reduces to Eq. (7.6) above. There is the unstable fixed point at the origin $a = 0$ and the stable limit cycle $a = e^{-i(\nu-\alpha)t}$ with amplitude 1 and frequency $|-\nu - \alpha|$.[7] As one can see from the general solution (7.9), the rotation velocity of the angle variable depends on the amplitude if $\alpha \neq 0$, and does not if $\alpha = 0$. We refer to these two situations as nonisochronous and isochronous cases, respectively.

### 7.2.2   Synchronization properties: isochronous case

Here we consider the case of isochronous autonomous oscillations, i.e., we take $\alpha = 0$:

$$\dot{a} = -i\nu a + a - |a|^2 a - ie. \tag{7.45}$$

Before going into details of the analysis of Eq. (7.45), we discuss an interpretation of the solutions in terms of the original variables $x \propto \text{Re}(a(t)e^{i\omega t})$, $y \propto \text{Im}(a(t)e^{i\omega t})$. If there is a stationary solution (fixed point) for $a$, the variables $x$, $y$ perform harmonic oscillations with the external frequency $\omega$. This regime can be characterized as perfect synchronization (phase locking): the only oscillations in the system are those with the external frequency. If there is a time-periodic solution for $a$, for the original variables one observes a quasiperiodic motion with two independent frequencies: one is the

---

[7] Apparently, this frequency depends on the detuning $\nu$ even if there is no forcing: this is due to our choice of the reference frame (7.37).

frequency of the external force and the other one is the frequency of the time-periodic solution of (7.45). Note that this latter frequency may vary with the parameters of the system.

We emphasize that the existence of the second (in addition to $\omega$) frequency does not necessarily mean desynchronization. Indeed, if we write $a(t) = R(t)e^{i\psi(t)}$ and $x(t) = \text{Re}(R(t)e^{i(\psi(t)+\omega t)})$, then the observed frequency of oscillations can be written as

$$\Omega = \langle \dot{\psi} \rangle + \omega. \tag{7.46}$$

(Note that $\psi$ is the correct difference between the phase of the oscillator and the phase of the external force, cf. (7.22).) The term $\langle \dot{\psi} \rangle$ depends on the trajectory of the system on the phase plane $(\text{Re}(a), \text{Im}(a))$. If this trajectory rotates around the origin, then $\langle \dot{\psi} \rangle \neq 0$, otherwise the variations of $\psi$ contribute only to modulation of the oscillations, but produce no frequency shift. Additionally, we note that Eq. (7.45) is invariant under the transformations $v \to -v$, $e \to -e$, $a \to a^*$ and $e \to -e$, $a \to -a$, therefore it is sufficient to consider the region $v > 0$, $e > 0$.

Investigations of Eq. (7.45) have a long history (see, e.g., [Appleton 1922; van der Pol 1927]), but a complete picture was established only recently by Holmes and Rand [1978] (see also [Argyris et al. 1994]). We refer the reader to these publications, and present here only the main features of the dynamics, leaving some petty aspects aside.

The (approximate) bifurcation diagram of Eq. (7.45) is shown in Fig. 7.6. We start the analysis with finding the stationary solutions (fixed points). Setting $\dot{a} = 0$, we obtain the following cubic equation for the square of the amplitude $R^2 = |a|^2$:

$$R^2(1 - R^2)^2 + v^2 R^2 = e^2.$$

This equation has three real roots provided

$$9v^2 + 1 - (1 - 3v^2)^{3/2} < \frac{27e^2}{2} < 9v^2 + 1 + (1 - 3v^2)^{3/2},$$

or one real root otherwise. Thus, Eq. (7.45) has either three or one fixed point(s). The region with three roots is denoted by A in Fig. 7.6. In the regions B, C and D there is only one fixed point.

To determine the stability of a fixed point, we have to linearize Eq. (7.45), which leads to the characteristic equation

$$\lambda^2 + (4R^2 - 2)\lambda + (1 - 3R^2)(1 - R^2) + v^2 = 0.$$

We see that the stability is determined by the value of the amplitude $R$, and, depending on the parameters, different stability types are possible (stable and unstable node, saddle, focus). The most important bifurcation here is the Hopf bifurcation at $4R^2 - 2 = 0$, which corresponds to the hyperbola

$$e^2 = v^2/2 + \frac{1}{8},$$

which separates regions B and D in Fig. 7.6.

These relations already yield a "coarse-grained" picture of possible transitions.[8] In regions A and B the only stable solution is a stable fixed point; this is the region of perfect synchronization, where the synchronous oscillations have a constant amplitude and a constant phase shift with respect to the external force (of course, those quantities are constants only in the approximation used). In regions C and D the global attractor in the system (7.45) is a limit cycle; here the motion of the forced weakly nonlinear oscillator is quasiperiodic.

It is instructive to look at how synchronization appears/disappears with the variation of the parameters $e$, $\nu$.[9] From Fig. 7.6 it is clear that the bifurcations are different for small and large values of $e$; we discuss these two cases separately.

[8] Here we do not describe the fine complex structure of bifurcations around the point $\nu = 0.6$, $e = 0.5$, see [Holmes and Rand 1978; Argyris *et al.* 1994] for details.

[9] These normalized parameters are related to the parameters of the original system (7.41) via Eqs. (7.44).

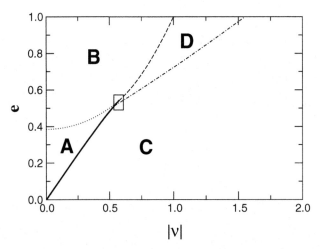

**Figure 7.6.** The bifurcation diagram for the isochronous forced weakly nonlinear oscillator (Eq. (7.45)): dependence on the frequency detuning $\nu$ and the (renormalized) forcing amplitude $e$. In regions A and B there is a stable fixed point, corresponding to a stable synchronized state (in A there is additionally a pair of unstable fixed points). In regions C and D there is an unstable fixed point and a stable limit cycle; the difference between C and D is illustrated in the text and in Fig. 7.9. The transition from A to C occurs via the saddle-node bifurcation (bold curve), it is illustrated in Figs. 7.7 and 7.8. The transition $B \rightarrow D \rightarrow C$ is illustrated in Figs. 7.9 and 7.10. The Hopf bifurcation line ($B \rightarrow D$) is shown by the dashed curve; the transition from the frequency locked to the nonsynchronous state ($D \rightarrow C$) is shown by the dashed-dotted curve. In the region shown by the box, complex bifurcations occur. From [Pikovsky *et al.* 2000].

### Synchronization transition at small amplitudes of the external force

Fixing the parameter $e$ at a small value ($\lesssim 0.5$) and varying $|v|$, we observe a transition
between regions A and C. In region A there are three fixed points: one unstable focus,
one stable node, and one unstable saddle. At the bifurcation the saddle and node collide
and a stable limit cycle appears, as shown in Figs. 7.7 and 7.8. This transition is very
similar to that occurring in the phase approximation as described in Section 7.1. This
is not surprising, as the theory of Section 7.1 should be universally valid for very small
forcing amplitudes.

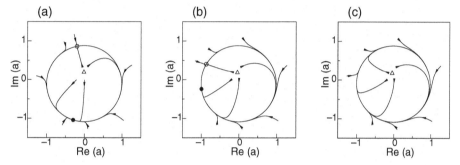

**Figure 7.7.** Loss of synchronization via a saddle-node bifurcation (the transition
A → C in the diagram Fig. 7.6). In the middle of the synchronization region there
exist an unstable focus (shown by triangle), a stable fixed point (filled circle) and an
unstable (open circle) fixed point (a). Stable and unstable fixed points come closer
near the border of synchronization (b). A stable limit cycle exists outside the
synchronization region (c); it is born from an invariant curve formed by the unstable
manifolds of the saddle. From [Pikovsky *et al.* 2000].

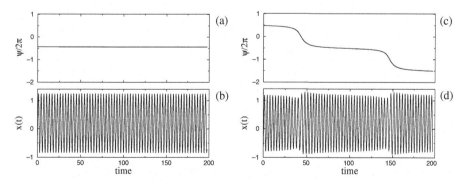

**Figure 7.8.** Oscillations in the forced weakly nonlinear oscillator at small forcing
amplitudes. Inside the synchronization region (A in Fig. 7.6) the amplitude and the
phase difference are constant (a,b). Just beyond the synchronization transition
(A → C in Fig. 7.6) the phase difference $\psi$ rotates nonuniformly, epochs of nearly
constant $\psi$ are intermingled with $2\pi$-slips (c); (d) the amplitude is slightly
modulated. From [Pikovsky *et al.* 2000].

### Synchronization transition at large amplitudes of the external force

Now we fix the parameter $e$ at a large value ($\gtrsim 0.5$) and vary $\nu$. The first transition B $\to$ D is the Hopf bifurcation (Fig. 7.9). In the middle of the synchronization region there is a stable node. When $|\nu|$ increases, it becomes a stable focus. At the transition point it loses stability, and a stable limit cycle appears. First, the amplitude of the limit cycle is small, so a point on it does not rotate around the origin. This means that the process $x(t)$ has both amplitude modulation and phase modulation, but the frequency of the oscillator remains the same as that of the external force (see Fig. 7.10a). The phase difference is not constant any more, but its mean growth rate $\Omega_\psi = \langle \dot{\psi} \rangle$ is still exactly zero. The situation changes if the cycle envelops the origin (transition D $\to$ C,

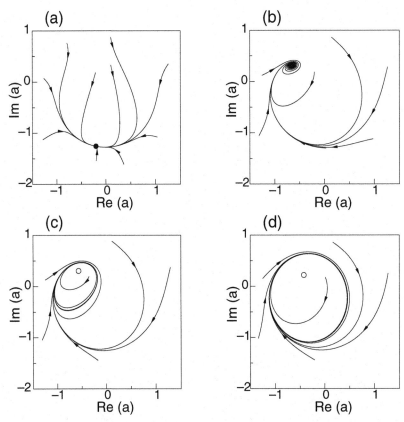

**Figure 7.9.** Transition to synchronization via a Hopf bifurcation (route B $\to$ D $\to$ C in Fig. 7.6). (a) Near the center of the synchronization region $\nu \approx 0$ all trajectories are attracted to a stable node. (b) Near the boundary of synchronization the fixed point is of focus type. (c) A stable limit cycle appears via a Hopf bifurcation, in region D, however, this cycle does not envelop the origin, so that the observed frequency is still that of the external force. (d) As the amplitude of the limit cycle grows, it envelops the origin and synchronization breaks down. From [Pikovsky *et al.* 2000].

Fig. 7.10b). Now the phase difference $\psi$ rotates and the observed frequency deviates from that of the external force.

The two types of transition to synchronization allow a clear physical interpretation. A small external force affects mainly the phase of oscillations, but not their amplitude. Therefore both in the synchronous and asynchronous state the amplitude is nearly constant, and the only change is in the phase dynamics: outside the synchronization region the phase difference rotates, and inside it is locked. Thus the transition to/from synchronization can be described as a phase locking/unlocking transition.

A strong force affects both the phase and the amplitude, and in the synchronization region the natural oscillations are totally suppressed: only oscillations with the forcing frequency are observed. If the frequency detuning grows, natural oscillations appear first in the form of small modulation of the forced regime; they become of the same order of magnitude as forced oscillations only for large mismatches. The differences in the properties of the synchronization transition for weak and strong forcing are schematically illustrated in Fig. 7.11.

The regimes of small and large forcing amplitude can be easily distinguished experimentally, if one observes (calculates) the power spectrum of the process (Figs. 7.12 and 7.13). In the synchronized state only a peak at the external frequency is present in the power spectrum. At the loss of synchronization a new frequency peak appears. For small forces (Fig. 7.12) this newly appearing mode has almost the same frequency as the external force, because near the saddle-node bifurcation the period of the emerging limit cycle tends to infinity (cf. Eq. (7.35)). For large amplitudes of the force (Fig. 7.13) the difference between the frequency of the newly appearing mode and the frequency of the external force is finite.

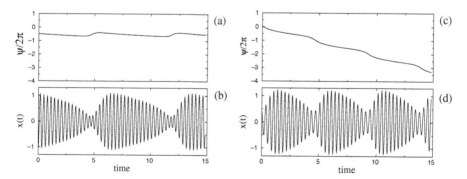

**Figure 7.10.** Oscillations in the forced oscillator at large forcing amplitudes. Inside the synchronization region (D in Fig. 7.6), but near the transition border, the phase difference is modulated, although bounded (a); the amplitude is modulated as well (b). After the transition (region C in Fig. 7.6), the phase difference $\psi$ rotates nonuniformly, but without epochs of "nearly synchronous" behavior (c); (cf. Fig. 7.8). The amplitude modulation is relatively strong (d). From [Pikovsky *et al.* 2000].

### 7.2.3   Synchronization properties: nonisochronous case

The bifurcation analysis of the full equation (7.43) is rather tiresome, and includes a number of codimension-2 and codimension-3 points. A detailed description is given by Levina and Nepomnyaschiy [1986] and Glendinning and Proctor [1993] (note that Glendinning and Proctor [1993] used a normalization of Eq. (7.43) that is different from ours). From their studies it follows that for small nonisochronity $\alpha^2 < 1/3$ the bifurcation structure remains qualitatively the same as in the isochronous case $\alpha = 0$; new bifurcations appear only for larger values of $\alpha$. We discuss here only the case of very small amplitude of the external force, when the phase approximation described in Section 7.1 remains valid. It is a simple exercise to apply the phase approximation to Eq. (7.43), cf. (7.16); the resulting equation for the phase reads

$$\dot{\phi} = -\nu - \alpha - e(\cos\phi + \alpha\sin\phi) = -\nu - \alpha - e\sqrt{1+\alpha^2}\cos(\phi - \phi_0) \qquad (7.47)$$

where $\tan\phi_0 = \alpha$. This phase equation defines the synchronization region

$$-\alpha - e\sqrt{1+\alpha^2} < \nu < -\alpha + e\sqrt{1+\alpha^2}.$$

We see that synchronization occurs in the vicinity of the natural frequency of the limit cycle, which differs from the "linear" frequency $\omega_0$ (see Eq. (7.36)). The natural

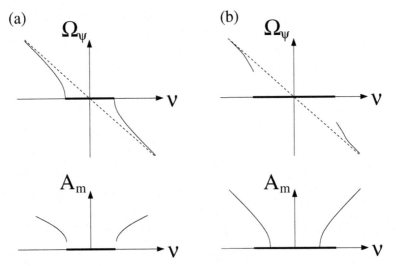

**Figure 7.11.** Schematic illustration of the synchronization transition for small (a) and large (b) forcing amplitudes. Beat frequency $\Omega_\psi = \langle\dot\psi\rangle$ and degree of the amplitude modulation $A_m$ are plotted vs. the detuning $\nu$. The domain of synchronization is shown by a bold horizontal bar. For weak forcing, the beat frequency at the border of synchronization obeys the square-root law (see Eq. 7.35); the amplitude of the beats appears with a finite value and remains relatively small. On the contrary, for strong forcing, the beat frequency appears as a finite value, whereas the amplitude modulation increases smoothly. Note that amplitude modulation is already observed within the synchronization region.

frequency is $\omega_0 - \alpha\mu$ in the original coordinates, and equals $-\alpha$ in the rotating frame in which Eq. (7.43) is written. Another interesting fact is that for nonisochronous oscillations the synchronization region is larger than for isochronous ones. This effect can be explained as follows: the external force has a twofold influence on the phase. First, it changes the phase directly, and this action is described as the $\alpha$-independent term $e \cos \phi$ in Eq. (7.47). Second, the force changes the amplitude, and this change results, due to nonisochronity, also in a phase shift; this effect is described with the term proportional to $e\alpha \sin \phi$ in Eq. (7.47). These two actions have different $\phi$-dependence, so that the resulting phase shift between the phase of the synchronized oscillations and the phase of the external force depends on the parameter $\alpha$.

## 7.3   The circle and annulus map

In previous sections we have used different approximations to describe the effect of an external force on self-sustained oscillators. Here we present a more general approach which is not restricted to small forces or to weak nonlinearities. We thus will be able to give a general description of synchronization, but this description will be only

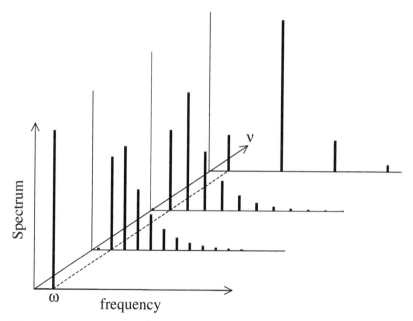

**Figure 7.12.** Evolution of the spectrum at loss of synchronization for a constant small forcing amplitude and variable detuning. The synchronized motion has only one peak at the external frequency $\omega$. At the transition a new peak (and its harmonics) appears with a very closed frequency; with increasing the frequency detuning this peak moves away from $\omega$. The horizontal axis is shifted arbitrarily (in particular, the origin of this axis does not correspond to zero frequency).

qualitative: it does not allow us to derive, say, borders of the synchronization region for given equations of motion.

This general approach is based on the construction of an **annulus map** that describes the dynamics of a periodically forced oscillator in the vicinity of the limit cycle. As the dynamics in the absence of the forcing are known, the structure of this map can be obtained just from continuity arguments. In some region of parameters, in particular for small forcing, the map possesses an attracting invariant curve so that one can consider the dynamics on this curve and obtain a one-dimensional **circle map**. The general properties of the circle map have been a subject of thorough mathematical consideration, resulting in a nice deep theory which is also important in other branches of nonlinear dynamics (Kolmogorov–Arnold–Moser theory, renormalization group, etc.). For large forcing the invariant curve is destroyed, and one has to analyze the full annulus map. This breakup is usually accompanied by the appearance of chaos, but a detailed description of these regimes is beyond the scope of the book.

In this section we outline the properties of the circle map dwelling upon those that are essential for synchronization. Next, we give an example of the annulus map and discuss how synchronization can be lost for very large forcings due to the appearance of chaos.

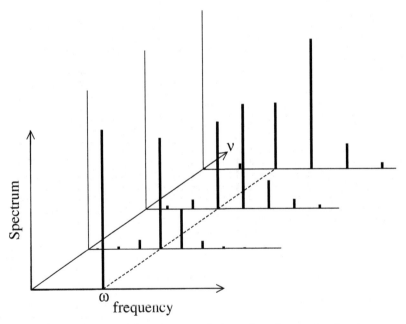

**Figure 7.13.** Evolution of the spectrum at the loss of synchronization for a large constant forcing amplitude and variable detuning. At the transition a new peak appears at a finite distance from the original peak at $\omega$; with increasing frequency detuning its amplitude grows and the amplitude of the peak at $\omega$ decreases. The layout of axes is as in Fig. 7.12.

## 7.3.1    The circle map: derivation and examples

In Section 7.1 we showed that the equations of motion of a general periodically forced dynamical system (7.5) in the vicinity of a limit cycle can be reduced, for small forcing amplitudes $\varepsilon$, to the phase equation (7.15)

$$\frac{d\phi}{dt} = \omega_0 + \varepsilon Q(\phi, t). \tag{7.48}$$

As an extension to the previous discussion we study this equation in the following without any further simplifications. Moreover, we will not use the smallness of the parameter $\varepsilon$, but will consider Eq. (7.48) in a rather general context. This will be justified in Section 7.3.3.

The r.h.s. of Eq. (7.48) is a $2\pi$-periodic function of the phase $\phi$ and a $T$-periodic function of time $t$. Thus, the phase space of the dynamical system (7.48) is a two-dimensional torus $0 \leq \phi < 2\pi$, $0 \leq t < T$. This two-dimensional system with continuous time can be reduced to a one-dimensional mapping. In the case of explicit time periodicity the construction of such a mapping is especially simple: one can take a stroboscopic mapping with time interval $T$. Fixing some phase of the external force by choosing $t = t_0$, we can find a one-to-one correspondence between the points $\phi(t_0)$ and $\phi(t_0 + T)$. In this way a smooth invertible *circle map* is defined:

$$\phi_{n+1} = \phi_n + \omega_0 T + \varepsilon F(\phi_n). \tag{7.49}$$

Several remarks are in order.

(i) The map is called a circle map because it is defined on the circle $0 \leq \phi < 2\pi$. Formally, one should also apply the operation modulo $2\pi$ to the r.h.s., but we will omit writing this operation for simplicity of presentation.

(ii) The map depends on the choice of the time moment $t_0$ (i.e., on the phase of the external force chosen for stroboscopic observations). Actually, this is not crucial: all maps for different choices of $t_0$ are equivalent, as they are connected by a smooth transformation $\phi(t_0) \rightarrow \phi(t'_0)$ via the solutions of (7.48).

(iii) If $\varepsilon = 0$, we get a *circle shift*. This describes stroboscopic observations of the motion on the limit cycle. The dynamics of the circle shift depend on the parameter $\omega_0 T$ and are trivial. If the quotient $T/T_0$ is rational, i.e., $\omega_0 T = 2\pi p/q$, each point of the circle is periodic with period $q$. If the ratio $T/T_0$ is irrational, then we have a quasiperiodic rotation on the circle. Note that we then have a slightly complicated description of unforced periodic oscillation due to the fact that we observe the system stroboscopically with a time interval $T$ which, generally, is not related to the period $T_0$. Our stroboscopic approach is thus appropriate for distinguishing between nonsynchronized (quasiperiodic) and synchronized (periodic with the period of the force or in a rational relation to this period) motions.

(iv) The form of the circle map (7.49) is not the most general one. From Eq. (7.48) it follows only that the map $\phi_{n+1} = \Phi(\phi_n)$ is a monotonically increasing function satisfying $\Phi(\phi + 2\pi) = \Phi(\phi) + 2\pi$. In (7.49) we have separated the shift term and the nonlinear function in order to stress the physical meaning of the parameters $\omega_0 T$ and $\varepsilon$: they correspond to the frequency and the amplitude of the external force, respectively. This separation is, strictly speaking, valid only for small $\varepsilon$, otherwise both the shift and the nonlinear term depend on both the amplitude and frequency of the force. The exact relation should be established in each particular problem.

Before proceeding with the discussion of the properties of the circle map, we shall present a simple example where the circle map can be written explicitly. Additionally, this example illustrates the second possibility of circle map construction: the phase of oscillations is not taken stroboscopically, but the phase of the external force is taken at some definite events of the oscillation (for the original dynamical system this represents another way of making a Poincaré section on the torus).

### An example: relaxation (integrate-and-fire) oscillator

A popular model of a relaxation self-sustained oscillator is the integrate-and-fire system. It is described by one scalar variable $x(t)$. The dynamics of $x$ consist of two parts:

(i) integration: $x$ grows linearly in time $x = (t - t_n)/T_0$, where $t_n$ is the time of the previous firing;

(ii) firing: as $x$ reaches the threshold $x_{\mathrm{up}} = 1$, the value of $x$ is instantaneously reduced to $x_{\mathrm{down}} = 0$.

Note that here we use the normalized variables, and the period of oscillations is $T_0$. The model can be easily generalized to a nonlinear growth of $x$ (see, e.g., [Mirollo and Strogatz 1990b]).

It should also be noted that we do not write the equations of motion of type (7.5) for this system. Indeed, such equations must be at least two-dimensional, describing both the slow motion (integration) and the fast one (jump). Generally, one can do this with Eqs. (7.5) with a large parameter describing the ratio of two time scales. The simplified description above is, strictly speaking, valid when this parameter tends to infinity, and the motion in the phase space is restricted to the so-called slow manifold (plus jumps). This quasi-one-dimensional character of the dynamics makes the description with a one-dimensional circle map exact.

There are several ways to force the relaxation oscillator, as shown in Fig 7.14.

(i) **Variation of the lower threshold.** The value of $x_{\mathrm{down}}$ is a periodic function of time, e.g., $x_{\mathrm{down}} = \varepsilon \sin \omega t$, see Fig. 7.14b. We denote the time of the $n$th firing event as $t_n$. Then the time of the next event can be calculated as

$t_{n+1} = t_n + T_0 - \varepsilon T_0 \sin \omega t_n$. Introducing the phase of the external force $\phi^{(e)} = \omega t$, we get the map defined on the circle $0 \leq \phi^{(e)} < 2\pi$:

$$\phi_{n+1}^{(e)} = \phi_n^{(e)} + \omega T_0 - \varepsilon \omega T_0 \sin \phi_n^{(e)}. \tag{7.50}$$

(ii) **Variation of the upper threshold.** If the value of $x_{up}$ is varied according to $x_{up} = 1 + \varepsilon \sin \omega t$ (see Fig. 7.14c), then again the times of the $n$th and $(n+1)$th firings are related by the equation $t_{n+1} = t_n + T_0 + \varepsilon T_0 \sin \omega t_{n+1}$. Unfortunately, this relation is implicit with respect to $t_{n+1}$. For small $\varepsilon$ we can approximately resolve it as $t_{n+1} = t_n + T_0 + \varepsilon T_0 \sin \omega (t_n + T_0)$, which leads to Eq. (7.50). However, for large $\varepsilon$ the relation between $t_{n+1}$ and $t_n$ becomes discontinuous.

(iii) **Pulse force.** If the integrate-and-fire oscillator is affected by another similar oscillator, the external force has the form of a sequence of pulses with period $T$. Let the amplitude of each pulse be $\varepsilon$, so that the state after the pulse is $x + \varepsilon$. Two situations are possible here: if $x + \varepsilon < x_{up}$, the integration continues; if $x + \varepsilon > x_{up}$, the oscillator fires and the variable $x$ is reset to $x_{down}$, see Fig. 7.14d. Let us denote the time interval between an external pulse and the next firing by $\tau_n$. Then for $\tau_{n+1}$ we get a discontinuous not a one-to-one circle map

$$\tau_{n+1} = \begin{cases} 0 & \text{if } 1 - \left\{ \dfrac{T - \tau_n}{T_0} \right\} < \varepsilon, \\[2ex] T_0(1 - \varepsilon) - T + \tau_n & \text{if } 1 - \left\{ \dfrac{T - \tau_n}{T_0} \right\} \geq \varepsilon, \end{cases}$$

where $\{\cdot\}$ denotes the fractional part.

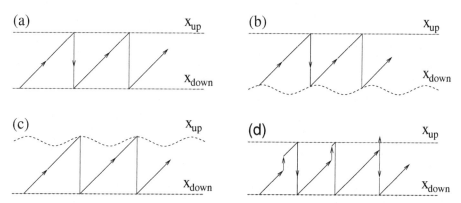

**Figure 7.14.** Different ways to force a relaxation oscillator. (a) An autonomous oscillator demonstrates purely periodic motion. (b) Variation of the lower threshold. (c) Variation of the upper threshold. (d) Forcing oscillator with a periodic sequence of pulses.

We see that, depending on the art of forcing, different types of circle map appear. The most common is a smooth circle map like (7.50); we will mainly concentrate our discussion on this case.

## 7.3.2   The circle map: properties

The circle map is one of the basic models of nonlinear dynamics, and is described in most books on nonlinear systems, including mathematically [Katok and Hasselblatt 1995] and physically oriented ones [Ott 1992; Argyris *et al.* 1994; Schuster 1988]. Here we outline some parts of the theory, focusing our attention on the features related to the synchronization problem. As a basic illustrative example we consider the sine circle map

$$\phi_{n+1} = \phi_n + \eta + \varepsilon \sin \phi_n. \tag{7.51}$$

This map depends on two parameters, their physical meaning follows directly from Eq. (7.49). The parameter $\eta = \omega_0 T = 2\pi T / T_0$ is proportional to the ratio of the period of the force and the period of self-sustained oscillations. It varies with the frequency of the external force. The parameter $\varepsilon$ is proportional to the amplitude of the external force. For $\varepsilon = 0$ we have linear rotation (the circle shift), and $\varepsilon$ governs the nonlinearity level[10] in the mapping (7.51). Our goal is to describe the dynamics of (7.51) on the plane of parameters $(\eta, \varepsilon)$; we do this in the form of the following propositions.

1. Because we consider the phase modulo $2\pi$, i.e., on the circle $0 \le \phi < 2\pi$, the dynamics do not change if we add $\pm 2\pi$ to the parameter $\eta$. This means that the state diagram is periodic in $\eta$ with period $2\pi$. Therefore, usually only the interval $0 \le \eta < 2\pi$ on the plane of parameters is considered in the literature. For our purposes, however, the case of the resonance $\eta \approx 2\pi$ (which means $T \approx T_0$) is most important. Note also that $2\pi(\eta - 1) = 2\pi(T - T_0)/T_0$ is equivalent to the parameter $\nu$ characterizing the frequency detuning (see Eq. (7.23) above).

2. The map (7.51) is invertible if $|\varepsilon| \le 1$, and noninvertible for $|\varepsilon| > 1$. (For a general map (7.49) the invertability fails when $\varepsilon F'(\phi) = -1$ at some point $\phi$.) The line $\varepsilon = 1$ is called the critical line: above it chaotic dynamics can occur. In fact, in the vicinity of the critical line the circle map no longer represents the real dynamics of the forced system (7.5), and the full $M$-dimensional annulus map should be considered instead. We discuss the borders of validity of the circle map in Section 7.3.4. In this section we describe invertible dynamics, assuming that $\varepsilon < 1$.

---

[10] Strictly speaking, the original self-sustained oscillator with $\varepsilon = 0$ is a nonlinear system, but in the context of the circle map it is convenient to consider the unforced case as a linear map. This is a consequence of the linearity of the phase equation (7.3).

3. For $\varepsilon = 0$ the dynamics of (7.51) are trivial linear rotation on the circle, which is periodic or quasiperiodic depending on whether the parameter $\eta/2\pi$ is rational or irrational. For $\varepsilon > 0$ one can also characterize the dynamics with a single parameter, called the *rotation number*. For a given initial point $\phi_0$ the rotation number is defined as the average phase shift per one iteration:

$$\rho(\phi_0) = \lim_{n \to \infty} \frac{\phi_n - \phi_0}{2\pi n}. \tag{7.52}$$

It is known (see, e.g., [Katok and Hasselblatt 1995]) that the rotation number does not depend on the initial point $\phi_0$ and is the same for forward $(n \to \infty)$ and backward $(n \to -\infty)$ iterations. This follows simply from the monotonic nature of the map: the iterations of different initial points cannot differ by more than $2\pi$, so that $\rho$ is $\phi_0$-independent. Thus $\rho$ depends on the parameters of the circle map only. It is clear that for $\varepsilon = 0$ the rotation number is $\rho = \eta/2\pi$.

There are only two types of dynamics: motions with rational and irrational rotation numbers. Before going into details, let us discuss the interpretation of the rotation number in terms of oscillations of the original system (7.5). Because $n$ measures time in units of the external period $T$, the rotation number can be rewritten as

$$\rho = \lim_{t \to \infty} \frac{T(\phi(t) - \phi(0))}{2\pi t} = \frac{\Omega}{\omega}, \tag{7.53}$$

where we introduce the observed frequency $\Omega$ as the average velocity of the phase rotation

$$\Omega = \lim_{t \to \infty} \frac{\phi(t) - \phi(0)}{t}.$$

Relation (7.53) shows that the rotation number is the ratio of the observed frequency of oscillations to that of the external force. (Note that this ratio should be inverted if the circle map is written for the phase of the external force, as for the forced relaxation oscillator considered above.)

4. The irrational rotation number corresponds to quasiperiodic dynamics of the phase. According to the Denjoy theorem [Denjoy 1932; Katok and Hasselblatt 1995], in the case of an irrational rotation number $\rho$, the nonlinear circle map (7.51) can be transformed with a suitable change of the variable $\phi = g(\theta)$ (with the obvious property $g(\theta + 2\pi) = 2\pi + g(\theta)$) to the circle shift

$$\theta_{n+1} = \theta_n + 2\pi\rho. \tag{7.54}$$

The solution of (7.54) is trivial, and the solution of the circle map is thus given by

$$\phi_n = g(\theta_0 + n2\pi\rho). \tag{7.55}$$

Any $2\pi$-periodic function of $\phi$ (e.g., any of the original phase space variables $x_k$) is therefore a quasiperiodic function of the discrete time $n$.

According to Eq. (7.53), an irrational value of $\rho$ means that the observed frequency and the frequency of the external force are incommensurate, and these are the two basic frequencies of the quasiperiodic motion.

5. If the circle map has a periodic orbit, the rotation number is rational. Indeed, on the periodic orbit the rotation number is obviously rational, and it is independent of the initial point. For linear rotation $\varepsilon = 0$, all points on the circle are periodic, but this degeneracy disappears in the nonlinear case $\varepsilon \neq 0$. Generally, periodic points of a nonlinear map are isolated. If a periodic orbit has the period $q$, and during $q$ iterations the phase $\phi$ makes $p$ rotations (so that $\phi_{n+q} = \phi_n + 2\pi p$), then the rotation number is $\rho = p/q$.

We now consider periodic orbits, starting with those having period one, i.e., with fixed points. They correspond to the main resonance $T \approx T_0$, therefore we set $\rho = 1$ and look at orbits with $q = p = 1$. These orbits satisfy

$$\phi + 2\pi = \phi + \omega_0 T + \varepsilon F(\phi)$$

and are solutions of the equation

$$\omega_0 T - 2\pi + \varepsilon F(\phi) = 0.$$

The solution exists if

$$-\varepsilon F_{\min} \leq \omega_0 T - 2\pi \leq \varepsilon F_{\max}, \tag{7.56}$$

moreover, in this region of parameters there are at least two solutions. It is easy to check that one is stable (with $d\phi_{n+1}/d\phi_n < 1$) and one unstable (with $d\phi_{n+1}/d\phi_n > 1$). For general functions $F$ having more than one minimum and maximum, there can be more than two solutions of the same period, but they always appear in pairs: a stable and an unstable. It is important to mention that relation (7.56) defines the synchronization region with rotation number one on the plane of parameters $(\eta, \varepsilon)$ (compare Fig. 7.4a and Fig. 7.15 later).

6. The properties of periodic orbits with longer periods are qualitatively the same as those of the orbit with period one. Indeed, iterating Eq. (7.49) $q$ times gives also a circle map of the type (7.49), with some more complex dependence on the parameters $\eta$, $\varepsilon$. A fixed point of this iterated map satisfying

$$\phi_q = \phi_0 + 2\pi p \tag{7.57}$$

is a periodic orbit of the map (7.49) with rotation number $\rho = p/q$.

The region of synchronization corresponding to the rotation number $p/q$ can be found in the first-order approximation in $\varepsilon$. Assuming that $\omega_0 T = 2\pi p/q + \varepsilon\kappa$ in (7.49) and keeping (by iteration of this mapping) only the terms $O(\varepsilon)$, we get

$$\phi_{n+q} = \phi_n + \varepsilon q\kappa + 2\pi pq \cdot \varepsilon \tilde{F}(\phi_n) \tag{7.58}$$

with a nonlinear function given by

$$\tilde{F}(\phi_n) = \frac{F(\phi_n) + F(\phi_n + 2\pi\frac{p}{q}) + \cdots + F(\phi_n + (q-1)2\pi\frac{p}{q})}{q}. \tag{7.59}$$

If we represent $F(\phi)$ as a Fourier series

$$F(\phi) = \sum_{k=-\infty}^{\infty} a_k e^{ik\phi},$$

then the function $\tilde{F}$ reads

$$\tilde{F}(\phi) = \sum_{l=-\infty}^{\infty} a_{lq} e^{ilq\phi}, \tag{7.60}$$

i.e., it consists only of the harmonics $0, \pm q, \pm 2q, \ldots$ of the initial $2\pi$-periodic function $F$. Using (7.58) and (7.57), we get the synchronization region (cf. (7.56))

$$-\varepsilon\tilde{F}_{\min} < \omega_0 T - 2\pi\frac{p}{q} < \varepsilon\tilde{F}_{\max}. \tag{7.61}$$

As the function $\tilde{F}$ is obtained from $F$ via effective averaging (7.59), its amplitude decreases with $q$ and therefore for large $q$, the synchronization interval is small. Moreover, as one can see from (7.60), if the initial nonlinear function $F$ does not contain the harmonics $\pm lq$, the synchronization region with the rotation number $p/q$ vanishes in the first-order approximation in $\varepsilon$. This is exactly the case for the sine circle map (7.51). Here one should consider higher order terms, and this leads to the following estimate of the width of the synchronization region [Arnold 1983]

$$\Delta\eta \sim \varepsilon^q. \tag{7.62}$$

7. As only two regimes are possible in the circle map, namely periodic and quasiperiodic, we can build the state diagram as shown in Fig. 7.15. All regions of synchronization have the form of vertical tongues [Arnold 1961], now called Arnold tongues. The tip of the tongue with the rotation number $\rho = p/q$ touches the point $\varepsilon = 0$, $\eta = 2\pi p/q$. All the space on the plane of the parameters between the tongues corresponds to quasiperiodic motions with irrational rotation numbers.

The Arnold tongues form vertical stripes on the plane of parameters, so the order of rational numbers on the line $\varepsilon = 0$ is lifted over the whole region

$0 < \varepsilon < 1$. This means that for each $\varepsilon$ we have an ordered sequence of synchronization intervals with all possible rational rotation numbers. In particular, these intervals are everywhere dense: between each two quasiperiodic regimes with different rotation numbers there exists a synchronization region.

From the topological viewpoint, the quasiperiodic regime is unstable and the synchronous state is stable: for fixed $\varepsilon$ and $\eta$ being varied, synchronization occurs on the intervals of $\eta$, while quasiperiodicity is observed on isolated points. This means that a quasiperiodic state can be destroyed with an arbitrary small perturbation. However, from the probabilistic viewpoint the quasiperiodic regimes are abundant:[11] as has been proven by Arnold [1961], for small $\varepsilon$ the Lebesgue measure of all synchronization intervals tends to zero as $\varepsilon \to 0$; this means that the perturbations that destroy quasiperiodicity are rather exceptional.

8. For fixed amplitude $\varepsilon$ the rotation number $\rho$ as a function of the parameter $\eta$ has constant values within the synchronization regions, i.e., this function takes distinct constant values on a dense set of subintervals. This function is also continuous and monotonic [Katok and Hasselblatt 1995], and is called a *devil's staircase* (Fig. 7.16). The measure of all rational subintervals vanishes for $\varepsilon \to 0$, but it is equal to the full measure on the critical line $\varepsilon = 1$. The devil's staircase with a positive measure of points between the constant subintervals is called incomplete, and the case when the measure of all constant subintervals is full (i.e., equal to the Lebesgue measure) is called a complete devil's staircase. The Cantor-type set of all irrational rotation numbers can be characterized by its fractal dimensions, see [Jensen *et al.* 1983] for details.

9. Transition to and from the synchronization occurs through the saddle-node bifurcation. This is exactly the same transition which we described in Section 7.1.7, so we shall not repeat the details here. The main result (7.35) can be directly applied to the dependence of the rotation number on the frequency

---

[11] Just as irrational numbers are abundant among real ones.

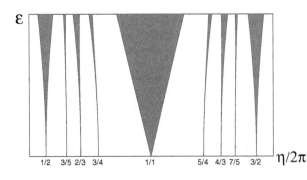

$\varepsilon$

1/2   3/5 2/3  3/4          1/1          5/4  4/3 7/5   3/2    $\eta/2\pi$

**Figure 7.15.** Major Arnold tongues in the sine circle map (Eq. (7.51)). The tips of the regions with rational rotation numbers touch the axis $\varepsilon = 0$ at rational values of $\eta/2\pi$. Note the symmetry $\eta \to 2\pi - \eta$.

detuning: near the borders of the synchronization region it follows the square-root law

$$|\rho - \rho_0| \sim |\eta - \eta_c|^{1/2}. \tag{7.63}$$

Note that the dependence (7.63) gives only an "envelope" for the rotation number, which as a function of $\eta$ in fact contains infinitely many small steps.

10. All the results above are valid for smooth one-to-one circle maps. As we have seen when considering relaxation oscillators, in some situations the circle map is not invertible and can have both flat regions and discontinuities. Some properties of such maps are close to the smooth case. The rotation number can be defined for monotonic maps with discontinuities, and it can be either rational or irrational. The main difference to the smooth case is that regimes with stable periodic orbits may be abundant, e.g., in mappings with finite flat regions and without vertical jumps quasiperiodic regimes may exist but they usually occupy a set of measure zero in the parameter space [Boyd 1985; Veerman 1989]. Here the transition to synchronization does not occur through the smooth saddle-node bifurcation, but may be abrupt.

11. When a quasiperiodic external force acts on a periodic oscillation, we can write instead of Eq. (7.48)

$$\frac{d\phi}{dt} = \omega_0 + \varepsilon Q(\phi, \omega_1 t, \omega_2 t),$$

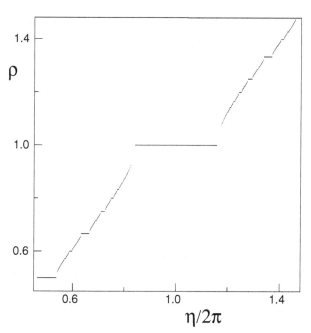

**Figure 7.16.** Devil's staircase: dependence of the rotation number on the parameter $\eta$ for the sine circle map (Eq. (7.51)) with $\varepsilon = 1$. The main plateaux correspond to rationals $1, 1/2, 2/3, \ldots$. Not all rational intervals are depicted, therefore the regions near the ends of large intervals look like gaps.

where $Q(\cdot, \cdot, \cdot)$ is $2\pi$-periodic in all its arguments. Here we have two external frequencies $\omega_1$, $\omega_2$, which are assumed to be incommensurate. This equation describes motion on a three-dimensional torus, and as the Poincaré map we obtain, instead of (7.51), a quasiperiodically forced circle map of the type (see [Ding *et al.* 1989; Glendinning *et al.* 2000] for details)

$$\phi_{n+1} = \phi_n + \eta + \varepsilon_1 \sin \phi_n + \varepsilon_2 \sin\left( 2\pi \frac{\omega_2}{\omega_1} n + \alpha \right).$$

As it follows from the results of Herman [1983], the rotation number (7.52) exists for this map as well. The dynamics can be either phase locked, if

$$\rho = \frac{p_2}{q_2} \frac{\omega_2}{\omega_1} + \frac{p_1}{q_1},$$

or nonsynchronized otherwise. In the synchronized state the observed frequency of oscillations is a rational combination of the two external frequencies; in the nonsynchronized state one has motion with three incommensurate frequencies. A quasiperiodic driving forcing can be also analyzed in the context of weakly nonlinear oscillators (see [Landa and Tarankova 1976; Landa 1980, 1996] for details).

The theory of the circle map has immediate applications for the synchronization properties of forced oscillators. First of all, it gives a complete qualitative description of the dynamics for small and moderate forcing. The main message (compared with the theory of Sections 7.1 and 7.2, where we have restricted ourselves to a first-order approximation in the forcing amplitude) is that there exists not only the main synchronization region, where the frequency of oscillations exactly coincides with that of the external force, but also regions of high-order synchronization of the type

$$\frac{\Omega}{\omega} = \frac{p}{q},$$

where the observed frequency is in a rational relation to the external frequency. Practically, the larger the numbers $p$ and $q$, the narrower is the phase locking region, and even in numerical experiments only synchronous regimes with small $p$ and $q$ can be observed.

### 7.3.3  The annulus map

In our derivation of the circle map (Section 7.3.1) we used the phase equation (7.48) which is valid for small forcing amplitudes $\varepsilon$ only. Now we demonstrate that the circle map has a much larger range of validity, and it properly describes the situation for moderate and large forcing as well. Here we cannot neglect the variations of the amplitude (i.e., any variable transverse to the limit cycle, see Section 7.1) in the full system (7.5). For simplicity of consideration (and especially of graphical presentation) we consider the case of two variables $\mathbf{x} = (x_1, x_2)$. Then a stroboscopic observation with the period of the external force provides a two-dimensional map

$$\mathbf{x}(t) \to \mathbf{x}(t + T).$$

Near the limit cycle of the unforced system this map has a simple structure: contraction in the transverse direction, and phase dynamics according to the circle map along the phase direction. The contraction in the amplitude direction means that we can restrict our attention to a strip around the limit cycle, which leads to the *annulus map*.

The dynamics of the circle map can be either periodic or quasiperiodic, the corresponding two regimes in the annulus map are shown in Fig. 7.17. In both cases all the points from the annulus are attracted to a closed invariant curve shown in Fig. 7.17 in bold. For vanishing forcing this is the limit cycle itself, and for small forcing this curve is slightly perturbed but remains invariant (i.e., it is mapped into itself).[12] In the quasiperiodic regime, rotation of the points on the invariant curve is topologically equivalent to the circle shift: all trajectories are dense and the dynamics are ergodic. In the periodic case there is both a stable and an unstable periodic orbit on the invariant curve (in Fig. 7.17 these are fixed points) and the stable orbit is the final minimal attractor. The invariant curve here is formed by unstable manifolds (separatrices) of the unstable periodic orbit. The existence of the invariant curve in the stroboscopic annulus map means that in the phase space of the original continuous-time system there is a two-dimensional invariant surface called an invariant torus (because the dynamics are periodic with respect to the phases of the system and of the external force).

[12] Note that invariance does not mean quasiperiodic dynamics. In the quasiperiodic case there are no invariant subsets on the curve, while for periodic dynamics the stable and the unstable orbits on the curve are also invariant.

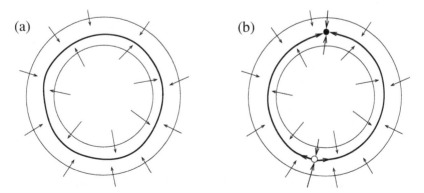

**Figure 7.17.** The structure of the annulus map for quasiperiodic rotation (a) and periodic dynamics (b). The arrows show the direction of contraction of the mapping (cf. Fig. 7.1); the stable and the unstable fixed points are depicted by a filled and open circle, respectively.

The existence of the stable (in the transverse direction) invariant curve justifies the validity of the circle map. Indeed, only the dynamics on this attracting curve are important asymptotically for large times, and on the curve we have exactly the map of the circle onto itself. For large forcing the invariant curve is destroyed, but before describing its fate we give an example of the annulus map.

### An example: kicked self-sustained oscillator

One simple way to obtain the stroboscopic map analytically is to consider a piecewise-solvable model. One can approximate the external force with a piecewise-constant function, solve the equations of motion in each time interval where the force is constant, and then obtain the annulus mapping by matching the solutions. We consider a particular simple example, namely a $T$-periodic sequence of $\delta$-pulses:

$$p(t) = \sum_{n=-\infty}^{\infty} \delta(t - nT). \tag{7.64}$$

Between the pulses we have an autonomous oscillator, so the whole problem can be divided into two tasks: solution of autonomous equations between pulses, and finding the action of the $\delta$-pulse. Both tasks are solvable if we consider the effect of the force (7.64) on the simple nonlinear oscillator (7.6). The equation has the form

$$\frac{dA}{dt} = (1 + i\eta) - (1 + i\alpha)|A|^2 A + i\varepsilon p(t). \tag{7.65}$$

Here the forcing is introduced in such a way that the pulse changes the variable $\text{Im}(A)$ by $\varepsilon$ and does not change the variable $\text{Re}(A)$: $\text{Im}(A_+) = \text{Im}(A_-) + \varepsilon$, $\text{Re}(A_+) = \text{Re}(A_-)$. The solution of Eq. (7.65) between the pulses is given by formulae (7.9) and the evolution during the time $T$ (starting from the point $\theta_0$, $R_0$) can be written as

$$R(T) = \left(1 + \frac{1 - R_0^2}{R_0^2} e^{-2T}\right)^{-1/2},$$

$$\theta(T) = \theta_0 + (\eta - \alpha)T - \frac{\alpha}{2}\ln(R_0^2 + (1 - R_0^2)e^{-2T}). \tag{7.66}$$

Representation of the change of the polar coordinates $R, \theta$ during the pulse as

$$R_+ = \sqrt{R_-^2 + 2R_-\varepsilon \sin\phi_- + \varepsilon^2}, \qquad \theta_+ = \tan^{-1}\left(\tan\theta_- + \frac{\varepsilon}{R_- \cos\theta_-}\right) \tag{7.67}$$

completes the construction of the mapping. Still, the formulae (7.66) and (7.67) can be significantly simplified if we consider small forcing $\varepsilon \ll 1$, and assume accordingly that the deviations of the amplitude $R$ from one (the limit cycle amplitude) are small.

We denote the variables just before the $n$th pulse as $R_n$, $\phi_n$. Then, for the variables just after the pulse we obtain from (7.67)

$$R_+ \approx R_n + \varepsilon \sin\phi_n, \qquad \phi_+ \approx \phi_n + \varepsilon \cos\phi_n.$$

Substituting this in the solution (7.66) we obtain the annulus map (cf. [Za-slavsky 1978])

$$R_{n+1} = 1 + (R_n - 1 + \varepsilon \sin \phi_n)e^{-2T},$$

$$\phi_{n+1} = \phi_n + (\eta - \alpha)T + \varepsilon \cos \phi_n - \alpha(R_n - 1 + \varepsilon \sin \phi_n)(1 - e^{-2T}).$$

If the period of the forcing is large, $T \gg 1$, then the annulus map is very squeezed in the $R$-direction. Neglecting variations of $R$ yields a circle map:

$$\phi_{n+1} = \phi_n + (\omega_0 - \alpha)T + \varepsilon \cos \phi_n - \alpha \varepsilon \sin \phi_n. \qquad (7.68)$$

Note that the two nonlinear terms in (7.68) have different contributions to the phase shift of oscillations with respect to the phase of the force. The term $\varepsilon \cos \phi_n$ describes the direct effect of the force on the phase, while the term $\alpha \varepsilon \sin \phi_n$ describes the effect of the amplitude perturbations: the force changes the amplitude, and because the oscillations are nonisochronous, an additional phase shift proportional to $\alpha$ occurs (cf. discussion in Section 7.2.3).

### 7.3.4    Large force and transition to chaos

As we have seen, for small forcing amplitude, the dynamics of the annulus map are simple (in fact, equivalent to the dynamics of the circle map): either quasiperiodic with irrational rotation number or synchronized, when the rotation number is rational. In both cases there exists an attracting invariant curve. For irrational rotation numbers, this curve is filled with trajectories and is ergodic; for a rational rotation number, there is a stable (node) and an unstable (saddle) periodic orbit on the curve (or several pairs of stable and unstable orbits), and the curve consists of unstable manifolds of the saddle. If the forcing is small, the only possibility for the synchronized regime to be destroyed is the saddle-node bifurcation.

For large amplitudes of the external force there are other transitions from synchronization, and they typically lead to chaotic behavior. These transitions have been described in [Aronson et al. 1982; Afraimovich and Shilnikov 1983], and we refer the reader to those papers for details. There are three main scenarios of a transition to chaos, as follows.

*Scenario I*

The first scenario is the transition to chaos through period-doublings of the stable periodic orbit (Fig. 7.18). Here a strange attractor appears continuously from the stable node and remains separated from the saddle. The motion becomes chaotic but nevertheless the rotation number is well-defined and remains rational, thus the appearing regime can be characterized as synchronized, albeit with a chaotic modulation. Note that in this scenario the invariant curve becomes nonsmooth prior to the first period-doubling, and does not exist as a smooth curve during further bifurcations. The

reason for this is that for period-doubling to occur, the multiplier of the periodic orbit
has to pass through $-1$, so it first must become complex. The node then becomes the
focus, and the invariant curve looks like that shown in Fig. 7.18a.

*Scenarios II, III*

At the second and the third scenarios the invariant curve can loose smoothness and
transform to a rather wild set as shown in Fig. 7.19. There is no attracting chaos
yet, but some of its precursors exist, in particular regions with local instability in
the phase space. These regions remain transient while the stable node exists, but
when the saddle-node bifurcation occurs, these regions can belong to the attractor
(if there are no other stable periodic orbits). One typically observes the transition
from synchronization to chaos via intermittency. The picture of the dynamics is close
to Fig. 7.5, with long laminar nearly synchronized epochs and phase slips; the only
difference is that the slips now occur at chaotic time intervals.

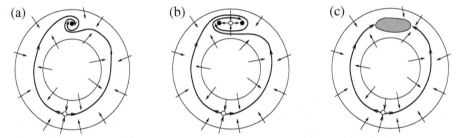

**Figure 7.18.** Destruction of invariant curve and period-doubling transition to chaos
(scenario I). (a) When the eigenvalues of the stable fixed point are complex or
negative, a smooth invariant curve does not exist. (b) After the first period-doubling
the unstable manifolds of the saddle wrap onto the period-two orbit. (c) After the
cascade of period-doublings a strange attractor appears. The rotation number
remains the same.

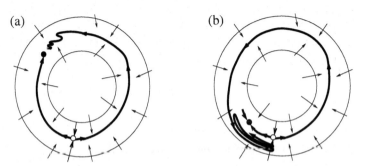

**Figure 7.19.** Two other scenarios of invariant curve destruction. (a) Scenario II: the
unstable manifold of the saddle makes a fold and does not form a smooth curve.
(b) Scenario III: the unstable manifold of the saddle crosses its stable manifold,
creating the homoclinic structure. In contrast to scenario II, an invariant
nonattracting chaotic set exists here.

Scenario I is typical for the center of the synchronization region, where the stable and unstable periodic orbits are well separated. Here, with an increase in the forcing amplitude, the described transition to chaos through period-doublings can occur. Scenarios II and III occur at the border of the synchronization region, usually as frequency detuning is increased (usually scenario II corresponds to lower amplitudes of forcing).

Apart from the possibility of chaotic behavior there are two other features of synchronization at large forcing amplitude that are different compared with small forcings. First, the dynamics are no longer uniquely determined by the rotation number. Different synchronization regions can overlap, thus leading to multistability, so that for certain parameters periodic oscillations with different rotation numbers (i.e., with different rational ratios between the observed and the forcing frequencies) coexist. This phenomenon has been experimentally observed by van der Pol and van der Mark [1927], see Fig. 3.29.[13] Another property observed at large forcing is that some synchronization regions can close off smoothly without transition to chaos, see [Aronson *et al.* 1986] for details.

## 7.4    Synchronization of rotators and Josephson junctions

In this chapter we have considered periodic self-sustained oscillators, and the effect of external force on them. At first glance, forced rotators do not belong to this class of systems, because they are not self-sustained. Nevertheless, if the rotators are driven by a constant force, they have the same features as self-sustained oscillators: in phase space there is a limit cycle, and one of the Lyapunov exponents is zero. Thus, driven rotations are similar to oscillations: they can be synchronized by a periodic external force. We shall first describe some examples of rotators, and then discuss their synchronization properties.

### 7.4.1    Dynamics of rotators and Josephson junctions

As the simplest example of a rotator let us consider a pendulum, driven by a constant torque $K$, see Fig. 7.20a. The equation of motion is

$$\frac{d^2\Psi}{dt^2} + \gamma\frac{d\Psi}{dt} + \kappa^2 \sin\Psi = \frac{K}{I}. \tag{7.69}$$

Here $\kappa$ is the frequency of small oscillations (it is rather irrelevant for the rotation regimes to be considered below, so we do not use our usual notation for frequency $\omega$), $\gamma > 0$ is the damping constant, $K$ is the torque, and $I$ is the moment of inertia.

---

[13] Remarkably, in trying to understand multistability of synchronization, Cartwright and Littlewood [1945] found some strange properties of the dynamics, which later enabled Smale [1980] to construct his famous chaotic horseshoe.

Equation (7.69) is dissipative, and two types of asymptotically stable regimes are possible here: a steady state (if the torque is small) or rotations (for large $K$).[14] The phase space of the dynamical system (7.69) is a cylinder $0 \leq \Psi < 2\pi$, $-\infty < \dot{\Psi} < \infty$ (we consider the positions of the pendulum differing by $2\pi$ as equivalent). A steady state is a fixed point on the cylinder, and rotations correspond to closed attracting trajectories on this cylinder. Although there is no physical mechanism for self-sustained oscillations here, a limit cycle appears due to the special structure of the phase space, i.e., to the periodicity in the angle variable $\Psi$.

Because the rotation is represented by a limit cycle, it has all the properties of the limit cycle in an autonomous system: it is an isolated closed orbit, and it has one zero and one negative Lyapunov exponent. Thus, we can treat constantly forced rotations as self-sustained oscillations. However, not all the theory developed for oscillations can be applied here: the particular cylindric structure of the phase space imposes some restrictions (e.g., a Hopf bifurcation of the cycle is impossible).

Remarkably, there is a quite different application of Eq. (7.69): it describes the dynamics of the Josephson junction. While referring to books by Barone and Paterno [1982] and Likharev [1991] for details, we sketch here the derivation. In the widely accepted resistively shunted junction model (Fig. 7.20b) it is assumed that the junction current consists of three components: a superconducting current $I_c \sin \Psi$, a resistive current $V/R$, and a capacitance current $\dot{V}C$. Here the dynamical variable $\Psi(t)$ is the phase difference between the macroscopic wave functions of superconductors on the two sides of the contact. In the context of the Josephson junction dynamics, it is often called "phase", but here we prefer to call it "angle", as the term "phase" is reserved for the variable on the limit cycle. The parameter $I_c$ is called the critical supercurrent of the junction. The resistance $R$ and the capacitance $C$ characterize the normal current. Finally, the voltage $V$ relates to the angle $\Psi$ according to the Josephson formula

$$\dot{\Psi} = \frac{2e}{\hbar} V, \qquad (7.70)$$

where the electron charge $e$ and the Planck constant reveal the underlying quantum nature of the phenomenon. Collecting all the terms, we obtain the equation of the

---

[14] In fact, there can be also a region of bistability of these two regimes.

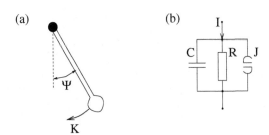

(a)     $\Psi$     $K$     (b)     I     C     R     J

**Figure 7.20.** (a) The pendulum is the simplest rotator. (b) The shunted model of the "electrical rotator", the Josephson junction. Both systems are described by the equivalent equations (7.69) and (7.71).

junction fed with the external current $I$

$$I = I_c \sin \Psi + \frac{\hbar}{2eR} \frac{d\Psi}{dt} + \frac{C\hbar}{2e} \frac{d^2\Psi}{dt^2}, \qquad (7.71)$$

which coincides with (7.69). The averaged frequency of rotation has a clear physical meaning for the Josephson junction: according to (7.70) it is proportional to the average (dc component) voltage across the junction. The steady state of the rotator (i.e., $\dot{\Psi} = 0$) corresponds to zero voltage, while for rotations the voltage is nonzero. Synchronized rotations correspond, as we will show below, to steps in the voltage–current characteristics of the junction.

## 7.4.2 Overdamped rotator in an external field

As we argued above, rotations are described by a limit cycle in the phase space, so that all the general results of Sections 7.1 and 7.3 can be applied here. Moreover, some synchronization phenomena can be studied in an even simpler context, due to the specific topological structure of the phase space. Indeed, usually periodic oscillations are impossible in one-dimensional dynamical systems (in order to have a limit cycle one needs at least two-dimensional phase space). But for rotations this is no longer true: if the phase space is a one-dimensional circle, rotations are possible.

For the rotator and the Josephson junction we obtain a one-dimensional system in the overdamped limit, when the coefficient of the second derivative tends to zero. In the case of the Josephson junction, this means vanishing capacitance. The equation of motion then reads

$$\frac{d\Psi}{dt} = \alpha(-\sin \Psi + \mathcal{I}), \qquad (7.72)$$

where $\alpha = 2eRI_c/\hbar$ and $\mathcal{I} = I/I_c$. This equation coincides with Eq. (7.29): the averaged phase equation for the forced oscillator is the same as the equation for an overdamped constantly driven rotator. The driving torque $\mathcal{I}$ corresponds to the frequency detuning, and the dependence of the rotation frequency $\Omega = \langle \dot{\Psi} \rangle$ on this parameter was shown in Fig. 7.4b. For the Josephson junction, $\Omega$ is proportional to voltage $V$, and $\mathcal{I}$ is proportional to the current, so that Fig. 7.4b represents the voltage–current characteristics of the contact.

We consider now the effect of periodic external force on the rotations. With an additional sinusoidal external current Eq. (7.72) reads

$$\frac{d\Psi}{dt} = \alpha(-\sin \Psi + \mathcal{I} + \mathcal{J}\cos(\omega t)). \qquad (7.73)$$

This equation is a particular form of Eq. (7.15), thus all the corresponding theory can be applied here. The Poincaré map for this equation is the circle map.

We follow here an analytical approach that allows us to find the synchronization regions if the external alternating current is large: $\mathcal{J} \gg 1$ (we remind the reader that the external current is normalized to the critical supercurrent of the junction). In this case we can first neglect the nonlinear term in (7.73), and then treat it as a perturbation. To find a synchronized solution, let us consider $\mathcal{J}^{-1}$ as a small parameter and expand the solution in powers of $\mathcal{J}^{-1}$:

$$\Psi = n\omega t + \mathcal{J}\Psi_{-1}(t) + \Psi_0(t) + \cdots, \qquad \langle \dot{\Psi}_{-1} \rangle = \langle \dot{\Psi}_0 \rangle = 0. \qquad (7.74)$$

In (7.74) we assume that the rotations have a frequency that is a multiple of the driving one. Substituting (7.74) in (7.73), we obtain for the terms $\sim \mathcal{J}$

$$\Psi_{-1}(t) = \alpha\omega^{-1} \sin(n\omega t) + \Psi_{-1}^0, \qquad \Psi_{-1}^0 = \text{constant}.$$

Substituting this in (7.73) and collecting the terms $\sim \mathcal{J}^0$ we obtain

$$\frac{d\Psi_0}{dt} = \alpha\mathcal{I} - n\omega - \alpha \sin[n\omega t + \alpha\mathcal{J}\omega^{-1} \sin(\omega t) + \mathcal{J}\Psi_{-1}^0].$$

Now using the condition $\langle \dot{\Psi}_0 \rangle = 0$ and performing the integration over the period $2\pi/\omega$ yields

$$0 = \alpha\mathcal{I} - n\omega - \alpha \sin \Psi_{-1}^0 J_n(-\alpha\mathcal{J}\omega^{-1}),$$

where $J_n$ is a Bessel function of the first kind. Because $\sin \Psi_{-1}^0$ lies between $-1$ and $1$, we obtain the width of the synchronization region:

$$\left| \mathcal{I} - n\frac{\omega}{\alpha} \right| < \left| J_n\left( -\frac{\alpha\mathcal{J}}{\omega} \right) \right|. \qquad (7.75)$$

In experiments with Josephson junctions, the synchronization regions (7.75) manifest themselves as steps on the voltage–current plots at $V_n = n\omega\hbar/2e$ known as Shapiro steps (cf. Section 4.1.8).

    The above theory shows that synchronization of rotators is completely analogous to synchronization of periodic oscillators. The phase of a rotator can be introduced using the angle as variable, and this leads to simple definitions of the period and frequency. Moreover, rotators are even simpler than oscillators: one does not encounter here amplitude-caused effects such as oscillation death.

## 7.5    Phase locked loops

The idea of a phase locked loop (PLL) is to use synchronization for effective and robust modulation/demodulation of signals in electronic communication systems. The information (e.g., an acoustic signal) in the frequency modulated radio signal is stored in the variations (modulations) of the frequency. In a receiver one has to extract these variations; this operation is called demodulation. Nowadays the PLL method is used in most radio receivers. Here we emphasize the principal concepts, connecting them to our physical understanding of synchronization as described above.

In a PLL one tries to synchronize a self-sustained oscillator with an external signal, but instead of using the signal as a forcing term (as in (7.5)), one constructs a circuit that ensures synchronization. The advantage of this method is that one can control different aspects of the phase dynamics separately; in fact one constructs a hardware device that simulates an equation for the phase with the desired synchronization properties.

A simple schematic diagram of a PLL is shown in Fig. 7.21. At the input we have a narrow-band signal $x(t)$ which can be considered as a sine function with slowly varying phase and amplitude:

$$x(t) = A(t) \sin \phi_e(t). \tag{7.76}$$

As the relevant information is contained in the variations of the phase, the output signal should be proportional to $\phi_e$.

The self-sustained oscillator to be synchronized is the voltage controlled oscillator (VCO), the frequency of which depends on the governing input signal $v(t)$. A good approximation is that the output of the VCO is sinusoidal with a constant amplitude and instantaneous frequency $\dot{\phi}$ that depends on $v(t)$ linearly:

$$y(t) = 2B \cos \phi(t), \qquad \frac{d\phi}{dt} = \omega_0 + K v(t). \tag{7.77}$$

Now the goal is to form the signal $v(t)$ proportional to the phase difference between the output of the VCO $y(t)$ and the input signal of the PLL $x(t)$. This is done in two steps. In the first step the phase detector forms a signal proportional to a function of the phase difference between $x$ and $y$. The simplest way to do this is to multiply these signals:

$$u(t) = x(t)y(t) = AB[\sin(\phi_e + \phi) + \sin(\phi_e - \phi)]. \tag{7.78}$$

Now it is necessary to separate the part of $u$ proportional to the phase difference, and this is done by virtue of the simplest low-band filter. It suppresses the term in (7.78) that is proportional to the sum of the phases (this term has the double frequency).

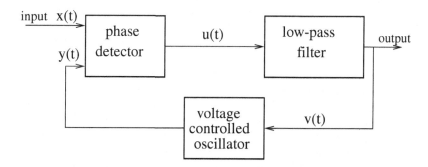

**Figure 7.21.** Block diagram of the phase locked loop.

Leaving only the term proportional to the difference between phases in the output of the filter, we can write

$$\frac{dv}{dt} + \frac{v}{\tau} = AB \sin(\phi_e - \phi), \tag{7.79}$$

where $\tau$ is the relaxation constant of the filter.

Introducing the phase difference $\psi = \phi - \phi_e$, we obtain from (7.77) and (7.79):

$$\frac{d^2\psi}{dt^2} + \frac{1}{\tau}\frac{d\psi}{dt} + KAB \sin\psi = -\frac{d^2\phi_e}{dt^2} - \frac{1}{\tau}\left(\frac{d\phi_e}{dt} - \omega_0\right). \tag{7.80}$$

This equation is similar to Eqs. (7.69) and (7.71). In the simplest case of constant external frequency $\dot{\phi}_e = \omega$, the r.h.s. of (7.80) is constant, and Eq. (7.80) has a stable fixed point (provided, of course, that we are in the synchronization region, i.e., $\omega \approx \omega_0$). The output signal $v$ is proportional to the phase difference $v \propto \sin\psi$. If the input phase $\phi_e$ is slowly modulated, the output signal changes according to this modulation. Thus the PLL generates a signal $y(t)$, having approximately the same (time-dependent) phase as the input signal $x(t)$; the demodulated output is provided by the signal $v(t)$. In physical language, the self-sustained oscillator (the VCO) is synchronized by the input signal $x(t)$.

The merit of the PLL is that the synchronized signal $y(t)$ is much cleaner than the input signal $x(t)$: all the noise in the amplitude $A(t)$ and in the high-frequency band disappears due to phase detection and filtering (see Fig. 7.22). There are many practical aspects of PLL dynamics, including schemes for digital processing (see [Best 1984; Lindsey and Chie 1985; Afraimovich et al. 1994] for details and further references).

(a)

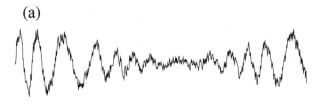

(b)

**Figure 7.22.** The PLL helps to reduce noise in the received signal $x(t)$ (a). The variations in amplitude, as well as high-frequency noisy components, do not appear in the oscillations of the VCO $y(t)$ (b). After Best [1984].

## 7.6    Bibliographic notes

The theory of synchronization by a periodic force is a classical problem in the theory of nonlinear oscillations. Nowadays it is mainly unusual behavior, like transition to chaos, that attracts the attention of researchers. Different examples have been studied numerically [Zaslavsky 1978; Gonzalez and Piro 1985; Aronson *et al.* 1986; Parlitz and Lauterborn 1987; Mettin *et al.* 1993; Coombes 1999; Coombes and Bressloff 1999] and experimentally [Martin and Martienssen 1986; Bryant and Jeffries 1987; Benford *et al.* 1989; Peinke *et al.* 1993]. In particular, Glass [1991] has reviewed synchronization of forced relaxation oscillators with application to cardiac arrhythmias and this work provides a rich bibliography; in the same issue the original paper by Arnold [1991] is reproduced. Several works have been devoted to the scaling structure of Arnold tongues and to characterization of the devil's staircase [Ostlund *et al.* 1983; Jensen *et al.* 1984; Alstrøm *et al.* 1990; Christiansen *et al.* 1990; Reichhardt and Nori 1999].

# Chapter 8

# Mutual synchronization of two interacting periodic oscillators

In this chapter we consider the effects of synchronization due to the interaction of two oscillating systems. This situation is intermediate between that of Chapter 7, where one oscillator was subject to a periodic external force, and the case of many interacting oscillators discussed later in Chapters 11 and 12. Indeed, the case of periodic forcing can be considered as a special case of two interacting oscillators when the coupling is unidirectional. However, two oscillators form an elementary building block for the case of many (more than two) mutually coupled systems. The problem can be formulated as follows: there are two nonlinear systems each exhibiting self-sustained periodic oscillations, generally with different amplitudes and frequencies. These two systems can interact, and the strength of the interaction is the main parameter. We are interested in the dynamics of the coupled system, with the main emphasis on the entrainment of phases and frequencies.

In Section 8.1 we develop a phase dynamics approach that is valid if the coupling is small – the problem here reduces to coupled equations for the phases only. Another approximation is used in Section 8.2, where the dynamics of weakly nonlinear oscillations are discussed. Finally, in Section 8.3 we describe synchronization of relaxation "integrate-and-fire" oscillators. No special attention is devoted to coupled rotators: their properties are very similar to those of oscillators.

## 8.1    Phase dynamics

If the coupling between two self-oscillating systems is small, one can derive, following Malkin [1956] and Kuramoto [1984], closed equations for the phases. The approach here is essentially the same as in Section 7.1. It is advisable to read that section first: below we use many ideas introduced there. Our basic model is a system of two interacting oscillators

$$\frac{d\mathbf{x}^{(1)}}{dt} = \mathbf{f}^{(1)}(\mathbf{x}^{(1)}) + \varepsilon \mathbf{p}^{(1)}(\mathbf{x}^{(1)}, \mathbf{x}^{(2)}),$$

$$\frac{d\mathbf{x}^{(2)}}{dt} = \mathbf{f}^{(2)}(\mathbf{x}^{(2)}) + \varepsilon \mathbf{p}^{(2)}(\mathbf{x}^{(2)}, \mathbf{x}^{(1)}). \tag{8.1}$$

Note that here we do not assume any similarity between the two interacting systems: they can be of different nature and have different dimensions. Also, the coupling can be asymmetric. We assume only that the autonomous dynamics (given by the functions $\mathbf{f}^{(1),(2)}$) can be separated from the interaction (described by generally different terms $\mathbf{p}^{(1),(2)}$), proportional to the coupling constant $\varepsilon$. This is motivated by the physical formulation of the problem: one has two independent oscillators that can operate separately, but which may also interact. We thus exclude a situation when two oscillating modes are observed in a complex system that cannot be decomposed into two parts.[1] Another case not covered by system (8.1) is that of a more complex coupling requiring additional dynamical variables.[2]

When the coupling constant $\varepsilon$ vanishes, each system has a stable limit cycle with frequencies $\omega_{1,2}$. Thus, as described in Section 7.1, we can introduce two phases on the cycles and in their vicinities,[3] (cf. Eq. (7.3))

$$\frac{d\phi_1}{dt} = \omega_1,$$

$$\frac{d\phi_2}{dt} = \omega_2. \tag{8.2}$$

In general, the frequencies $\omega_1, \omega_2$ are incommensurate, therefore the motion of the uncoupled oscillators is quasiperiodic.

In the first approximation we can write the equations for the phases in the coupled system, analogous to (7.14), as

$$\frac{d\phi_1(\mathbf{x}^{(1)})}{dt} = \omega_1 + \varepsilon \sum_k \frac{\partial \phi_1}{\partial x_k^{(1)}} p_k^{(1)}(\mathbf{x}^{(1)}, \mathbf{x}^{(2)}),$$

$$\frac{d\phi_2(\mathbf{x}^{(2)})}{dt} = \omega_2 + \varepsilon \sum_k \frac{\partial \phi_2}{\partial x_k^{(2)}} p_k^{(2)}(\mathbf{x}^{(2)}, \mathbf{x}^{(1)}). \tag{8.3}$$

Assuming that for small coupling the perturbations of the amplitudes are small, on the r.h.s. we can substitute the values of the variables $\mathbf{x}^{(1)}, \mathbf{x}^{(2)}$ on the cycles, where these

---

[1] Nevertheless, some high-dimensional systems (e.g., lasers) can generate two independent modes that can be considered in the framework of the model.

[2] In electronics, this difference corresponds to that between a coupling with resistors (no new equations) and a coupling with reactive elements such as capacitors and inductances (which requires additional equations). One such example will be considered in Section 12.3.

[3] Compared to Section 7.1, we omit here the subscript "0" when denoting the natural frequencies; instead we use a subscript corresponding to the oscillator number.

variables are unique functions of the phases only. We thus obtain closed equations for the phases

$$\frac{d\phi_1}{dt} = \omega_1 + \varepsilon Q_1(\phi_1, \phi_2),$$

$$\frac{d\phi_2}{dt} = \omega_2 + \varepsilon Q_2(\phi_2, \phi_1),$$

(8.4)

with $2\pi$-periodic (in both arguments) functions $Q_{1,2}$.

The possibility of writing closed equations for the phase variables means that in the high-dimensional phase space of variables $(\mathbf{x}^{(1)}, \mathbf{x}^{(2)})$ there exists a two-dimensional invariant surface parameterized with the phases $\phi_1, \phi_2$. Moreover, this surface is a torus as the $2\pi$ shift in each phase leads to the same point in the phase space. This two-dimensional torus is completely analogous to the invariant torus for the nonautonomous system described in Section 7.3. There are two ways to characterize the dynamics on an invariant torus. First, we can use the smallness of the parameter $\varepsilon$ and average Eqs. (8.4). Another approach is based on the construction of the circle map.

## 8.1.1   Averaged equations for the phase

The $2\pi$-periodic functions $Q_{1,2}$ in Eqs. (8.4) can be represented as double Fourier series

$$Q_1(\phi_1, \phi_2) = \sum_{k,l} a_1^{k,l} e^{ik\phi_1 + il\phi_2}, \qquad Q_2(\phi_2, \phi_1) = \sum_{k,l} a_2^{l,k} e^{ik\phi_1 + il\phi_2}.$$

In the zero approximation the phases rotate with the unperturbed (natural) frequencies

$$\phi_1 = \omega_1 t, \qquad \phi_2 = \omega_2 t,$$

and in the functions $Q_{1,2}$ all the terms correspond to fast rotations except for those satisfying the resonance condition

$$k\omega_1 + l\omega_2 \approx 0.$$

Let us assume that the two natural frequencies $\omega_{1,2}$ are nearly in resonance:

$$\frac{\omega_1}{\omega_2} \approx \frac{m}{n}.$$

Then all the terms in the Fourier series with $k = nj$, $l = -mj$ are resonant and contribute to the averaged equations. As a result we obtain

$$\frac{d\phi_1}{dt} = \omega_1 + \varepsilon q_1(n\phi_1 - m\phi_2),$$

$$\frac{d\phi_2}{dt} = \omega_2 + \varepsilon q_2(m\phi_2 - n\phi_1),$$

(8.5)

where

$$q_1(n\phi_1 - m\phi_2) = \sum_j a_1^{nj,-mj} e^{ij(n\phi_1 - m\phi_2)},$$

$$q_2(m\phi_2 - n\phi_1) = \sum_j a_2^{mj,-nj} e^{ij(m\phi_2 - n\phi_1)}.$$

For the difference between the phases of two oscillators $\psi = n\phi_1 - m\phi_2$ we obtain from (8.5)

$$\frac{d\psi}{dt} = -\nu + \varepsilon q(\psi), \tag{8.6}$$

where

$$\nu = m\omega_2 - n\omega_1, \qquad q(\psi) = nq_1(\psi) - mq_2(-\psi). \tag{8.7}$$

One can see that Eq. (8.6) has exactly the same form as Eq. (7.24) in Section 7.1.6, and we do not need to repeat its analysis here. In the case of synchronization, Eq. (8.6) has a stable fixed point $\psi_0$ and the observed frequencies of the oscillators are

$$\Omega_{1,2} = \langle \dot{\phi}_{1,2} \rangle = \omega_{1,2} + \varepsilon q_{1,2}(\pm\psi_0).$$

It is easy to see that the relation between these frequencies is constant inside the synchronization region:

$$\frac{\Omega_1}{\Omega_2} = \frac{m}{n}.$$

Let us focus on the simplest case of $1 : 1$ resonance, i.e., on the case when the natural frequencies of the oscillators nearly coincide $\omega_1 \approx \omega_2$. Here we should put $m = n = 1$ in the formulae above. Further, we assume that the coupling is symmetric, i.e., $q_1(\psi) = q_2(\psi)$; then, according to (8.7), we get an antisymmetric coupling function in (8.6) $q(\psi) = -q(-\psi)$. The simplest and the most natural antisymmetric $2\pi$-periodic function is sine, and the corresponding model for the interaction of two oscillators reads

$$\frac{d\psi}{dt} = -\nu + \varepsilon \sin \psi. \tag{8.8}$$

Depending on the sign of $\varepsilon$ there are two cases, called attractive and repulsive interactions.[4] If $\varepsilon < 0$, then the stable value of the phase difference $\psi$ lies in the region $-\pi/2 < \psi < \pi/2$, and, in particular, for zero frequency detuning $\nu$ the stable phase difference is zero. One can say that the phases "attract" each other. If $\varepsilon > 0$, the stable phase difference is in the interval $\pi/2 < \psi < 3\pi/2$, and for coinciding natural frequencies it is equal to $\pi$; this is the "repulsive" case. The two types of synchronous motion are sometimes called "in-phase" and "anti-phase" (or "out-of-phase").[5] Remarkably, quantitative properties of synchronization (in particular, the

---

[4] Or, equivalently, one can say that $\varepsilon$ is positive, but the coupling function changes sign $\sin \psi \rightarrow -\sin \psi$.

[5] We remind the reader that Huygens described the "out-of-phase" synchronization of pendulum clocks in his first observation of the phenomenon.

width of the synchronization region) are the same for both cases. It is worth noting that attraction and repulsion between the phases may not correspond to attraction and repulsion between some original variables $x_k^{(1,2)}$, due to possibly nontrivial forms of isochrones in the vicinity of the limit cycle (see [Han et al. 1995, 1997; Postnov et al. 1999a] for such an example).

In the averaged description, synchronization appears as the perfect phase locking: the stable fixed point of Eq. (8.8) $\psi_0$ means not only that both oscillators have the same frequency, but also that there is a constant phase shift between the phases of the oscillators $\phi_1 = \phi_2 + \psi_0$. The latter property is no longer valid if one considers the full system (8.4): due to nonresonant terms the phases are not perfectly locked, but oscillate around the trajectory of the averaged system (8.5). These oscillations can be especially large if the oscillators are close to relaxation ones, i.e., if the coupling functions $Q_{1,2}$ have many harmonics.

## 8.1.2    Circle map

The right-hand sides of Eqs. (8.4) are $2\pi$-periodic in both variables; therefore the flow on the two-dimensional phase plane $(\phi_1, \phi_2)$ is equivalent to the flow on a two-dimensional torus $0 \le \phi_1 < 2\pi$, $0 \le \phi_2 < 2\pi$. This two-dimensional flow can be reduced to an invertible circle map.

We take the line $\phi_2 = 0$ as a secant line. Starting a trajectory at $\phi_1(0)$, $\phi_2(0) = 0$ and following it until the point $\phi_1(t)$, $\phi_2(t) = 2\pi$, we obtain the mapping $\phi_1(0) \to \phi_1(t)$, which we write introducing the discrete time $n$ as

$$\phi_1(n+1) = F(\phi_1(n)), \tag{8.9}$$

where the function $F$ is such that $F(x + 2\pi) = 2\pi + F(x)$.[6] For noninteracting systems, this mapping reduces to the linear circle shift

$$\phi_1(n+1) = \phi_1(n) + 2\pi \frac{\omega_1}{\omega_2}.$$

For the circle map (8.9) we can define the rotation number $\rho$ according to Eq. (7.52); this gives the ratio between the two observed frequencies

$$\rho = \frac{\Omega_1}{\Omega_2}.$$

We note that there is an equivalent way of obtaining the circle map: one can take the line $\phi_1 = 0$ as a section and obtain a mapping $\phi_2 \to \tilde{F}(\phi_2)$; the new rotation number will be the inverse of the old one.

The full theory of the circle map (Section 7.3) can be applied here. In particular, synchronization is destroyed via a saddle-node bifurcation, as is described in Sections 7.1 and 7.3.

---

[6]  In fact, we use here the smallness of the interaction: the flow on the torus is not arbitrary, but close to rotations in both coordinates. This ensures the absence of fixed points and closed trajectories not wrapping the torus, and therefore the existence of the Poincaré map.

## 8.2    Weakly nonlinear oscillators

If the coupling between two oscillators is relatively large, it affects not only the phases but also the amplitudes. In general, the features of strong interaction are nonuniversal, but in the case of weakly nonlinear self-sustained oscillators, one can use the method of averaging and obtain universal equations depending on a few essential parameters. We will follow here the approach of Aronson *et al.* [1990]. As we have already outlined the method of averaging in Section 7.2, here we simply modify the equations to account for the bidirectional coupling.

### 8.2.1    General equations

We take two oscillators, generally different, and couple them linearly

$$\ddot{x}_1 + \omega_1^2 x_1 = f_1(x_1, \dot{x}_1) + D_1(x_2 - x_1) + B_1(\dot{x}_2 - \dot{x}_1), \tag{8.10}$$

$$\ddot{x}_2 + \omega_2^2 x_2 = f_2(x_2, \dot{x}_2) + D_2(x_1 - x_2) + B_2(\dot{x}_1 - \dot{x}_2). \tag{8.11}$$

Here $\omega_{1,2}$ are the frequencies of the linear uncoupled oscillators. Some comments are in order.

(i) We consider the coupling to be linear in the variables $x_{1,2}$, $\dot{x}_{1,2}$. This is justified if the natural frequencies $\omega_1$ and $\omega_2$ are close, which corresponds to the resonance 1 : 1. Indeed, the main terms on the r.h.s. are those having frequencies $\omega_{1,2}$, and such terms are linear. If the linear terms vanish, one has to consider higher-order terms; the synchronization will be weaker in this case.

(ii) The coupling terms are chosen to be proportional to differences of the variables and their derivatives. This coupling vanishes if the states of the two systems coincide $x_1 = x_2$, $\dot{x}_1 = \dot{x}_2$. Aronson *et al.* [1990] call this coupling "diffusive". Another possibility could be "direct" coupling, where, e.g., Eq. (8.10) is modified to

$$\ddot{x}_1 + \omega_1^2 x_1 = f_1(x_1, \dot{x}_1) + D_1 x_2 + B_1 \dot{x}_2.$$

The difference between "direct" and "diffusive" coupling will be important in considering the "oscillation death" phenomenon, otherwise the properties of synchronization are similar for both coupling types.[7]

As is usually done in the method of averaging, we look for a solution oscillating with some common (yet unknown) frequency $\omega$, with slowly varying complex amplitudes $A_{1,2}$. Using the ansatz

$$x_{1,2}(t) = \tfrac{1}{2}(A_{1,2}(t)e^{i\omega t} + c.c.), \qquad y_{1,2}(t) = \tfrac{1}{2}(i\omega A_{1,2}(t)e^{i\omega t} + c.c.)$$

---

[7] We discuss the difference between the coupling terms proportional to $B_{1,2}$ and $D_{1,2}$ in Section 8.2.3.

we obtain the following general equations for the slowly varying complex amplitudes $A_{1,2}$ (cf. Eq. (7.41))

$$\dot{A}_1 = -i\Delta_1 A_1 + \mu_1 A_1 - (\gamma_1 + i\alpha_1)|A_1|^2 A_1 + (\beta_1 + i\delta_1)(A_2 - A_1),$$
$$\dot{A}_2 = -i\Delta_2 A_2 + \mu_2 A_2 - (\gamma_2 + i\alpha_2)|A_2|^2 A_2 + (\beta_2 + i\delta_2)(A_1 - A_2).$$

(8.12)

Here the frequency detuning parameters can, in the first approximation, be represented as

$$\Delta_{1,2} = \omega_{1,2} - \omega.$$

The coupling parameters $\beta_{1,2}, \delta_{1,2}$ are proportional to the coupling constants $B_{1,2}, D_{1,2}$. The other parameters $\mu_{1,2}, \gamma_{1,2}, \alpha_{1,2}$ are the same as in Eq. (7.41). Introducing the real amplitudes and the phases according to $A_{1,2} = R_{1,2}e^{i\phi_{1,2}}$ we obtain a system of four real equations:

$$\dot{R}_1 = \mu_1 R_1(1 - \gamma_1 R_1^2) + \beta_1(R_2 \cos(\phi_2 - \phi_1) - R_1) - \delta_1 R_2 \sin(\phi_2 - \phi_1),$$
$$\dot{\phi}_1 = -\Delta_1 - \mu_1\alpha_1 R_1^2 + \delta_1\left(\frac{R_2}{R_1}\cos(\phi_2 - \phi_1) - 1\right) + \beta_1\frac{R_2}{R_1}\sin(\phi_2 - \phi_1),$$
$$\dot{R}_2 = \mu_2 R_2(1 - \gamma_2 R_2^2) + \beta_2(R_1 \cos(\phi_1 - \phi_2) - R_2) - \delta_2 R_1 \sin(\phi_1 - \phi_2),$$
$$\dot{\phi}_2 = -\Delta_2 - \mu_2\alpha_2 R_2^2 + \delta_2\left(\frac{R_1}{R_2}\cos(\phi_1 - \phi_2) - 1\right) + \beta_2\frac{R_1}{R_2}\sin(\phi_1 - \phi_2).$$

(8.13)

Remarkably, the coupling terms depend on the phase difference only, so that we can reduce the number of equations by introducing the phase difference $\psi = \phi_2 - \phi_1$. With this variable the system (8.13) takes the form

$$\dot{R}_1 = \mu_1 R_1(1 - \gamma_1 R_1^2) + \beta_1(R_2 \cos\psi - R_1) - \delta_1 R_2 \sin\psi,$$

$$\dot{R}_2 = \mu_2 R_2(1 - \gamma_2 R_2^2) + \beta_2(R_1 \cos\psi - R_2) + \delta_2 R_1 \sin\psi,$$

$$\dot{\psi} = -\nu + \mu_1\alpha_1 R_1^2 - \mu_2\alpha_2 R_2^2 + \left(-\delta_1\frac{R_2}{R_1} + \delta_2\frac{R_1}{R_2}\right)\cos\psi + \delta_1 - \delta_2$$

$$-\left(\beta_1\frac{R_2}{R_1} + \beta_2\frac{R_1}{R_2}\right)\sin\psi.$$

Here $\nu = \omega_2 - \omega_1$ is the detuning (mismatch) of the natural frequencies.

The above equations are rather general, and the analysis of all possible cases is too cumbersome. We can reduce the number of parameters if we assume that the oscillators are identical except for their linear frequencies, i.e., $\mu_1 = \mu_2 = \mu$, etc. Additionally, we can normalize the time by $\mu$ and the amplitudes by $\sqrt{\gamma/\mu}$, to get rid of two parameters. Then the remaining coefficients $\beta, \delta$ should be considered as normalized to $\mu$, and $\alpha$ as normalized to $\gamma/\mu$; for the sake of simplicity we use, however, the same notation and rewrite the system as

$$\dot{R}_1 = R_1(1 - R_1^2) + \beta(R_2 \cos\psi - R_1) - \delta R_2 \sin\psi, \tag{8.14}$$

$$\dot{R}_2 = R_2(1 - R_2^2) + \beta(R_1 \cos\psi - R_2) + \delta R_1 \sin\psi, \tag{8.15}$$

$$\dot{\psi} = -v + \alpha(R_1^2 - R_2^2) + \delta\left(-\frac{R_2}{R_1} + \frac{R_1}{R_2}\right)\cos\psi - \beta\left(\frac{R_2}{R_1} + \frac{R_1}{R_2}\right)\sin\psi. \tag{8.16}$$

These equations have been studied in detail by Aronson *et al.* [1990]. Here, we do not want to present all their results; rather we discuss the most important physical effects.

Before proceeding, we recall the physical meaning of the parameters in Eqs. (8.14)–(8.16). The parameter $\alpha$ describes the nonlinear frequency shift of a single oscillator; isochronous oscillations correspond to $\alpha = 0$. The parameter $v$ is the detuning of the natural frequencies; when the frequencies coincide, $v = 0$. The parameters $\delta$ and $\beta$ are the coupling constants, we will classify them below.

If the oscillators are isochronous ($\alpha = 0$), the transition to synchronization typically occurs via the saddle-node bifurcation at which a limit cycle appears, similar to the scenario described in Section 8.1. A more complex structure of bifurcations is observed for the nonisochronous case $\alpha \neq 0$.

## 8.2.2  Oscillation death, or quenching

An interesting phenomenon – oscillation death (quenching) – can be observed if the coupling is of a diffusive type, it has no analogy in the case of periodic forcing and direct coupling. Here, for sufficiently large coupling $\beta$ and frequency detuning $v$, the origin $R_1 = R_2 = 0$ becomes stable and the oscillations in both systems die out due to coupling.

To demonstrate this, let us write linearized equations (8.12), where for simplicity we consider all the parameters to be equal. Moreover, we assume the coupling to be purely dissipative (see discussion below), $\delta = 0$. Finally, by choosing the frequency $\omega = (\omega_1 + \omega_2)/2$ we can set $\Delta_1 = -\Delta_2 = \Delta$ and obtain

$$\dot{A}_1 = (i\Delta + \mu)A_1 + \beta(A_2 - A_1),$$

$$\dot{A}_2 = (-i\Delta + \mu)A_2 + \beta(A_1 - A_2).$$

Linear stability analysis yields the eigenvalues

$$\lambda = \mu - \beta \pm \sqrt{\beta^2 - \Delta^2}.$$

The steady state $A_1 = A_2 = 0$ is thus stable if $\mu < \beta < (\mu^2 + \Delta^2)/2\mu$. Physically, this coupling-induced stability can be easily understood if we note that the diffusive coupling brings additional dissipation in each system and that this dissipation cannot be compensated for by forcing from the other oscillator if the detuning is large.

### 8.2.3    Attractive and repulsive interaction

Let us reduce the system (8.14)–(8.16) to a single equation for the phase. We can do this if the coupling is small, i.e., if the parameters $\beta$ and $\delta$ can be considered small. In fact, we could obtain this with the phase dynamics approximation outlined in Section 8.1, but, because we already have averaged Eqs. (8.14)–(8.16), it is easier to derive the phase equation directly from them. In the first approximation the amplitudes $R_{1,2}$ deviate slightly from their unperturbed values $R_{1,2} = 1$:

$$R_{1,2} \approx 1 + r_{1,2}, \qquad r_{1,2} \ll 1.$$

Inserting this in Eqs. (8.14) and (8.15) yields in the first approximation

$$\dot{r}_{1,2} = -2r_{1,2} + \beta(\cos\psi - 1) \mp \delta \sin\psi.$$

We see that the perturbations of the amplitude are strongly damped, so we can assume $\dot{r}_{1,2} \approx 0$ to obtain

$$R_{1,2} = 1 + \frac{\beta}{2}(\cos\psi - 1) \mp \frac{\delta}{2}\sin\psi.$$

Substituting this in (8.16) we obtain

$$\dot{\psi} = -\nu - 2(\beta + \alpha\delta)\sin\psi. \tag{8.17}$$

This equation coincides with Eq. (8.8), the effective coupling constant is $\varepsilon = -2(\beta + \alpha\delta)$.

Let us first consider identical oscillators, $\nu = 0$. It is clear from Eq. (8.17) that the stable phase difference between the phases of two oscillators depends on the sign of the coefficient $\beta + \alpha\delta$. If it is positive, then the stable phase difference is 0 and we have attraction of the phases; if it is negative, the stable phase difference is $\pi$ and a repulsive interaction is observed. It is instructive to discuss the physical meaning of these two types of interaction.

First of all, we should distinguish between dissipative and reactive coupling. In the system (8.10) and (8.11) the terms proportional to $D_{1,2}$ are reactive and the terms proportional to $B_{1,2}$ are dissipative. Indeed, let us neglect for a minute the nonlinear and dissipative terms (i.e., set $f_{1,2} = 0$), and consider the linear conservative oscillators. The result of coupling is then comprehensible: the coupling terms proportional to $D_{1,2}$ only shift the eigenfrequencies, while the terms proportional to $B_{1,2}$ bring dissipation.[8] These effects exhibit themselves in the nonlinear case as well. In the terms used by Aronson *et al.* [1990] the two coupling terms correspond to scalar ($B$) and nonscalar ($D$) coupling. To understand the origin of these notions, let us rewrite Eq. (8.10) as a system of two first-order equations

$$\dot{x}_1 = y_1,$$

[8] One can see that the divergence of the phase volume is given by $-(B_1 + B_2)$, cf. [Schmidt and Chernikov 1999].

$$\dot{y}_1 = -\omega_1^2 x_1 + f_1(x_1, y_1) + D_1(x_2 - x_1) + B_1(y_2 - y_1).$$

We see that the term proportional to $B_1$ describes linear coupling through the variable $y$ in the equation for $y$, whereas the term proportional to $D_1$ describes linear coupling through the variable $x$ in the equation for $y$. In general, if an oscillating system is written as a system of first-order ordinary differential equations, the scalar term couples similar variables, while the nonscalar term describes cross-coupling.

Physically, the dissipative coupling proportional to $\beta$ drives the two interacting systems to a more homogeneous regime where their states coincide (provided, of course, that $\beta > 0$). As a result, this coupling directly favors "in phase" synchronization of the oscillators, in accordance with Eq. (8.17). Contrary to this, the influence of the reactive coupling is not clear *a priori*. To describe the influence of the different types of coupling on the phase dynamics, let us look at the sketches in Figs. 8.1 and 8.2.

In Fig. 8.1 we illustrate the case of isochronous oscillators ($\alpha = 0$). Panel (a) shows the interaction due to dissipative (scalar) coupling: only coefficients $B$ in Eqs. (8.10) and (8.11) are present. The coupling acts as a force in the $y$-direction, and this force is proportional to the difference of $y$-variables on the two limit cycles. So the coupling of the phases is attractive.[9] If the phases are close, it acts not over the whole period of oscillations, but only when the $y$-variables are sufficiently different, i.e., when $x$ is near to the minimum or to the maximum. Panel (b) shows the case of reactive (nonscalar) coupling, here we assume that the coefficients $B$ vanish. Now the force acts also in the $y$-direction, but it is proportional to the difference of $x$-variables. Thus the force acts when $x$ is nearly zero, and it neither brings the phases together nor pulls them apart. Hence, the reactive coupling changes only the amplitudes of the isochronous oscillators, not the phases. The phases can be affected

[9] Strictly speaking, this conclusion is valid only for weakly nonlinear oscillations. In the case of strongly nonlinear oscillations even diffusive coupling of state variables may lead to repulsion of phases [Han *et al.* 1995, 1997; Postnov *et al.* 1999a].

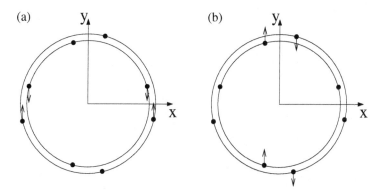

**Figure 8.1.** Sketch of coupling forces of two interacting isochronous oscillators for the case of dissipative (a) and reactive (b) coupling. For clarity of presentation the amplitudes of the two limit cycles are drawn slightly differently.

indirectly, only if they depend on the amplitudes, i.e., if the oscillations are non-isochronous (in the isochronous case the phases do not depend on the amplitudes; the isochrones are the radial lines on the phase plane). We conclude that reactive coupling has no effect on isochronous oscillators. This corresponds to the fact that the coefficient of reactive coupling $\delta$ appears in Eq. (8.17) multiplied by the isochronity constant $\alpha$.

The situation changes if the reactive force couples nonisochronous oscillations, i.e., if $\alpha \neq 0$. This case is illustrated in Fig. 8.2. The force increases the amplitude of one oscillator and decreases the amplitude of the other, and in the case of the nonisochronity of oscillations this leads to a phase shift, as the frequency is amplitude-dependent. Depending on the sign of the product $\alpha\delta$ this interaction can be either repulsive or attractive. As a result, the stable phase shift will be 0 or $\pi$. This follows also from Eq. (8.17). The same mechanism is responsible for phase instability in an oscillatory medium, discussed later in Chapter 11.

One remark is in order here. We have illustrated above only the role of the leading terms in the phase dynamics. If these terms vanish, Eq. (8.17) is no longer valid. An account of the higher-order terms is then required to describe the phase interaction leading to synchronization.

## 8.3     Relaxation oscillators

There is no universal model for relaxation oscillators, so we would like to present one example here: the so-called integrate-and-fire oscillator. It is not described by a system of differential equations, rather it is defined by a separate description of slow and fast motions. The oscillator is described by a single variable $x$ that grows from 0 to 1 according to a given dynamical law (the "integrating" stage; this can be an ordinary

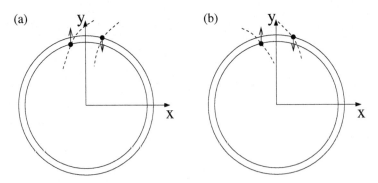

**Figure 8.2.** Sketch of the reactive coupling of two nonisochronous oscillators. The isochrones (lines of constant phase) are depicted with dashed lines, their form depends on the sign of $\alpha$. Correspondingly, one has either repulsive (a) or attractive (b) coupling depending on the sign of the product $\delta\alpha$.

differential equation, or just a given function of time). As the threshold $x = 1$ is reached, the oscillator instantly jumps ("fires") to $x = 0$. The interaction of two such oscillators is supposed to take place only during the firing events. When one oscillator $x_1$ fires, it acts on the second one pulling its variable $x_2$ by an amount $\varepsilon$. If $x_2 + \varepsilon$ exceeds the threshold (i.e., $x_2 + \varepsilon > 1$), the second oscillator fires as well, but no back action on the first oscillator occurs (the oscillator in the firing mode is insensitive to external action).

The dynamics of this particular coupling are very dissipative. Indeed, if the phases of the two oscillators happen to be close to each other, then the first oscillator that fires causes the firing of the other one, so that they fire simultaneously. After this event the phases of the oscillators are identical. If the natural frequencies of oscillators are close, they continue to fire together. Perfect synchronization, with coinciding firing events, will be observed, and the period will be the smallest one of the two natural periods.

We now proceed to an analytical treatment of the problem, following the approach of Mirollo and Strogatz [1990b]. We assume first that the oscillators are identical, with a natural frequency $\omega_0$. The slow motion is given by a function $x = f(\phi)$, where $\phi$ is the phase, obeying $\dot{\phi} = \omega_0$. The slow motion corresponds to growth of the phase from 0 to $2\pi$ and at $\phi = 2\pi$ the firing occurs.

For two oscillators, we have a flow on a two-dimensional torus (Fig. 8.3); this can be reduced to a one-dimensional Poincaré map. Let us choose the line $\phi_1 = 0$ as a secant, which means that we look for the phase of the second oscillator at the moments of time when the first oscillator fires. The construction of the Poincaré map $\phi_2^{(0)} \rightarrow F(\phi_2^{(0)})$ is illustrated in Fig. 8.3. We start from a point 0 with coordinates $\phi_1 = 0$, $\phi_2 = \phi_2^{(0)}$. At the first stage of slow motion, point 1 with $\phi_2^{(1)} = 2\pi$, $\phi_1^{(1)} = 2\pi - \phi_2^{(0)}$ is reached. At this moment oscillator 2 fires and the phase of the first one changes as well. As we assumed that the variable $x$ changes by $\varepsilon$, the new phase of

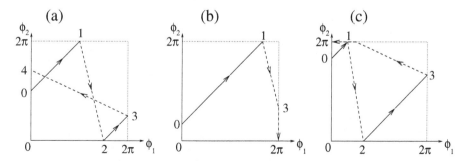

**Figure 8.3.** Construction of the Poincaré map for coupled integrate-and-fire oscillators. Solid lines, slow motions ("integrate"); dashed lines, fast jumps ("fire"). Three possible types of trajectory starting at $\phi_1 = 0$ are shown. In (a) the oscillators do not fire simultaneously. In (b) the firing of oscillator 2 induces the firing of oscillator 1. In (c) the firing of oscillator 1 induces the firing of oscillator 2.

the first oscillator is given by $\phi_1^{(2)} = f^{-1}(f(\phi_1^{(1)}) + \varepsilon)$, where $f^{-1}$ is the function inverse to $f$. Two cases are now possible. If $\phi_1^{(2)} \geq 2\pi$, the first oscillator fires and the phases of both oscillators jump to zero (Fig. 8.3b), so $F(\phi_2^{(0)}) = 0$. Otherwise there is another piece of slow motion to the point $\phi_1^{(3)} = 2\pi$, $\phi_2^{(3)} = 2\pi - \phi_1^{(2)}$, followed by the firing of oscillator 1. At this firing the phase $\phi_2$ changes, and again either a firing of oscillator 2 is induced (Fig. 8.3c) and the phase $\phi_2$ jumps to $2\pi$,[10] or the firing is not induced and the end point of the Poincaré mapping is $\phi_2^{(4)} = f^{-1}(f(\phi_2^{(3)}) + \varepsilon)$ (Fig. 8.3a).

It is easy to see that, due to symmetry, the Poincaré map $\phi_2^4 = F(\phi_2^0)$ can be written as a double iteration of a map $h$:

$$F(\phi) = h(h(\phi)), \qquad h(\phi) = f^{-1}(f(2\pi - \phi) + \varepsilon). \tag{8.18}$$

The map $h$ is called a **firing map**. The resulting Poincaré map has two plateaux where $F(\phi) = 0$ or $F(\phi) = 2\pi$, and a smooth region between them, see Fig. 8.4. The smooth region is given by Eq. (8.18), and depends on the form of oscillations $f$. Mirollo and Strogatz [1990b] demonstrated that if $f$ is monotonic and concave down (i.e., $f' > 0$, $f'' < 0$), then the smooth part of the mapping $F$ is strictly expanding (i.e., the derivative is larger than one). This means that no other attractors of the Poincaré map are possible, except for $\phi = 0$. The attracting point $\phi = 0$ corresponds exactly to the state of perfect synchrony, when two oscillators fire simultaneously.

The above construction can be easily generalized to the case of different (in particular, having different natural frequencies) oscillators. The slow motion in Fig. 8.3 will now be a straight line not parallel to the diagonal, but having a slope $\omega_1/\omega_2$. The resulting Poincaré mapping has a smooth part (which, similar to Eq. (8.18), is

---

[10] Not to zero, to show that here we have two firings of oscillator 2 during one firing of oscillator 1.

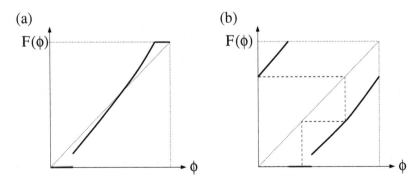

**Figure 8.4.** The nonsmooth circle map describing synchronization of relaxation oscillators. (a) The oscillators are identical; the only attractor is the fixed point $\phi = 0$. (b) The oscillators have different natural frequencies. On the attracting periodic orbit (dashed line) only one firing event occurs simultaneously in both oscillators; other firings do not coincide.

represented by a superposition of now different firing mappings) and a part where the new phase (modulo $2\pi$) is identically zero, see Fig. 8.4b. The stable periodic orbit of this map passes necessarily through the point $\phi_1 = \phi_2 = 0$, i.e., there is exactly one event of coinciding firings. An example of $2:3$ synchronization in a system with the ratio of natural frequencies $\omega_1/\omega_2 = 1.55$ is shown in Fig. 8.5.

The properties of this nonsmooth Poincaré map are similar to that of the smooth one (see Section 7.3). The main difference is in the structure of the devil's staircase and the Arnold tongues. As in the smooth case, all rational and irrational rotation numbers are possible, but the measure of irrational numbers (in the space of parameters) is now zero: a situation with a quasiperiodic regime in the mapping Fig. 8.4 is exceptional, typically periodic orbits will be observed [Boyd 1985; Veerman 1989]. This is a consequence of strong dissipation in the considered integrate-and-fire model, as discussed above.

## 8.4    Bibliographic notes

There is a vast literature on the dynamics of coupled systems. Nowadays the main emphasis is on complex behavior and the appearance of chaos due to interaction. In theoretical [Waller and Kapral 1984; Pastor-Diáz *et al.* 1993; Volkov and Romanov 1994; Pastor-Diáz and López-Fraguas 1995; Tass 1995; Kurrer 1997; Lopez-Ruiz and Pomeau 1997; Reddy *et al.* 1999] and experimental [Bondarenko *et al.* 1989; Thornburg *et al.* 1997] papers the interested reader can find further references. Coupled rotators have been intensively studied in the context of Josephson junction arrays [Jain *et al.* 1984; Saitoh and Nishino 1991; Valkering *et al.* 2000]. Finally, we mention some recent papers where different generalizations of "integrate-and-fire" oscillators have been considered [Kirk and Stone 1997; Ernst *et al.* 1998; Coombes and Bressloff 1999; S. H. Park *et al.* 1999b].

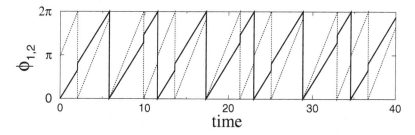

**Figure 8.5.** The phase locking of nonidentical relaxation oscillators with the ratio of natural frequencies $\omega_1/\omega_2 = 1.55$. The $2:3$ synchronization is observed, and this regime corresponds to the periodic trajectory of the map shown in Fig. 8.4b.

# Chapter 9

# Synchronization in the presence of noise

In previous chapters we considered synchronization in purely deterministic systems, neglecting all irregularities and fluctuations. Here we discuss how the latter effects can be incorporated in the picture of phase locking. We start with a discussion of the effect of noise on autonomous self-sustained oscillations. We show that noise causes phase diffusion, thus spoiling perfect time-periodicity. Next, we consider synchronization by an external periodic force in the presence of noise. Finally, we discuss mutual synchronization of two noisy oscillators.

## 9.1    Self-sustained oscillator in the presence of noise

No oscillator is perfectly periodic: all clocks have to be adjusted from time to time, some even rather often. There are many factors causing irregularity of self-sustained oscillators, for simplicity we will call them all noise. A detailed analysis of noisy oscillators must include a thorough mathematical description of the problem, where fluctuations of different nature (e.g., technical, thermal) should be taken into account. This has been done for different types of oscillators (see, e.g., [Malakhov 1968]); here we want to discuss the basic phenomena only.

As the first model, we consider a self-sustained oscillator subject to a noisy external force. In revising the basic equations of forced oscillators of Chapter 7, one can see that only the approximation of phase dynamics is valid in the case of a fluctuating force as well, since we do not assume any regularity of the force in the derivation of Eq. (7.15). Thus, we can utilize this equation for noisy forcing as well:

$$\frac{d\phi}{dt} = \omega_0 + \varepsilon Q(\phi, t), \tag{9.1}$$

where $Q$ is a $2\pi$-periodic function of $\phi$ and an arbitrary random function of time.

The simplest case is the situation when the stochastic term in the phase equation (9.1) does not depend on the phase at all, so that we can write

$$\frac{d\phi}{dt} = \omega_0 + \xi(t) \qquad (9.2)$$

with a stationary random process $\xi(t)$. Because the instantaneous frequency $\dot\phi$ (velocity of the phase rotation) is a random function of time, the phase undergoes a random walk, or diffusive motion. The solution of (9.2) is

$$\phi(t) = \phi_0 + \omega_0 t + \int_0^t \xi(\tau)\,d\tau, \qquad (9.3)$$

and from this solution one can easily find statistical characteristics of the phase diffusion. We assume that the average of the noise vanishes (if not, one can absorb the mean value in the frequency $\omega_0$), then the ensemble averaged value of the phase at time $t$ is $\phi_0 + \omega_0 t$. The variance can be obtained by averaging the square of (9.3); for large times it obeys the usual Green–Kubo relation for diffusion processes (see, e.g., [van Kampen 1992])

$$\langle(\phi(t) - \phi_0 - \omega_0 t)^2\rangle \propto tD, \qquad D = \int_{-\infty}^{\infty} K(t)\,dt, \qquad (9.4)$$

where $K(t) = \langle\xi(\tau)\xi(\tau + t)\rangle$ is the correlation function of the noise.

The diffusion of the phase implies that the oscillations are no longer perfectly periodic, and the diffusion constant $D$ measures the quality of the self-sustained oscillations, which is an important characteristic of clocks and electronic generators. In the case of $\delta$-correlated Gaussian noise

$$K(t) = \langle\xi(\tau)\xi(\tau + t)\rangle = 2\sigma^2\delta(t), \qquad (9.5)$$

the distribution of the phase is also Gaussian with variance $2\sigma^2 t$ (so that the diffusion constant is equal to the noise intensity $D = 2\sigma^2$). This allows one to calculate the autocorrelation function of a natural observable $x(t) = \cos\phi$. Simple calculations give exponentially decaying correlations

$$\langle x(t)x(t + \tau)\rangle = \tfrac{1}{2}\exp[-\tfrac{1}{2}\tau\sigma^2]\cos\omega_0\tau,$$

which corresponds to the Lorentzian shape of the power spectrum centered at the mean frequency $\omega_0$. The width of the spectral peak is $\sigma^2$, i.e., it is proportional to the diffusion constant of the phase.

## 9.2 Synchronization in the presence of noise

### 9.2.1 Qualitative picture of the Langevin dynamics

As we have seen in Section 7.1, the basic features of the phase dynamics are captured by the averaged equation (cf. Eq. (7.24) and (8.6)):

$$\frac{d\psi}{dt} = -v + \varepsilon q(\psi), \qquad (9.6)$$

where $\psi$ is the difference between the phases of the oscillator and the external force. A natural way to describe the effect of noise is to include a noise term in the r.h.s. and to consider the Langevin equation

$$\frac{d\psi}{dt} = -v + \varepsilon q(\psi) + \xi(t) \qquad (9.7)$$

with an additive noise term $\xi(t)$. Equation (9.7) thus describes a situation when a self-sustained oscillator is forced simultaneously by a periodic and a stochastic driving signal.

Physically, it is convenient to interpret the Langevin dynamics (9.7) as a random walk of a "particle" in a one-dimensional potential (see Fig. 9.2 later). Indeed, the deterministic force on the r.h.s. of (9.7) can be written as

$$-v + \varepsilon q(\psi) = -\frac{dV}{d\psi}, \qquad V(\psi) = v\psi - \varepsilon \int^{\psi} q(x)\,dx. \qquad (9.8)$$

Moreover, the motion is overdamped because the phase $\psi$ has no inertia.[1] Without noise, the "particle" either is trapped in a minimum of the inclined potential, or slides down along it (Fig. 9.1). A potential with minima yields stable steady states[2] and describes synchronization, while a monotonic potential corresponds to a quasiperiodic state with rotating phase (see also the qualitative discussion in Section 3.4).

Noise has little influence on the quasiperiodic state: here the mean velocity of the "particle" motion is nonzero and it only slightly changes because of the noise. On the contrary, the influence of noise on the synchronous state can be dramatic. Indeed, it can kick the "particle" out of the stable positions and, if the noisy force is large, it can drive the "particle" from one equilibrium to a neighboring one. The phase then changes by $\pm 2\pi$ and this event is called **phase slip**. We illustrate this process in Fig. 9.2. For the phase slip to occur, the "particle" should overcome the potential

---

[1] A particle with damping usually obeys an equation $m\ddot{x} + \gamma\dot{x} + dV/dx = 0$ (cf. Eqs. (7.69) and (7.71)), but when the damping is very large ($\gamma \to \infty$) one can neglect the second derivative and obtain an equation of the type (9.6). The same equation can be obtained in the limit $m \to 0$.

[2] For simplicity of presentation we consider only a situation with one minimum of the potential in the interval $[0, 2\pi)$.

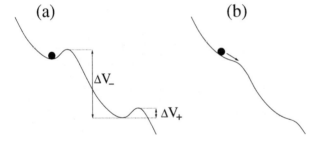

(a)                    (b)

$\Delta V_-$

$\Delta V_+$

**Figure 9.1.** Phase as a "particle" in an inclined potential $V(\psi)$. (a) The case of synchronization: the potential has a minimum and the particle stays there. (b) Outside the synchronization region the "particle" slides down. From [Pikovsky *et al.* 2000].

barrier $\Delta V_{\pm}$, so that the probability of slips can be small. In general, the probability of a slip grows with the noise intensity and decreases with the barrier height. Thus, if $v \neq 0$, the probabilities of slips $+2\pi$ and $-2\pi$ are different, so that on average the "particle" moves in one direction and the observed frequency difference $\Omega_\psi = \langle \dot{\psi} \rangle$ is nonzero.[3] The time evolution of the phase resembles that of the noise-free system near the synchronization transition (compare Fig. 9.2 with Fig. 7.5), only now the phase slips appear irregularly.

Here we should, however, distinguish the cases of bounded and unbounded noise. In the case of unbounded (e.g., Gaussian) noise very large kicks can occur and slips are possible even if the barriers $\Delta V$ are high. In this situation the slips appear for any nonzero noise intensity, and for any nonzero $v$ the probabilities of right and left slips are different. Thus, the frequency difference $\Omega_\psi$ is a monotononically decreasing function of $v$ and the synchronization region (the region where $\Omega_\psi = 0$) disappears. Another scenario occurs if the noise is bounded. Now, for small (with respect to the

---

[3] For convenience, we introduce the notation $\Omega_\psi = \langle \dot{\psi} \rangle$; this corresponds to the difference between the observed frequency of the oscillator and the frequency of the external force, or between the two observed frequencies of the coupled oscillators.

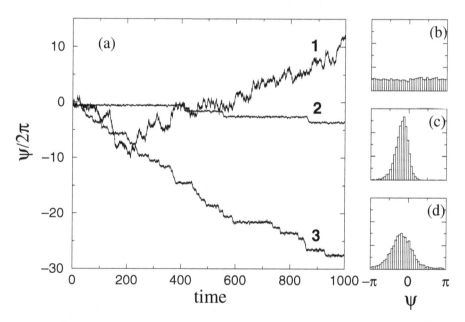

**Figure 9.2.** (a) Pictures of Langevin dynamics with different noise amplitude $\sigma^2$ and mismatch $v$. Curve 1: diffusion of the phase in the "free" noisy oscillator ($\varepsilon = 0$). Curve 2: small noise and moderate mismatch; the slips are rare. Curve 3: the same mismatch as for curve 2, but stronger noise; the slips appear rather often and the phase rotates faster. In (b)–(d) the histograms of the phase taken modulo $2\pi$ are shown for curves 1–3, respectively. Free diffusion yields the homogeneous distribution. Otherwise, the distribution of the phase has a maximum near the stable phase position of the noise-free system. From [Pikovsky *et al.* 2000].

barrier height) noise the slips are impossible and the observed frequency remains exactly zero. Only when the barriers are small enough do slips occur and synchronization is, strictly speaking, destroyed. We illustrate these two possibilities in Fig. 9.3.

At this point we would like to discuss the definition of synchronization in noisy systems. If one prefers to define synchronization as a perfect frequency entrainment, without phase slips, then the situation of Fig. 9.3b does not satisfy this definition. It seems to us reasonable, for noisy systems, to loosen the requirement of exact coincidence of frequencies, and to describe the process as in Figs. 9.2 and 9.3b, where the frequencies nearly adjust, as synchronized or nearly synchronized.

We can also adopt another viewpoint on Langevin dynamics, focusing on the question: how the periodic force influences the noisy phase dynamics. From the qualitative picture of Fig. 9.1 it is clear that the force suppresses not only the drift velocity of the particle, but also the diffusion. The diffusion constant can vanish for bounded noise; but even for unbounded Gaussian noise it can become exponentially small in the center of the synchronization region, as we demonstrate below for the case of white noise.

### 9.2.2   Quantitative description for white noise

Now we would like to present a quantitative description of synchronization in the presence of noise. We can do this if we assume that the noise has good statistical properties. If it is Gaussian and $\delta$-correlated, a powerful theory, based on the Fokker–Planck equation, can be applied [Stratonovich 1963; Risken 1989; Gardiner 1990]. Therefore we use this assumption in the rest of this section. The mean value of the noise $\xi$ is assumed to be zero (otherwise it can be absorbed in the frequency detuning parameter $\nu$: $\nu \to \nu - \langle \xi \rangle$), and its intensity is characterized with the single parameter $\sigma^2$, see Eq. (9.5). We follow here the Stratonovich interpretation of the stochastic differential equation (9.7). Then the probability distribution density of the phase $P(\psi, t)$ obeys the Fokker–Planck equation (FPE) [Stratonovich 1963; Risken 1989; Gardiner 1990]:

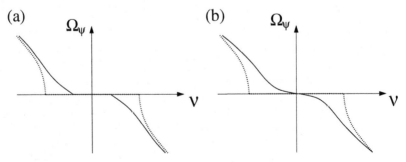

**Figure 9.3.** Frequency difference vs. mismatch for bounded (a) and unbounded (b) noise. The plateau (synchronization region) is retained under small bounded noise, but disappears in the case of unbounded noise.

$$\frac{\partial P}{\partial t} = -\frac{\partial[(-\nu + \varepsilon q(\psi))P]}{\partial \psi} + \sigma^2 \frac{\partial^2 P}{\partial \psi^2}. \tag{9.9}$$

Equivalently, we can write

$$\frac{\partial P}{\partial t} + \frac{\partial S}{\partial \psi} = 0,$$

where the probability flux $S$ is defined as

$$S = -P\frac{dV}{d\psi} - \sigma^2 \frac{\partial P}{\partial \psi}, \tag{9.10}$$

and $V$ is the potential given by Eq. (9.8).

We look for a stationary (time-independent) probability density. The stationary solution of the FPE (9.9) must be $2\pi$-periodic in $\psi$, and it is easy to check that the following expression satisfies both Eq. (9.9) and this condition:

$$\bar{P}(\psi) = \frac{1}{C} \int_{\psi}^{\psi+2\pi} \exp\left(\frac{V(\psi') - V(\psi)}{\sigma^2}\right) d\psi'. \tag{9.11}$$

The constant $C$ can be found from the normalization condition $\int_0^{2\pi} \bar{P}(\psi) d\psi = 1$. Using the solution (9.11), one can find the constant probability flux $S$

$$S = \frac{\sigma^2}{C}(1 - e^{2\pi\nu\sigma^{-2}}),$$

which is directly related to the mean frequency difference $\Omega_\psi$. Indeed, calculating $\langle \dot{\psi} \rangle$ from the Langevin equation (9.7) and using (9.10) we get

$$\Omega_\psi = \langle \dot{\psi} \rangle = \langle -\nu + \varepsilon q(\psi) \rangle = \int_0^{2\pi} -\frac{dV}{d\psi} \bar{P}(\psi) d\psi = 2\pi S.$$

To proceed, we consider a typical form of the coupling term $q(\psi) = \sin\psi$. In this case the constants $S, C$ can be expressed through Bessel functions of complex order [Stratonovich 1963], but for practical numerical purposes the continuous fraction representation of the solution described by Risken [1989] is more convenient. Representing $\bar{P}(\psi)$ as a Fourier series

$$\bar{P} = \sum_{-\infty}^{+\infty} P_n e^{in\psi}$$

and substituting this in (9.10) yields the infinite system

$$-(in\sigma^2 + \nu)P_n + \frac{\varepsilon}{2i}(P_{n-1} - P_{n+1}) = S\delta_{n,0}. \tag{9.12}$$

The normalization condition implies $P_0 = 1/2\pi$ and the reality of $\bar{P}$ implies $P_{-n} = P_n^*$. Rewriting (9.12) for $n > 0$ as

$$\frac{P_n}{P_{n-1}} = \frac{1}{(\nu + in\sigma^2)\frac{2i}{\varepsilon} + \frac{P_{n+1}}{P_n}}$$

and iterating this relation starting from $n = 1$, we can represent the solution as a rapidly converging (and thus numerically pleasant) continuous fraction

$$P_1 = \cfrac{(2\pi)^{-1}}{2\dfrac{iv - \sigma^2}{\varepsilon} + \cfrac{1}{2\dfrac{iv - 2\sigma^2}{\varepsilon} + \cfrac{1}{2\dfrac{iv - 3\sigma^2}{\varepsilon} + \cdots}}} \qquad (9.13)$$

From Eq. (9.12) at $n = 0$ we get a relation between the first harmonic $P_1$ and the flux $S$:

$$S = -\frac{v}{2\pi} - \varepsilon \operatorname{Im}(P_1).$$

Thus the average frequency difference is

$$\Omega_\psi = -v - 2\pi\varepsilon \operatorname{Im}(P_1).$$

It is interesting that the real part of $P_1$ gives the Lyapunov exponent of the phase dynamics. Indeed, the linearized Eq. (9.7) (in the case $q(\psi) = \sin\psi$) reads

$$\frac{d\delta\psi}{dt} = \varepsilon \cos\psi \cdot \delta\psi,$$

and the average logarithm of the perturbation grows as

$$\lambda = \left\langle \frac{d \ln \delta\psi}{dt} \right\rangle = \varepsilon \langle \cos\psi \rangle = 2\pi \operatorname{Re}(P_1).$$

The Lyapunov exponent characterizes the stability of the phase. For vanishing noise, $\sigma^2 = 0$, only the synchronized state is stable, while the quasiperiodic state has a zero Lyapunov exponent. In the case $\sigma^2 > 0$, the qualitative difference disappears: the Lyapunov exponent is negative for all mismatches $v$.

We show the dependencies of the average frequency difference and of the Lyapunov exponent on the system parameters in Fig. 9.4. (One can see from expression (9.13) that the quantity $P_1$ depends in fact on the two parameters $v/\varepsilon$ and $\sigma^2/\varepsilon$; here we fix $\varepsilon$ and present in Fig. 9.4 a one-parameter family of curves.) In the limit $\sigma^2 \to 0$ we obtain the results for the purely deterministic case, considered in Chapter 7. The effect of noise is to smear the plateau in the dependence of the frequency difference $\Omega_\psi$ on the frequency detuning $v$, although for small noise the deviations in the center of the synchronization region are exponentially small.

The Lyapunov exponent is always negative, both in the synchronization region around $v = 0$ and in the state where the deterministic regime is quasiperiodic. Thus, the phase dynamics are stable with respect to perturbations of initial conditions. This results in the possibility of synchronizing two identical oscillators by driving them with the same noise (see Section 15.2).

A rather complete statistical picture of the phase dynamics can be formulated in the case of small noise [Stratonovich 1963]. Here the phase slips are rare and well separated events, of duration much less than the characteristic time interval between them (see curve 2 in Fig. 9.2). The dynamics of the phase can be thus represented as a sequence of independent $+2\pi$ and $-2\pi$ jumps. As the phase quickly "forgets" its previous jump, the whole process can be approximately considered Poissonian, characterized by a single parameter, the jump rate. If the frequency detuning $\nu$ is nonzero, the jump rates in the positive and the negative directions, $G_+$ and $G_-$, are different. In terms of these rates one can express the mean frequency difference as the mean phase drift velocity

$$\Omega_\psi = 2\pi(G_+ - G_-),$$

and the variance of the phase distribution

$$\langle(\psi - \langle\psi\rangle)^2\rangle = (2\pi)^2(G_+ + G_-)t$$

defines the diffusion constant. The formulae above result from the observation that the numbers of positive (negative) jumps during time interval $t$ are random numbers with mean values $G_\pm t$ and variances $G_\pm t$ (the process is Poissonian!); since the positive and negative jumps are statistically independent, the mean values and variances can be simply added. The exact expressions for $G_+$ are given by Stratonovich [1963], they are obtained there as mean times for a Brownian particle to overcome the barriers of the potential. Diffusion over a barrier in the case of small noise is described with the

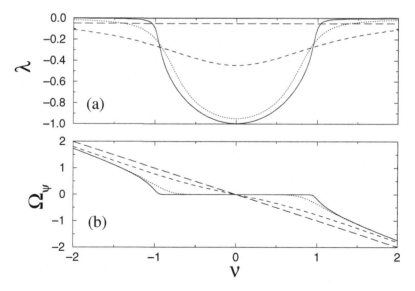

**Figure 9.4.** The Lyapunov exponent (a) and the averaged frequency (b) of the noisy periodic oscillator under external force, for $\varepsilon = 1$ and different noise amplitudes (solid line: $\sigma^2 = 0.01$; dotted line: $\sigma^2 = 0.1$; dashed line: $\sigma^2 = 1$; long-dashed line: $\sigma^2 = 10$).

well-known Kramers' formula [Risken 1989; Gardiner 1990], according to which the probability exponentially depends on the barrier height (see Fig. 9.1):

$$G_\pm \propto \exp\left(\frac{-\Delta V_\pm}{\sigma^2}\right).$$

Typically, one of the probabilities is much larger than the other, and we observe a sequence of phase slips for which the mean frequency difference and the phase diffusion constant are of the same order of magnitude. Only in the center of the synchronization region, where the barriers of the potential $\Delta V_\pm$ are equal and the mean frequency difference $\Omega_\psi$ vanishes, a random walk exhibits positive and negative jumps with equal probability. The diffusion constant in the latter case is exponentially small for vanishing noise.

### 9.2.3 Synchronization by a quasiharmonic fluctuating force

Here we discuss the action of a quasiharmonic (narrow-band) external stochastic force on a periodic oscillator. Such a problem appears naturally, e.g., in the theory of phase locked loops (Section 7.5). Indeed, if a signal carrying some information has to be demodulated using the effect of phase locking, then it is clear that such a signal cannot be considered as purely periodic, but rather as a signal with slow variations of frequency and of amplitude. As is usual in the theory of communication, these variations should be considered as random, albeit relatively slow (compared to the period of oscillations). This is exactly the problem of phase locking by a narrow-band random force.

Because modulations of the frequency and amplitude of the force are slow, we can still use the basic equation for phase locking (9.6), but consider the detuning $\nu$ and the amplitude $\varepsilon$ as random functions of time:

$$\frac{d\psi}{dt} = -\nu(t) + \varepsilon(t)q(\psi). \tag{9.14}$$

To study the influence of fluctuations on the phase locking, we assume the variations of $\nu$ and $\varepsilon$ to be small. So we write

$$\nu = \nu_0 + \Delta\nu(t), \qquad \varepsilon = \varepsilon_0 + \Delta\varepsilon(t), \qquad \psi = \psi_0 + \Delta\psi(t).$$

Assuming furthermore the simplest form of the nonlinear term $q(\psi) = -\sin\psi$ we obtain, after linearization,

$$\frac{d\Delta\psi}{dt} = -\sqrt{\varepsilon_0^2 - \nu_0^2}\,\Delta\psi + \Delta\nu(t) + \frac{\nu_0}{\varepsilon_0}\Delta\varepsilon(t).$$

From this linear relation, assuming independence of the fluctuations of the frequency and of the amplitude, we can represent the power spectrum of the process $\Delta\psi(t)$ through the spectra of $\nu(t)$ and $\varepsilon(t)$:

$$S_\psi(\varsigma) = \frac{S_\nu(\varsigma) + v_0^2 \varepsilon_0^{-2} S_\varepsilon(\varsigma)}{\varepsilon_0^2 - v_0^2 + \varsigma^2}.$$

This formula (cf. [Landa 1980]) shows that the fluctuations of the phase remain bounded in the center of the synchronization region $|v_0| < |\varepsilon_0|$, and diverge on its ends. Therefore, for given statistical characteristics of the modulation, there may be a region where there are no phase slips, similar to the situation shown in Fig. 9.3a. In the context of the phase locked loop (PLL), this means that the loop almost perfectly demodulates the input signal. Near the borders of the synchronization region slips are unavoidable; here the PLL works erroneously.

We remind the reader that the variable $\psi$ is the difference between the phase of the oscillator and that of the external force (cf. Eq. (7.22)). Therefore phase locking by a narrow-band force does not mean the absence of phase diffusion: the phase of the oscillator just follows the random variations of the forcing phase.

### 9.2.4 Mutual synchronization of noisy oscillators

We have considered the effect of noise on synchronization by a periodic force. As already mentioned in Chapter 8, the equations for the phase difference of two mutually coupled oscillators are the same as for a periodically forced oscillator. Thus, the dynamics of the phase difference of two coupled noisy oscillators are completely described by the theory presented above. However, the phase difference is not the only quantity of interest for coupled oscillators. In many applications (e.g., using oscillators as clocks) the quality of the oscillations is important, i.e., the diffusion coefficients of individual phases.

Here we demonstrate that the diffusion of the phases can be reduced due to coupling. We consider, following Malakhov [1968], synchronization of two mutually coupled noisy oscillators in the phase approximation:

$$\dot{\phi}_1 = \omega_1 + \varepsilon_1 \sin(\phi_2 - \phi_1) + \xi_1,$$
$$\dot{\phi}_2 = \omega_2 + \varepsilon_2 \sin(\phi_1 - \phi_2) + \xi_2.$$

Below, the coupling constants are assumed to be different, as are the noise intensities

$$\langle \xi_i(t)\xi_j(t') \rangle = 2\sigma_i^2 \delta_{ij} \delta(t - t'),$$

while the frequency detuning is supposed to be zero: $\omega_1 = \omega_2 = \omega$. For the phase difference $\psi = \phi_1 - \phi_2$ and the "sum" of the phases $\theta = \varepsilon_2 \phi_1 + \varepsilon_1 \phi_2$ we find

$$\dot{\psi} = -(\varepsilon_1 + \varepsilon_2) \sin \psi + \xi_1 - \xi_2,$$
$$\dot{\theta} = \varepsilon_2 \xi_1 + \varepsilon_1 \xi_2 + \omega(\varepsilon_1 + \varepsilon_2).$$

The equation for $\psi$ describes mutual synchronization of the oscillators, it is equivalent to Eq. (9.7). This means that for the phase difference the whole theory developed in Section 9.2 above holds. We do not repeat it here, but mention only that the diffusion of the phase difference $\psi$ is exponentially small for large coupling. Contrary to this, there is no restoring force for the variable $\theta$, which therefore exhibits strong diffusion with a diffusion constant

$$D_\theta = 2(\varepsilon_2^2 \sigma_1^2 + \varepsilon_1^2 \sigma_2^2),$$

to which both fluctuating terms contribute.

Representing the phases of the oscillators as

$$\phi_1 = \frac{\varepsilon_1 \psi + \theta}{\varepsilon_1 + \varepsilon_2}, \qquad \phi_2 = \frac{-\varepsilon_2 \psi + \theta}{\varepsilon_1 + \varepsilon_2},$$

and neglecting the phase difference $\psi$ compared with the phase sum $\theta$ (we can do this because of the difference in their diffusion constants), we obtain for the individual phases of the oscillators the equal diffusion constants

$$D_1 = D_2 = D = 2(\varepsilon_1^2 \sigma_2^2 + \varepsilon_2^2 \sigma_1^2)(\varepsilon_1 + \varepsilon_2)^{-2}. \tag{9.15}$$

This common diffusion constant should be compared with the diffusion constants of the noninteracting oscillators

$$D_1^0 = 2\sigma_1^2, \qquad D_2^0 = 2\sigma_1^2.$$

Consider first the simplest case of unidirectional coupling: $\varepsilon_1 = 0$. Then $D = 2\sigma_1^2 = D_1^0$: the quality of the synchronized state is that of the driving signal. The fluctuations caused in the second oscillator by the noisy term $\xi_2$ are suppressed due to synchronization. So if the first oscillator has better quality, it can improve the quality of the second one: this effect can be used to obtain high-quality large-power oscillations. Examining formula (9.15), we can also see that, even in the case of mutual coupling, improvement in quality is possible if the influence of the best oscillator (i.e., that with the lesser noise intensity $\sigma^2$) is larger than that of the worst one. For example, if the first oscillator is of better quality (i.e., $\sigma_1^2 < \sigma_2^2$) and its influence is stronger (i.e., $\varepsilon_2 > \varepsilon_1$), then the resulting diffusion constant decreases and the quality of oscillations increases.

## 9.3   Bibliographic notes

The theory of synchronization in noisy oscillators was already established in the 1960s and has been presented in books by Stratonovich [1963], Malakhov [1968], Landa [1980] and Risken [1989]. Recent relevant works mainly deal with chaotic oscillations or with ensembles; we give the corresponding references in Chapters 10 and 12.

# Chapter 10

# Phase synchronization of chaotic systems

So far we have considered synchronization of periodic oscillators, also in the presence of noise, and have described phase locking and frequency entrainment. In this chapter we discuss similar effects for chaotic systems. The main idea is that (at least for some systems) chaotic signals can be regarded as oscillations with chaotically modulated amplitude and with more or less uniformly rotating phase. The mean velocity of this rotation determines the characteristic time scale of the chaotic system that can be adjusted by weak forcing or due to weak coupling with another oscillator. Thus, we expect to observe phase locking and frequency entrainment for this class of systems as well. It is important that the amplitude dynamics remain chaotic and almost unaffected by the forcing/coupling.

The effects discussed in this chapter were initially termed "phase synchronization" to distinguish them from the other phenomena in coupled (forced) chaotic systems (these phenomena are presented in Part III). As these effects are a natural extension of the theory presented in previous sections, here we refer to them simply as synchronization.

In the following sections we first introduce the notion of phase for chaotic oscillators. Next, we present the effect of phase synchronization of chaotic oscillators and discuss it from statistical and topological viewpoints. We consider both entrainment of a chaotic oscillator by a periodic force and mutual synchronization of two nonidentical chaotic systems. Our presentation assumes that the reader is familiar with the basic aspects of the theory of chaos.

## 10.1    Phase of a chaotic oscillator

### 10.1.1    Notion of the phase

The first problem in extending the basic ideas of synchronization theory to cover chaotic self-sustained oscillators is how to introduce the phase for this case. For periodic oscillations we defined phase as a variable parameterizing the motion along the limit cycle and growing proportionally with time (see Section 7.1). Introduced in this way, the phase rotates uniformly and corresponds to the neutrally stable (i.e., having zero Lyapunov exponent) direction in the phase space. Our goal is to extend this definition in such a way that:

(i) it gives the correct phase of a periodic orbit (and, hence, fulfils the obvious requirement to include the old case) and produces reasonable results for the multiple periodic orbits embedded in chaotic attractors;[1]

(ii) it provides the phase that corresponds to the shift of the phase point along the flow and therefore corresponds to a zero Lyapunov exponent;

(iii) it has clear physical meaning and can be easily computed.

There seems to be no general and rigorous way to meet these demands for an arbitrary chaotic system. We present here an approach that at least partially complies with the conditions formulated above.

Suppose that we can construct a Poincaré map for our autonomous continuous-time chaotic system, i.e., we can choose a surface of section that is crossed transversally by all trajectories of the chaotic (strange) attractor. Then, for each piece of a trajectory between two cross-sections with this surface, we define the phase as a linear function of time, so that the phase gains $2\pi$ with each return to the surface of section:

$$\phi(t) = 2\pi \frac{t - t_n}{t_{n+1} - t_n} + 2\pi n, \qquad t_n \leq t < t_{n+1}. \qquad (10.1)$$

Here $t_n$ is the time of the $n$th crossing of the secant surface. Obviously, the definition is ambiguous because it crucially depends on the choice of the Poincaré surface. Nevertheless, introduced in this way, the phase meets the conditions listed above as follows.

(i) For a periodic trajectory that crosses the surface of section only once, the phase (10.1) completely agrees with Eq. (7.3). This is not the case for periodic orbits that have several cross-sections with the surface; for these trajectories the phase (10.1) is a piece-wise linear function of time. For any particular periodic orbit one can choose a small piece of the surface of section around it and define the

---

[1] We remind the reader that a chaotic motion can be represented via an expansion in the motion along the unstable periodic orbits (UPOs) embedded in the strange attractor; a discussion and citations are given in Section 10.2. The properties of the phase for the set of UPOs are considered in Section 10.2.3.

phase that satisfies Eq. (7.3), but it is impossible to do this globally, i.e., for all periodic orbits. It is important to note that for a periodic orbit with period $T$, the increment of the phase according to Eq. (10.1) is $\phi(T) - \phi(0) = 2\pi\mathcal{N}$, where $\mathcal{N}$ is the number of cross-sections of this orbit with the surface of section; hence, the above definition of the phase provides the correct value of the frequency (and that is why we consider this definition as reasonable). Note that $\mathcal{N}$ is nothing else but the period of the corresponding orbit in the Poincaré map; we call it the topological period (this notion becomes important in Section 10.2).

(ii) The phase (10.1) is locally proportional to time, and thus its perturbation neither grows nor decays. Hence, the phase (10.1) corresponds to zero Lyapunov exponent.

(iii) The phase (10.1) counts "rotations" of the chaotic oscillator, where rotations are defined according to the choice of the Poincaré map: each iteration of the mapping is equivalent to one cycle of the continuous-time system. The phase is easy to calculate; moreover, if we are only interested in characterizing the dynamics in terms of frequency locking, we can simply compute the (mean) frequency as the number of returns to the Poincaré surface (i.e., number of cycles) per time unit.

The phase is an important ingredient of a (commonly used in nonlinear dynamics) reduction of a continuous-time dynamical system to a discrete Poincaré map (see, e.g., Eqs. (10.7) and (10.8) later). In mathematical literature one often goes the opposite way: starting with a mapping, one builds a continuous-time flow using a special flow construction [Cornfeld et al. 1982].

In order not to be overly abstract, we illustrate this approach with two prototypic models of nonlinear dynamics, namely the Rössler [Rössler 1976] and the Lorenz [Lorenz 1963] oscillators. Both are three-dimensional dissipative systems with chaotic attractors.

## A simple case: the Rössler system

The dynamics of the Rössler system

$$
\begin{aligned}
\dot{x} &= -y - z, \\
\dot{y} &= x + 0.15y, \\
\dot{z} &= 0.4 + z(x - 8.5),
\end{aligned}
$$

(10.2)

are shown in Fig. 10.1. The time dependencies $x(t)$ and $y(t)$ can be viewed as "nearly sinusoidal" chaotically modulated oscillations, while $z(t)$ is a sequence of chaotic pulses. A proper choice of the Poincaré surface for the computation of the phase according to Eq. (10.1) can be, e.g., the halfplane $y = 0$, $x < 0$. A projection of the phase portrait on the plane $(x, y)$ looks like a smeared limit cycle with well-defined rotations around the origin. In this and similar cases one can also introduce the angle variable $\theta$ via transformation to the polar coordinates (with the origin coinciding with the center of rotation, cf. [Goryachev and Kapral 1996; Pikovsky *et al.* 1996]):

$$\theta = \tan^{-1}(y/x). \tag{10.3}$$

We can consider the angle variable $\theta$ as an easily computable estimate of the phase $\phi$; for this particular example the difference between $\theta$ and $\phi$ is negligible [Pikovsky *et al.* 1996].

Note that although the phase $\phi$ and the phase-like variable $\theta$ do not coincide microscopically (i.e., on a time scale less than one characteristic period of oscillation, see Fig. 10.3 below), they provide equal mean frequencies: the mean frequency defined as the time average $\langle d\theta/dt \rangle$ coincides with a straightforward determination of the mean frequency as the average number of crossings of the Poincaré surface per time unit. Alternatively, the phase can be estimated from an "oscillatory" observable ($x$ and $y$ are equally suitable for this purpose) with the help of the analytic signal approach based on the Hilbert transform; this method is especially useful for experimental applications (see Chapter 6 and Appendix A2 for details).

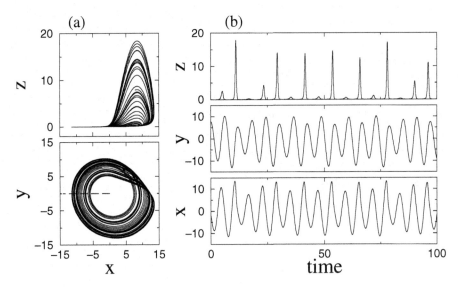

**Figure 10.1.** The dynamics of the Rössler system (Eqs. (10.2)). (a) Projections of the strange attractor onto the planes $(x, y)$ and $(x, z)$. The dashed line presents the Poincaré section used for computation in Fig. 10.5 below. (b) The time series of the variables $x$, $y$, $z$. The processes $x(t)$ and $y(t)$ can be well represented as oscillations with chaotically modulated amplitude.

### A nontrivial case: the Lorenz system

The strange attractor of the Lorenz system

$$\dot{x} = 10(y - x),$$
$$\dot{y} = 28x - y - xz, \tag{10.4}$$
$$\dot{z} = -8/3 \cdot z + xy,$$

shown in Fig. 10.2, is not topologically similar to the Rössler attractor. The variable $z(t)$ demonstrates characteristic chaotically modulated oscillations, but the processes $x(t)$ and $y(t)$ show additionally switchings due to evident symmetry $(x, y) \rightarrow (-x, -y)$ of the Lorenz equations. While the oscillations of $z$ are rather regular, the switchings are not. To overcome the complications due to this mixture of oscillations and switchings, we can introduce a symmetrized observable $u(t) = \sqrt{x^2 + y^2}$ and project the phase portrait on the plane $(u, z)$, Fig. 10.3. In this projection the phase portrait resembles that of the Rössler attractor, and the phase can be introduced in a similar way. A possible Poincaré section is, e.g., $z = 27$, $u > 12$. Alternatively, one can define an angle variable $\theta(t)$ choosing the point $u_0 = 12$, $z_0 = 27$ on the plane Fig. 10.3 as the origin and calculating

$$\theta = \tan^{-1}((z - z_0)/(u - u_0)). \tag{10.5}$$

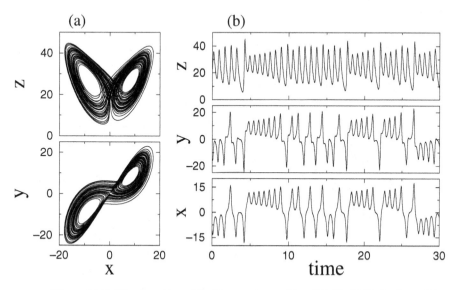

**Figure 10.2.** The dynamics of the Lorenz system (Eqs. (10.4)). (a) Projections of the phase portrait onto planes $(x, y)$ and $(x, z)$. None of these projections shows a rotation around a unique center. (b) The time series of the variables $x$, $y$, $z$. Only $z(t)$ looks like modulated oscillations; in $x(t)$ and $y(t)$ one sees a combination of oscillations and jumps accompanied by the sign change.

Again, this angle variable gives the same mean frequency as the phase based on the Poincaré map.

## Limitations of the approach

At this point we would like to stress that the above approach is rather restricted, and for many chaotic systems we cannot define the phase or a phase-like variable in a proper way. As an example we take again the Rössler system, but this time we choose a set of parameters different from those used in (10.2); the equations are:

$$
\begin{aligned}
\dot{x} &= -y - z, \\
\dot{y} &= x + 0.3y, \\
\dot{z} &= 0.4 + z(x - 7.5).
\end{aligned}
\tag{10.6}
$$

For these parameters the systems possess the so-called funnel attractor shown in Fig. 10.4. On the plane $(x, y)$ it is difficult to find a line that is transversally crossed by all the trajectories. Correspondingly, in the time series $x(t)$, $y(t)$ we see small and large oscillations, and it is difficult to decide which waveform should be considered as a cycle and assigned to the $2\pi$ increase in phase. The funnel Rössler attractor is probably the simplest example of a system with ill-defined phase. In general we expect that the concept of phase is hardly applicable for high-dimensional chaos with many different time scales of oscillations.

Another remark is in order here. We want to introduce the phase as a variable corresponding to the zero Lyapunov exponent of the system. The zero exponent exists in all autonomous continuous-time dynamical systems (if the attractor is not a fixed point), and in all such systems there is the possibility of making a small or a large

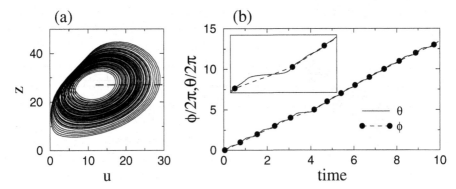

**Figure 10.3.** (a) The dynamics of the Lorenz system in the variables $u$, $z$ looks like a smeared limit cycle with rotations around the unstable fixed point of the system. The dashed line shows the surface of section $z = 27$, $u > 12$. (b) The evolution of the phase $\phi$ based on the Poincaré map (Eq. (10.1), dashed line) and the angle variable $\theta$ (Eq. (10.5), solid line). They coincide at the points (filled circles) where the Poincaré surface is crossed, and differ slightly on the time scale smaller than a characteristic return time (see inset in (b)).

shift along the trajectory, so that this perturbation neither grows nor decays. Exactly this property is important for synchronization, as it makes possible the adjustment of the phases of two systems (or of one oscillator and the force). Dynamical systems with discrete time do not generically have a zero Lyapunov exponent, regardless of the attractor type (periodic or chaotic). Indeed, for discrete-time orbits we are not able to perform small shifts along the trajectory, only large ones. Hence, the notion of phase is not applicable here. The same is valid for continuous-time but nonautonomous systems. For example, chaotic behavior can be exhibited by a periodically forced nonlinear oscillator of the type

$$\ddot{x} + \delta\dot{x} + F(x) = \varepsilon \cos \omega t.$$

Here again, the neutrally stable direction generally does not exist and all Lyapunov exponents are nonzero. One can say that the forced oscillations $x(t)$ are adjusted to the phase of the external force and are not invariant to small shifts of time. We should remember, however, that for a noisy force, the property of having neutral perturbations can exist in a statistical sense; see the discussion of stochastic resonance in Part I (Section 3.6.3).

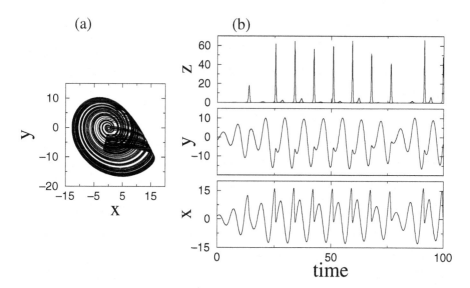

**Figure 10.4.** The dynamics of the Rössler system with a funnel attractor (10.6). (a) Projection of the phase portrait on the plane $(x, y)$. Two characteristic loops corresponding to large and small oscillations are seen. (b) The time series of the variables $x$, $y$, $z$. The oscillations of $x$ and $y$ are complex, so that the phase and the amplitude cannot be defined unambiguously.

### 10.1.2   Phase dynamics of chaotic oscillators

In contrast to the case of periodic oscillations, the growth of the phase of a chaotic system cannot be expected to be uniform. Instead, the instantaneous frequency depends in general on the amplitude. Let us stick to the phase definition based on the Poincaré map (Eq. (10.1)); the phase dynamics can therefore be described as

$$A_{n+1} = \mathcal{M}(A_n), \tag{10.7}$$

$$\frac{d\phi}{dt} = \omega(A_n) \equiv \omega_0 + F(A_n). \tag{10.8}$$

As the amplitude $A_n$ we take the set of coordinates of the point on the secant surface; it does not change during the growth of the phase from $2\pi n$ to $2\pi(n+1)$ and can be considered as a discrete variable; the transformation $\mathcal{M}$ defines the Poincaré map. The phase evolves according to (10.1); the "instantaneous" frequency $\omega(A_n) = 2\pi/T_n$ is determined by the Poincaré return time $T_n = t_{n+1} - t_n$ and depends in general on the position on the secant surface, i.e., on the amplitude. Assuming chaotic behavior of the amplitudes, we can consider the term $\omega(A_n)$ as the sum of the averaged frequency $\omega_0$ and of some effective "noise" $F(A)$ (although this irregular term is of purely deterministic origin); in exceptional cases $F(A)$ may vanish.

The crucial observation is that Eq. (10.8) is similar to Eq. (9.2), describing the evolution of the phase of a periodic oscillator in the presence of external noise. Thus, the dynamics of the phase are generally diffusive (cf. Eq. (9.4)), and the phase performs a random walk. The diffusion constant $D$ (see Eq. (9.4)) determines the phase coherence of the chaotic oscillations. Roughly speaking, $D$ is proportional to the width of the peak at $\omega_0$ in the power spectrum of a typical observable of the chaotic system. For the Rössler attractor, the diffusion constant is extremely small ($D < 10^{-4}$), which corresponds to an extremely sharp peak in the spectrum; this oscillator is therefore often called phase-coherent. For the Lorenz attractor, the spectral peak is essentially broader, and the diffusion constant is not very small, $D \approx 0.2$. Therefore, we anticipate that the phase dynamics of the Rössler system are rather close to those of a periodic oscillator, whereas the effective noise in the Lorenz system is not negligible, and the latter should behave like a periodic system perturbed by a relatively strong noise.

We illustrate the coherence properties of the Rössler and Lorenz attractors in Fig. 10.5, where we show the return times $T_n$, or the "periods" of rotation. For the Rössler oscillator, the variation of $T_n$ is comparatively small, while for the Lorenz oscillator the return time $T_n$ can be arbitrarily large (this corresponds to the slow motion in the vicinity of the saddle at $x = y = z = 0$). As we show below, this feature determines essentially different synchronization properties of these two systems.

In conclusion, we expect that the synchronization phenomena for chaotic systems are similar to those in noisy periodic oscillations. However, one should be aware that the "noisy" term $F(A)$ can hardly be calculated explicitly, and definitely cannot be considered as a Gaussian $\delta$-correlated noise as is commonly assumed in statistical treatments of noisy oscillators (see Chapter 9).

## 10.2 Synchronization of chaotic oscillators

In this section we describe synchronization properties of chaotic systems for which the phase is well-defined and can be directly computed. Hence, the synchronization (called *phase synchronization* to distinguish it from other types of synchronizations of chaotic systems considered in Part III) can be characterized in a straightforward way in terms of phase and frequency locking. The mean observed frequency of the oscillator can thus be easily calculated as

$$\Omega = \lim_{t \to \infty} 2\pi \frac{N_t}{t}, \tag{10.9}$$

where $N_t$ is the number of intersections of the phase trajectory with the Poincaré section during the observation time $t$. This method can also be applied to time series; in the simplest case one can, e.g., take for $N_t$ the number of maxima of an appropriate oscillatory observable ($x(t)$ for the Rössler system and $z(t)$ for the Lorenz system).

Similarly to the case of periodic oscillations, we describe here synchronization by a periodic external force as well as mutual synchronization of coupled systems. Furthermore, we discuss how synchronization of an externally forced oscillator can be characterized indirectly, i.e., without implicit computation of the phase. This characterization, which is independent of the phase definition, is also suitable for the study of chaotic systems with ill-defined phase.

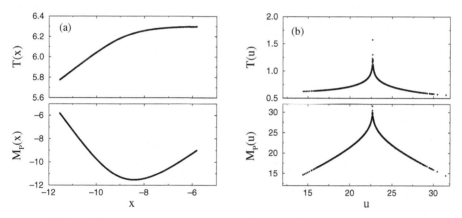

**Figure 10.5.** The return times and the Poincaré maps for the attractors of (a) the Rössler (Eqs. (10.2)) and (b) the Lorenz (Eqs. (10.4)) systems (the surfaces of section are depicted in Figs. 10.1 and 10.3). The graphs look like functions of one variable, but actually they are projections of two-dimensional functions (the internal Cantor structure is very thin and can be hardly seen; therefore we preserve the same notation $\mathcal{M}(\cdot)$ for the mapping). The return time of the Lorenz system has a logarithmic singularity at $u \approx 23$.

### 10.2.1  Phase synchronization by external force

**The Rössler system**

We start with the Rössler model (10.2) and add a periodic external force to the equation for $x$:

$$\dot{x} = -y - z + \varepsilon \cos \omega t,$$
$$\dot{y} = x + 0.15y, \tag{10.10}$$
$$\dot{z} = 0.4 + z(x - 8.5).$$

Calculating the observed frequency $\Omega$ in dependence of the parameters of the external force, we get (shown in Fig. 10.6) a plateau where $\Omega = \omega$. This picture is very similar to the usual picture of the main synchronization region for periodic oscillators. It is remarkable that a relatively small force is able to lock the frequency without a large influence on the amplitude. To illustrate this, we show in Fig. 10.7 stroboscopic (taken with the period of the force) plots of the phase plane $(x, y)$. In the synchronization region the points are concentrated in phase and distributed in amplitude; in the non-synchronous case broad distributions in both phase and amplitude are observed.

The results presented in Figs. 10.6 and 10.7 show that even a weak periodic force can entrain the phase of a chaotic oscillator in the same way it entrains the phase of a periodic one. The effect on the amplitude is relatively small: the force does not suppress chaos. This can also be seen from the calculations of the Lyapunov exponents. The largest Lyapunov exponent remains mainly positive (except for non-avoidable periodic windows) in the whole range of parameters of Fig. 10.6. Outside the synchronization region the second Lyapunov exponent is practically zero, while inside it is negative. The second Lyapunov exponent has, therefore, the same properties as the largest exponent in the system with periodic oscillations. This demonstrates the relative independence of the dynamics of the amplitude and the phase for small external forces.

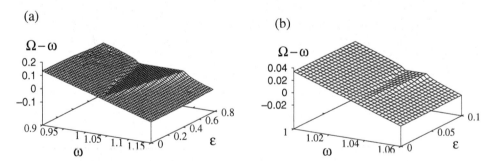

**Figure 10.6.** Synchronization in the Rössler system (10.10). (a) The observed frequency as a function of the amplitude and frequency of the external force. (b) Magnification of the domain of small forcing amplitudes demonstrates that the synchronization threshold is very small; this means that the influence of chaotic amplitudes on the phase dynamics (the effective noise) is weak.

### The Lorenz system

Now let us consider the Lorenz system with a periodic external force acting on the oscillatory variable $z$:

$$\dot{x} = 10(y - x),$$
$$\dot{y} = 28x - y - xz, \qquad\qquad\qquad (10.11)$$
$$\dot{z} = -8/3 \cdot z + xy + \varepsilon \cos \omega t.$$

The effective noise in the Lorenz system is larger than in the Rössler system, and the phase synchronization is not perfect. The dependence of the difference between the observed frequency and the frequency of the external force, $\Omega - \omega$, on $\omega$ (Fig. 10.8) shows a plateau where $\Omega \approx \omega$, but we do not observe exact frequency locking $\Omega = \omega$.

To illustrate the fine features of synchronization of the Lorenz system (see [E.-H. Park et al. 1999; Zaks et al. 1999]) we show in Fig. 10.9a the time dependence of the angle variable $\theta$ defined according to Eq. (10.5). For large time intervals we observe locking by the external force, but these intervals are intermingled with the phase slips. In this way the phase dynamics of the Lorenz system resemble those of a periodic oscillator disturbed with unbounded noise: the slips can be observed for any detuning.[2] The distinction from the case of Gaussian noise is that the probability of having a positive slip is larger than for a negative slip, and this results in the deviation of the observed frequency $\Omega$ from $\omega$. This specific feature of the Lorenz system follows from

---

[2] We note that the phase dynamics of the Rössler system can be regarded as analogous to those of a periodic oscillator with the bounded noise: the irregular in time phase slips are observed at the border of the synchronization region and do not appear in the middle of it.

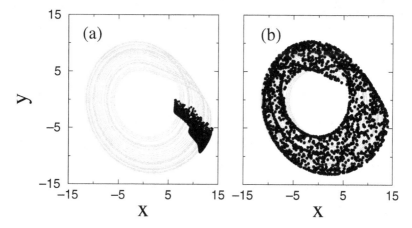

**Figure 10.7.** Stroboscopic plots of the forced Rössler system. (a) Inside the synchronization region ($\varepsilon = 0.16$, $\omega = 1.04$) the phases are concentrated in a small domain. (b) In the nonsynchronized state ($\varepsilon = 0.16$, $\omega = 1.1$) the phases are spread. The autonomous attractor is shown in the background in gray. Equivalently, the plot can be viewed as a distribution of the states in an ensemble of identical Rössler systems driven by a common force.

the Poincaré map and the distribution of the return times shown in Fig. 10.5b. One can see that a long sequence of iterations with small $u$ ($u \approx 15$) is possible, at which the return times are small. During these iterations the oscillator phase is ahead of the force; as a result the observed frequency is slightly larger than the driving one (see details in [Zaks *et al.* 1999]).

In Fig. 10.9b the phase dynamics of the forced system are illustrated by the stroboscopic portrait: the majority of points are concentrated in a small part of the phase space, but there is also a tail of points that lags. Eventually, some of the points make additional rotations with respect to the external force, and some of them miss one or several; the points spread over the attractor although their distribution remains nonuniform.

## 10.2.2  Indirect characterization of synchronization

Up to now we have characterized synchronization by means of direct computation of the frequency. This is possible for systems with a well-defined phase, like the Rössler (Eqs. (10.2)) or the Lorenz (Eqs. (10.4)) systems, but may be difficult and ambiguous if the phase is ill-defined, as in the case of a Rössler system with a funnel attractor (see Eqs. (10.6) and Fig. 10.4). In this respect it is useful to portray synchronization without any reference to a particular definition of the phase. The main idea is that the onset of synchronization means the appearance of coherence between the chaotic process and the external force. The coherence can be seen in the power spectrum of oscillations and in the collective dynamics of an ensemble of identical systems driven by the same force. This suggests two approaches to indirect characterization of entrainment of a chaotic oscillator by an external force.

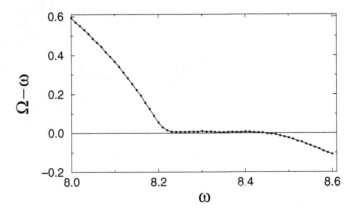

**Figure 10.8.** Synchronization in the Lorenz system: the observed frequency vs. the external frequency for the relatively large forcing amplitude $\varepsilon = 10$. In the plateau $8.2 \lesssim \omega \lesssim 8.5$ there is small discrepancy $\Omega - \omega \approx 0.005$ due to phase slips (see Fig. 10.9).

Typically, the power spectrum of a chaotic system has a broad-band component and a peak at the mean frequency of oscillations;[3] this peak is very narrow for the "good" Rössler system (10.2) and relatively wide for the Lorenz attractor. With periodic forcing, an additional $\delta$-peak appears in the spectrum at frequency $\omega$ of the force (and at its harmonics $n\omega$). Outside the synchronization region, both peaks exist and the intensity of the $\delta$-peak is roughly of the order of the forcing amplitude. Inside the synchronization region, the $\delta$-peak absorbs the whole intensity of the peak at the frequency of the autonomous oscillations. Thus, synchronization appears as a strong increase of the intensity $S$ of the spectral component at the driving frequency. This increase is due to the appearing coherence of chaotic oscillations. Indeed, the periodic force suppresses the phase diffusion, so that the phases at the times $T, 2T, 3T, \ldots$ have almost the same value. The autocorrelation function, therefore, has nondecreasing maxima at time lags $T, 2T, 3T, \ldots$, where $T = 2\pi/\omega$, which corresponds in the spectral domain to the $\delta$-peak at the frequency $\omega$ of the external force.

Another possibility to reveal coherence is to build an ensemble of *replicas* of the chaotic system. Without forcing, the mixing property of chaos leads to decoherence in the ensemble: starting from different initial conditions, the systems eventually spread over the chaotic attractor according to the natural invariant measure. The same happens outside the synchronization region; see Fig. 10.7 for an illustration of spreading. In the synchronous regime, all the systems tend to have approximately the same phase, as can be seen from Figs. 10.7 and 10.9b. Now the small cloud of points rotates as a stable object and does not spread. The simplest way to quantify coherence in an ensemble of $N$ identical systems is to calculate the mean field $X = N^{-1} \sum_1^N x_k$, where $x_k$ is

---

[3] Strictly speaking, the spectrum depends on the observable; we assume that a general nondegenerate oscillatory observable is used for its computation (e.g., $x(t)$ for the Rössler system and $z(t)$ for the Lorenz system).

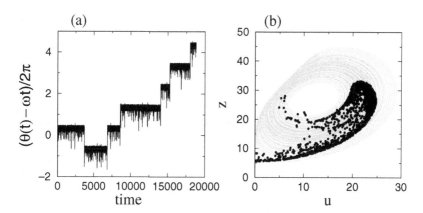

**Figure 10.9.** Imperfect synchronization of the Lorenz system for $\varepsilon = 10$, $\omega = 8.35$. (a) The phase difference as a function of time: large locked intervals are intermingled with slips. (b) The stroboscopic portrait of the forced system. The autonomous attractor is shown in the background in gray.

the observable of a $k$th replica system.[4] If synchronization is absent, the points in the phase space are spread over the whole attractor, and the mean field is very small and nearly constant in time. Contrary to this, in the case of synchronization the mean field is an oscillating (with frequency $\omega$) function of time with a large amplitude.

We emphasize again that computation of both the mean field $X$ and the intensity of the discrete spectral component $S$ does not require determination of the phase and is therefore independent of its definition. It should be noted that application of these indicators is not restricted to the case of chaotic oscillators; they can also be used to characterize the dynamics of noisy systems. We illustrate usage of indirect indicators of synchronization in Fig. 10.10. The two indirect approaches are, of course, interrelated. Because of the mixing, the ensemble of identical systems can be constructed in the other way, by taking observations of a single system at times $0, nT, 2nT, \ldots$ with $n$ large enough. Thus the coherence in time of one system is just the same as the coherence in an ensemble of independent identical systems driven by a common force. The intensity of the $\delta$-peak in the power spectrum is thus proportional to the intensity of the mean field oscillations (see [Pikovsky *et al.* 1997b] for details).

### 10.2.3  Synchronization in terms of unstable periodic orbits

Another approach to phase synchronization of chaotic oscillators is based on consideration of the properties of individual unstable periodic orbits (UPOs) embedded in chaotic attractors. These orbits are known to build a kind of "skeleton" of a chaotic set (see, e.g., [Ott 1992; Katok and Hasselblatt 1995]). In particular, the autonomous systems (10.2) and (10.4) possess an infinite number of unstable periodic solutions.

---

[4] Like the power spectrum, the mean field depends on the observable; see the previous footnote.

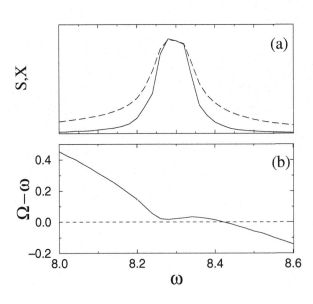

**Figure 10.10.** The resonance-like dependencies of the intensity of the discrete spectral component $S$ (solid curve) and the variance of the mean field in a large ensemble $X$ (dashed curve), both in arbitrary units, on the forcing frequency in the Lorenz system (10.11), for forcing amplitude $\varepsilon = 5$. For comparison, the observed frequency is also plotted. Both indirect characteristics, calculated for the observable $z(t)$, have a maximum in the synchronization region.

Let us pick one of these solutions and consider the influence of a small periodic force upon it. With the exception that the cycle is now unstable, we come to the standard problem of periodically forced oscillations described in Section 7.3. Thus we can use the results of the theory, and characterize synchronization of a particular periodic orbit with a rotation number. An irrational rotation number means quasiperiodic motion on the invariant torus that evolves from the periodic orbit of the unforced system. (Note that this torus inherits the instability of the periodic orbit.) A rational rotation number means that there are two periodic orbits on the invariant torus, one "phase-stable" and one "phase-unstable". The regions of rational rotation numbers are the usual Arnold tongues (see Fig. 7.15). On the border of the locking region the two closed orbits on the torus coalesce and disappear via the saddle-node bifurcation. Outside the tongues the motion corresponding to this particular periodic orbit is not synchronized and the trajectories are dense on the torus.

On the plane of the parameters $\omega$ and $\varepsilon$ the tip of the main Arnold tongue of a periodic orbit lies at the point $\varepsilon = 0$, $\omega = \omega_0^{(i)}$, where $\omega_0^{(i)}$ is the mean frequency of the $i$th autonomous orbit. This frequency differs from the formally defined frequency of the periodic solution $2\pi / T_i$, where $T_i$ is the period of the orbit: we also take into account here the topological period $\mathcal{N}_i$ (the number of rotations of the orbit, see Section 10.1) and write $\omega_0^{(i)} = 2\pi \mathcal{N}_i / T_i$. Generally, the values of $\omega_0^{(i)}$ vary for different periodic orbits: the frequencies $\omega_0^{(i)}$ can be close to each other or widespread, depending on the properties of the return times. If the frequencies $\omega_0^{(i)}$ are not very different, the Arnold tongues overlap (Fig. 10.11), and one can find a parameter region where the

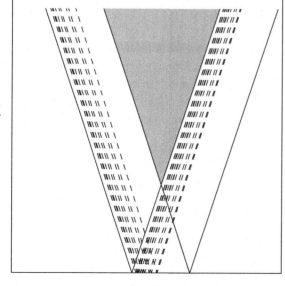

$\varepsilon$

$\omega$

**Figure 10.11.** Schematic view of Arnold tongues for unstable periodic orbits in a chaotic system. Generally, the autonomous orbits have different frequencies $\omega_0^{(i)}$, therefore the tongues touch the $\omega$-axis at different points. The leftmost and the rightmost tongues corresponding to the orbits with the minimal and maximal $\omega_0^{(i)}$ are shown by solid lines. In the shadowed region all the cycles are synchronized and the mean frequency of oscillations virtually coincides with the forcing frequency.

motion along all periodic orbits is locked by the external force. If the forcing remains moderate, this is the overlapping region for the leftmost and the rightmost Arnold tongues which correspond to the periodic orbits of the autonomous system with the smallest and the largest values of $\omega_0^{(i)}$, respectively. Inside this region, the chaotic trajectories repeatedly visit the neighborhoods of the tori; moving along the surface of a torus they approach the phase-stable solution and remain there for a certain time before the "transverse" (amplitude) instability bounces them to another torus. Since all periodic motions are locked, the phase remains localized within the bounded domain: one observes phase synchronization. Outside the region where all the tongues overlap, synchronization cannot be perfect: for some time intervals a trajectory follows the locked cycles and the phase follows the external force, but for other time epochs the trajectory comes close to nonlocked cycles and performs a phase slip. The latter situation is observed in the Lorenz system (Fig. 10.9).

### 10.2.4  Mutual synchronization of two coupled oscillators

We demonstrate this effect with a system of two diffusively coupled Rössler oscillators (10.2):

$$
\begin{aligned}
\dot{x}_1 &= -(1+v)y_1 - z_1+ & \dot{x}_2 &= -(1-v)y_2 - z_2+ \\
&\quad \varepsilon(x_2 - x_1), & &\quad \varepsilon(x_1 - x_2), \\
\dot{y}_1 &= (1+v)x_1 + 0.15y_1, & \dot{y}_2 &= (1-v)x_2 + 0.15y_2, \\
\dot{z}_1 &= 0.2 + z_1(x_1 - 10), & \dot{z}_2 &= 0.2 + z_2(x_2 - 10).
\end{aligned}
\tag{10.12}
$$

Note that the oscillators are not identical; the additionally introduced parameter $v$ governs the detuning of the natural mean frequencies. The coupling is diffusive and proportional to the coupling strength $\varepsilon$.

The observed frequencies of both oscillators can be calculated directly, by counting the number of oscillations per time unit. The resulting dependence of the frequency difference (Fig. 10.12) gives a typical picture of the synchronization region. The phase diagram of different regimes (as functions of the coupling $\varepsilon$ and the frequency mismatch $v$) exhibits three regions of qualitatively different behavior as follows.

(I) The synchronization region, where the frequencies are locked, $\Omega_1 = \Omega_2$. It is important to note that there is almost no threshold of synchronization; this is a particular feature of the phase-coherent Rössler attractor.

(II) The region of nonsynchronized oscillations, where $|\Omega_1 - \Omega_2| = |\Omega_b| > 0$. In analogy to the case of periodic oscillators, this frequency $\Omega_b$ can be considered as the "beat frequency".

(III) In this region oscillations in both systems disappear due to diffusive coupling. This effect is known for periodic systems as oscillation death (or quenching) (see Chapter 8).

It is instructive to characterize the synchronization transition by means of the Lyapunov exponents (see Fig. 10.13). The six-order dynamical system (10.12) has six Lyapunov exponents. For zero coupling we have a degenerate situation of two independent systems, each of them having one positive, one zero, and one negative exponent. The two zero exponents correspond to the two independent phases. With coupling, the phases become dependent and degeneracy is removed: only one Lyapunov exponent should remain exactly zero. We observe, however, that for small coupling the second zero Lyapunov exponent also remains extremely small (in fact, numerically indistinguishable from zero); this corresponds to the "quasiperiodic" dynamics of the phases, where both of them can be shifted. Only at relatively strong coupling, when synchronization sets in, the second Lyapunov exponent becomes negative: now the phases are locked and the relation between them is stable. Note that the two largest exponents remain positive in the whole range of couplings, meaning that the amplitudes are chaotic and independent: the coupled system remains in a hyperchaotic state.

## 10.3   **Bibliographic notes**

In our presentation we have followed the papers [Rosenblum *et al.* 1996; Pikovsky *et al.* 1997b] where different definitions of the phase for chaotic systems as well as direct and indirect characterization of synchronization were discussed. Phase synchronization of chaotic oscillators by a periodic external force was described by

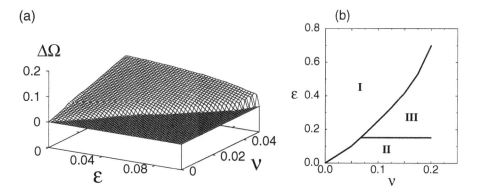

**Figure 10.12.** (a) The synchronization region for two coupled Rössler oscillators (10.12): the plot of the difference in observed frequencies $\Delta\Omega = \langle\dot\theta_1 - \dot\theta_2\rangle$ as a function of the coupling $\varepsilon$ and mismatch $\nu$ exhibits a domain where $\Delta\Omega$ vanishes. (b) A schematic diagram showing the regions of nonsynchronous (II) and synchronous (I) motion, and of oscillation death (III). The diagram is approximate, the windows of periodic behavior in regions I and II are not shown. Panel (b) is reprinted from Osipov *et al.*, *Physical Review* E, Vol. 55, 1997, pp. 2353–2361. Copyright 1997 by the American Physical Society.

Pikovsky [1985] who observed an increase of the discrete spectral component in the power spectrum of a driven system, by Stone [1992] who stroboscopically observed a periodically kicked Rössler system, and by Vadivasova *et al.* [1999]. Similar effects in coupled oscillators, in particular the overlapping of spectral peaks due to interaction, were discussed by Landa and Perminov [1985], Anishchenko *et al.* [1991], Landa and Rosenblum [1992], Anischenko *et al.* [1992] and Landa and Rosenblum [1993].

Phase synchronization in terms of periodic orbits was considered by Pikovsky *et al.* [1997c], Zaks *et al.* [1999] and S. H. Park *et al.* [1999a]. The synchronization transition and the scaling properties of intermittency were described by Pikovsky *et al.* [1997a], Rosa Jr. *et al.* [1998] and Lee *et al.* [1998]. Phase synchronization effects can be also observed in large ensembles, lattices, and chaotically oscillating media, see [Brunnet *et al.* 1994; Goryachev and Kapral 1996; Pikovsky *et al.* 1996; Osipov *et al.* 1997; Brunnet and Chaté 1998; Goryachev *et al.* 1998; Blasius *et al.* 1999] and Section 12.3. Synchronization of a chaotic system by a stochastic force was considered in [Pikovsky *et al.* 1997b].

Experimental observations of phase synchronization of chaos were reported for electronic circuits [Pikovsky 1985; Parlitz *et al.* 1996; Rulkov 1996], plasmas [Rosa Jr. *et al.* 2000], and lasers [Tang *et al.* 1998a,c]. See also the special issue on this topic [Kurths (ed.) 2000].

For coupled nonidentical chaotic oscillators, phase synchronization is observed even for a small coupling strength. For strong coupling the tendency to complete

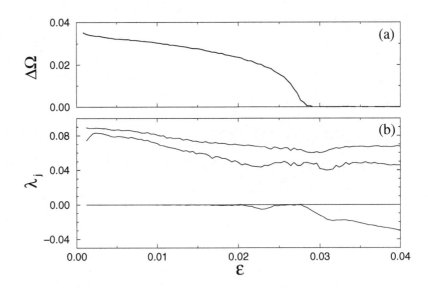

**Figure 10.13.** The four largest Lyapunov exponents and the frequency difference vs. coupling $\varepsilon$ in the coupled Rössler oscillators (10.12); $\nu = 0.015$. For small couplings there are two positive and two nearly zero Lyapunov exponents. Transition to phase synchronization occurs at $\varepsilon \approx 0.028$; at this value of the coupling one of the zero Lyapunov exponents becomes negative.

synchronization is observed (for a detailed discussion of complete synchronization see Part III). For intermediate coupling strengths, an interesting state can be observed: the states of the two interacting systems nearly coincide if one is shifted in time, $\mathbf{x}_1(t) \approx \mathbf{x}_2(t - \tau)$. This state, called *lag synchronization*, was studied theoretically by Rosenblum *et al.* [1997b] and Sosnovtseva *et al.* [1999] and experimentally by Taherion and Lai [1999].

Postnov *et al.* [1999b] described multistability of phase locked states for coupled oscillators with multiband attractors. Finally, we would like to mention that Fujigaki *et al.* [1996] and Fujigaki and Shimada [1997] use the term *phase synchronization* in a different sense.

# Chapter 11

# Synchronization in oscillatory media

An oscillatory medium is an extended system, where each site (element) performs self-sustained oscillations. A good physical (in fact, chemical) example is an oscillatory reaction (e.g., the Belousov–Zhabotinsky reaction) in a large container, where different sites can oscillate with different periods and phases. Typically, the reaction is accompanied by variation of the color of the medium. Hence, the phase profile is clearly seen in such systems: different phases correspond to different colors.

In our presentation we start with a description of the phase dynamics in lattices and spatially continuous systems. Next we consider a medium of weakly nonlinear oscillators that demonstrates a rich variety of behaviors; we describe some of them in Section 11.3.

## 11.1  Oscillator lattices

A lattice of oscillators is a natural generalization of the system of two coupled systems described in Chapter 8. We start by considering a one-dimensional chain of oscillators (numbered with subscript $k$) having different natural frequencies $\omega_k$ and assume that only nearest neighbors interact. If the coupling is weak, the approximation of phase dynamics can be used and the equations can be written as an obvious generalization of Eq. (8.5):

$$\frac{d\phi_k}{dt} = \omega_k + \varepsilon q(\phi_{k-1} - \phi_k) + \varepsilon q(\phi_{k+1} - \phi_k), \qquad k = 1, \dots, N. \quad (11.1)$$

Here we assume for simplicity that the oscillators differ only by their natural frequencies $\omega_k$ and the coupling terms are identical for all pairs. The boundary conditions for (11.1) are taken as $\phi_0 = \phi_1$, $\phi_{N+1} = \phi_N$.

To get a general impression of what can happen in the lattice, let us look at the limiting cases. If the coupling vanishes ($\varepsilon = 0$), the phase of each oscillator rotates with its own natural frequency and in the lattice of $N$ oscillators one observes quasiperiodic motion with $N$ different frequencies. In the other limiting case, when the coupling is very large $\varepsilon \gg |\omega_k|$, the differences in the natural frequencies are negligible, and hence all oscillators eventually synchronize. In between these situations we expect to find regimes with partial synchronization, where several (less than $N$) different frequencies are present. As the coupling tends to synchronize nearest neighbors, *clusters* of synchronized oscillators are observed. We illustrate this qualitative picture with numerical results for a lattice of five oscillators (Fig. 11.1).

The transition from independent oscillators to the fully synchronized state depends on the distribution of frequencies $\omega_k$. Two models have been considered in the literature: a random distribution of natural frequencies and a regular linear distribution. While many-cluster states are difficult to describe, the transition from complete synchronization to the state with two clusters can be treated analytically. As the first step we introduce the phase difference between the neighboring sites, $\psi_k = \phi_{k+1} - \phi_k$, and obtain from (11.1) a set of $N - 1$ equations

$$\frac{d\psi_k}{dt} = \Delta_k + \varepsilon[q(\psi_{k-1}) + q(\psi_{k+1}) - 2q(\psi_k)], \quad k = 1, \ldots, N - 1. \tag{11.2}$$

Here $\Delta_k = \omega_{k+1} - \omega_k$ are the frequency differences. For a stationary state $\dot{\psi}_k = 0$ we obtain a linear tridiagonal algebraic system for the unknown variables $u_k = q(\psi_k)$; it has a unique solution. The problem is in inverting this relation and finding

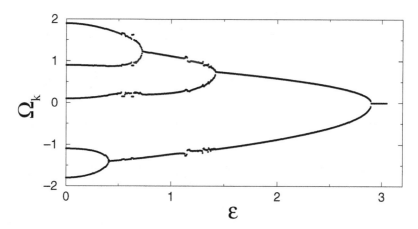

**Figure 11.1.** Dependence of the observed frequencies $\Omega_k = \langle \dot{\phi}_k \rangle$ on the coupling constant $\varepsilon$ in a lattice of five oscillators (11.1). The natural frequencies are $-1.8$, $-1.1$, $0.1$, $0.9$, $1.9$, and the coupling function is $q(x) = \sin x$. With an increase of the coupling strength, first the oscillators 1 and 2 form a cluster at $\varepsilon \approx 0.4$. Then, at $\varepsilon \approx 0.6$ a cluster of oscillators 4 and 5 appears. Oscillator 3 joins it at $\varepsilon \approx 1.4$. Finally, at $\varepsilon \approx 3$ all oscillators become synchronized.

$\psi_k = q^{-1}(u_k)$. As the coupling function $q(\psi)$ is periodic, there are many inverse
solutions provided all the components $u_k$ of the solution lie in the interval $(q_{\min}, q_{\max})$.
As it has been proved by Ermentrout and Kopell [1984], if the coupling function has
one minimum and one maximum, from all possible $2^{N-1}$ roots only one solution
is stable, whereas other fixed points are saddles and unstable nodes. At the critical
coupling, at which, for some $l$, $u_l = q_{\min}$ or $u_l = q_{\max}$, the stable fixed point
disappears via a saddle-node bifurcation and a periodic orbit appears. The phase space
of the system (11.2) is an $(N-1)$-dimensional torus, and the appearing periodic
trajectory rotates in the direction of the variable $\psi_l$. As a result, $\langle \dot{\psi}_k \rangle = 0$ for all
$k$ except for $k = l$, where $\langle \dot{\psi}_l \rangle \neq 0$. In terms of the phases $\phi_k$, this means that
all oscillators $1, \ldots, l$ have the same observed frequency $\Omega_1$, whereas the frequency
$\Omega_2$ of all oscillators $l + 1, \ldots, N$ has a different value. Thus, two clusters of synchronized oscillators appear. Note that the phase differences $\psi_k$ are not constants,
but oscillate, because generally all $\psi_k$ are periodic functions of time. Further decrease of the coupling constant leads to bifurcations at which clusters become split,
etc. For a large lattice and random natural frequencies a typical picture is as in
Fig. 11.2.

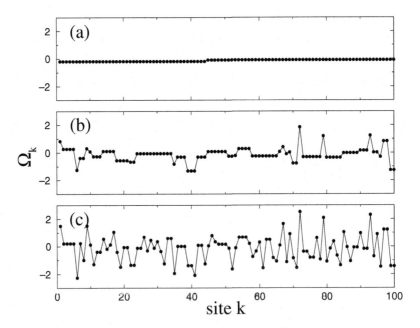

**Figure 11.2.** Clusters in a lattice (11.1) of 100 phase oscillators with random natural
frequencies (normally distributed with unit variance) and the coupling function
$q(x) = \sin x$. (a) A two-cluster state at $\varepsilon = 4$. (b) Many relatively large clusters at
$\varepsilon = 1$. (c) A few small clusters plus many nonsynchronized oscillators at $\varepsilon = 0.2$.

## 11.2    Spatially continuous phase profiles

In many cases the oscillators cannot be considered as discrete units but rather as a continuous oscillatory medium. A typical example is a periodic chemical reaction in a vessel, where at each point one observes oscillations, and these oscillators are coupled via diffusion. Here the phase is a function of spatial coordinates and time, and our first goal is to derive a partial differential equation describing its evolution.

One possible approach, based on the phase approximation to homogeneous oscillatory solutions of partial differential equations, will be outlined in Section 11.3. Here we start from the lattice equations (11.1) and consider their continuous limit. In this limit, the spacing between neighboring sites tends to zero and the coupling constant tends to infinity. If we set $\varepsilon = \tilde{\varepsilon}(\Delta x)^{-2}$ and expand $q$ in a Taylor series, we obtain

$$\frac{d\phi_k}{dt} = \omega_k + \tilde{\varepsilon}q'(0)\frac{\phi_{k-1} - 2\phi_k + \phi_{k+1}}{(\Delta x)^2}$$

$$+ \tilde{\varepsilon}q''(0)\frac{(\phi_{k+1} - \phi_k)^2 + (\phi_k - \phi_{k-1})^2}{2(\Delta x)^2} + \cdots. \tag{11.3}$$

In the continuous limit we have to set $\phi_{k+1} - \phi_k = O(\Delta x)$, thus the second and the third terms on the r.h.s. for $\Delta x \to 0$ converge to the second derivative and the square of the first derivative of the phase. It is easy to check that all other terms in the expansion vanish in this limit. As a result we obtain

$$\frac{\partial\phi(x,t)}{\partial t} = \omega(x) + \alpha\nabla^2\phi(x,t) + \beta(\nabla\phi(x,t))^2. \tag{11.4}$$

This equation is valid if $\alpha > 0$; otherwise, the anti-phase oscillations in the lattice are stable, and their description in a continuous limit requires special analysis.

Equation (11.4) is generally valid if spatial variations of the phase are slow; more precisely, if the characteristic length scale $L$ tends to infinity. Indeed, a partial differential equation for the phase can contain only its derivatives – not the phase itself (because the dynamics are invariant under phase shifts). Next, the symmetry $x \to -x$ imposes the condition that the total power of the derivative should be even. This means that all possible additional terms on the r.h.s. of Eq. (11.4) have either higher derivatives or higher nonlinearities (e.g., $\nabla^4\phi$, $(\nabla\phi)^2\nabla^2\phi$). Assuming $\nabla\phi \sim L^{-1}$, we can estimate these additional terms to be of order $L^{-4}$, $L^{-6}$, ..., i.e., much smaller than the terms of order $L^{-2}$ that are presented in (11.4). We will see in Section 11.3, however, that if the characteristic length scale of phase variations is finite, one should include higher-order terms in (11.4).

### 11.2.1    Plane waves and targets

We first consider a medium of identical oscillators, i.e., a medium with $\omega(x) = $ constant. In this case Eq. (11.4) has plane wave solutions

$$\phi(x,t) = \mathcal{K}x + (\omega + \beta\mathcal{K}^2)t + \phi_0. \tag{11.5}$$

The oscillations in the medium are synchronous, but in general the frequency deviates from the natural one, and the phase shift between different points is nonzero. The sign of coefficient $\beta$ determines the dispersion of waves, i.e., whether their frequency grows or decreases with the wave number (the former case is more typical). Which solution of the family (11.5) is realized depends mainly on boundary conditions. For zero-gradient boundary conditions $\nabla\phi|_B = 0$, the only possible value of the wave number $\mathcal{K}$ is zero. Thus a spatially homogeneous profile of the phase is established, with perfect synchronization between all points of the medium (i.e., all the phases are equal). The last stage of this process is described by the linearized version of (11.4) which is the diffusion equation, so that the characteristic time of the onset of synchronization is $L^{-2}$, where $L$ is the length of the system.

In an infinite one-dimensional medium more complex structures are possible, due to nonlinearity of (11.4). In particular, two plane waves can form a stationary junction [Kuramoto 1984]:

$$\phi(x,t) = \omega t + \beta(a^2 + b^2)t + ax + \frac{\alpha}{\beta}\ln\cosh\frac{b\beta}{\alpha}(x + 2a\beta t). \tag{11.6}$$

This solution depends on two parameters $a$ and $b$ that define the plane waves asymptotics at $x \to \pm\infty$. Assuming for definiteness $\beta b > 0$, we have for large $|x|$

$$\phi \sim \mathcal{K}_\pm x + (\omega + \beta\mathcal{K}_\pm^2)t, \qquad \mathcal{K}_\pm = a \pm b. \tag{11.7}$$

The position of the junction moves with velocity $-2a\beta$. Sometimes the junction is called the domain wall, it can also be a source or a sink of waves. Analysis of the relations (11.7) shows that for $\beta > 0$ the wave pattern with larger frequency wins: the junction moves in the direction of the smaller frequency.

For periodic boundary conditions, plane waves (11.5) with all $\mathcal{K}$ satisfying $\mathcal{K}L = m2\pi$ are possible. One can define the "topological charge" (or the "spatial rotation number") of a phase profile as

$$Q = \frac{1}{2\pi}\int_0^L \nabla\phi \, d\phi,$$

which can have integer values only. This quantity measures the phase shift along the medium. It is clear that the topological charge is conserved during the evolution, therefore any initial phase profile having charge $Q$ eventually evolves to the plane wave (11.5) with the wave number $\mathcal{K} = Q2\pi/L$.

When the profile of natural frequencies $\omega(x)$ is nontrivial, the following Hopf–Cole transformation

$$\phi = \frac{\alpha}{\beta}\ln u \tag{11.8}$$

helps to convert the nonlinear Eq. (11.4) to the linear equation

$$\frac{\partial u}{\partial t} = \frac{\beta}{\alpha}\omega(x)u + \alpha\nabla^2 u. \tag{11.9}$$

This can be considered as the imaginary-time Schrödinger equation with a potential $-(\beta/\alpha)\omega(x)$. The asymptotic ($t \to \infty$) solution is dominated by the first mode having the slowest decay rate. The largest eigenvalue of

$$\lambda u = \frac{\beta}{\alpha}\omega(x)u + \alpha\nabla^2 u \tag{11.10}$$

thus gives the frequency of stationary oscillations. In determining this frequency one should carefully account for boundary conditions; due to the nonlinear transformation (11.8) they depend on the spatial rotation number $Q$.

Suppose that the frequency profile is asymptotically homogeneous: $\omega \to \omega_0$ as $x \to \pm\infty$, but has a local maximum (we assume that $Q = 0$, thus $u \to 0$ at $x \to \pm\infty$). This corresponds to the Schrödinger equation with a local minimum (maximum) of the potential, depending on the sign of $\beta$. It is known from elementary quantum mechanics that, in one dimension, a well-type potential always has at least one discrete eigenvalue $\lambda_1$ in the interval $\omega_{max} > \lambda > \omega_0$. Thus the leading eigenfunction has a form of oscillations with frequency $\lambda_1$: a small region with locally enlarged (for positive $\beta$) frequency dominates the whole medium. If the local inhomogeneity corresponds to a potential hill, the eigenvalue problem (11.10) has no discrete eigenvalues and the observed frequency is that of homogeneous regions. A similar situation is observed in two- and three-dimensional versions of the problem: a local inhomogeneity having a larger/smaller frequency than the environment emits concentric waves and gives rise to a target pattern, depending on the sign of $\beta$.

## 11.2.2 Effect of noise: roughening vs. synchronization

Even a small noise can spoil synchronization in a large system. Let us model a homogeneous ($\omega = $ constant) noisy oscillatory medium with the Langevin approach, i.e., add a fluctuating term to the r.h.s. of (11.4):

$$\frac{\partial\phi(x,t)}{\partial t} = \omega + \alpha\nabla^2\phi(x,t) + \beta(\nabla\phi(x,t))^2 + \xi(x,t). \tag{11.11}$$

This equation is well-known in the theory of roughening interfaces as the Kardar–Parisi–Zhang equation (see, e.g., [Barabási and Stanley 1995; Halpin-Healy and Zhang 1995]). In our context the interface is the phase profile $\phi(x,t)$, and the roughening means developing a rough function of $x$ (with large deviations from its mean value) from an initially flat profile.

To demonstrate the roughening, let us consider the linearized version of (11.11), where we neglect the term $\sim(\nabla\phi)^2$:

$$\frac{\partial\phi(x,t)}{\partial t} = \omega + \alpha\nabla^2\phi(x,t) + \xi(x,t). \tag{11.12}$$

In the theory of roughening interfaces Eq. (11.12) is called the Edwards–Wilkinson equation. Performing the Fourier transform in space we can rewrite (11.12) as a set of linear independent equations for Fourier modes

$$\frac{d\phi_\mathcal{K}}{dt} = \omega\delta_{\mathcal{K},0} - \alpha\mathcal{K}^2\phi_\mathcal{K} + \xi_\mathcal{K}(t).$$

If we assume the noise to be Gaussian $\delta$-correlated in space and time, then all the Fourier components $\xi_\mathcal{K}$ are independent $\delta$-correlated in time processes having the same intensity $\langle\xi_\mathcal{K}(t)\xi_{\mathcal{K}'}(t')\rangle = 2\sigma^2\delta_{\mathcal{K}\mathcal{K}'}\delta(t-t')$. Thus the spectral components $\phi_\mathcal{K}(t)$ are also independent processes. Writing for each $\phi_\mathcal{K}$ the Fokker–Planck equation, we obtain the Gaussian stationary distribution with variance $Var(\phi_\mathcal{K}) = \sigma^2\alpha^{-1}\mathcal{K}^{-2}$. Thus, the spatial spectrum of the phase profile $\phi(x,t)$ is proportional to $\mathcal{K}^{-2}$, and one can show (see, e.g., [Halpin-Healy and Zhang 1995]) that the same is true for the nonlinear Kardar–Parisi–Zhang equation (11.11) as well.

The variance of the instantaneous phase profile $Var(\phi) = \langle(\phi - \langle\phi\rangle)^2\rangle$ can be calculated as the integral over the spatial spectrum. As the integral diverges at $\mathcal{K} \to 0$, we should take into account the cutoff at the smallest wave number $\mathcal{K}_0$ corresponding to the system length $L$: $\mathcal{K}_0 \sim L^{-1}$ (another cutoff at small scales is necessary to get rid of the ultraviolet divergence). At this point the result starts to depend on the dimension of the system $d$:

$$Var(\phi) \sim \int_{L^{-1}}^{C} \mathcal{K}^{d-1}\,\mathcal{K}^{-2}\,d\mathcal{K} \sim \begin{cases} L & d = 1 \\ \ln L & d = 2 \\ \text{constant} & d \geq 3. \end{cases} \tag{11.13}$$

The roughness is the property of indefinite growth of the "interface" width with the system size. As follows from Eq. (11.13), roughness essentially depends on the dimensionality of the system, i.e., on whether the oscillatory medium is one-, two-, or three-dimensional. The spatial spectrum can be integrated without divergence in dimensions larger than three, so in this case there is no roughening: the variance of the interface is finite even for very large systems. Contrary to this, the phase profile is rough in dimensions $d = 1, 2$.

For the phase dynamics the transition roughening–nonroughening can be interpreted as the decoherence–coherence transition (see [Gallas *et al.* 1992; Grinstein *et al.* 1993]). Let us start with the roughened one-dimensional case. Note first that due to spatial homogeneity the observed frequencies in (11.11) are the same at all points (which is not surprising because the natural frequencies are also equal). Thus, in the sense of coincidence of the observed frequencies the oscillations are synchronized. However, they are not coherent. At each moment of time the phase profile can be seen as a curve of random-walk-type (one can see this from the spatial spectrum $\sim\mathcal{K}^{-2}$). Thus, locally in space the phases are not very much different and on small spatial scales one can consider the oscillations as synchronized. However, for a larger length scale the characteristic phase difference exceeds $2\pi$ and no correlation between the phases taken modulo $2\pi$ can be seen. If we look at some phase-dependent observable, e.g., $\sin\phi$, its average over the medium will be a constant in time as in the case of fully independent oscillations. In this sense the roughening means decoherence in large systems. Conversely, if the phase profile is not rough, i.e., if the variations of the phase along the whole system are less than $2\pi$, then not only do the frequencies of all oscillations coincide, but also the phases are correlated and the averaging of a phase-dependent quantity over the whole medium will give oscillations with the common frequency. Chaté and Manneville [1992] have described different examples of such behavior.

## 11.3    Weakly nonlinear oscillatory medium

In Section 11.2 we characterized an oscillating medium with a single phase variable. This is possible if the deviations of all other variables from the limit cycle describing homogeneous periodic oscillations are small. Otherwise, one has to consider full partial differential equations for the state variables. The situation is simplified if the oscillations are weakly nonlinear. In this case one can introduce a complex amplitude $A$, depending on space and time, and represent a state variable $u(x, t)$ as $u = \text{Re}(A(x, t)e^{i\omega t})$. Here $\omega$ is the frequency of natural oscillations. An equation for $A$ can be derived for a particular problem with the help of the averaging method or its variations (see, e.g., [Kuramoto 1984; Haken 1993; Bohr *et al.* 1998]). Here we use the same approach as in Section 11.2: we start with a lattice of weakly nonlinear self-sustained oscillators and consider its continuous limit.

### 11.3.1    Complex Ginzburg–Landau equation

A one-dimensional lattice of weakly coupled nonlinear oscillators is described by a generalization of Eqs. (8.12):

$$\frac{d A_k}{dt} = \mu A_k - (\gamma + i\alpha)|A_k|^2 A_k + (\beta + i\delta)(A_{k+1} + A_{k-1} - 2A_k). \tag{11.14}$$

Here we assume that all oscillators have the same parameters. A transition to a continuous medium assumes that the difference $A_{k+1} - A_k$ is of order $\Delta x$; correspondingly the interaction constants $\beta$ and $\delta$ are large. Setting $\beta = \tilde{\beta}(\Delta x)^{-2}$ and $\delta = \tilde{\delta}(\Delta x)^{-2}$, we get

$$\frac{\partial A}{\partial t} = \mu A - (\gamma + i\alpha)|A|^2 A + (\tilde{\beta} + i\tilde{\delta})\nabla^2 A.$$

It is convenient to use here the same scaling as in Section 8.2, i.e., normalize time by $\mu$ and the amplitude by $\sqrt{\gamma/\mu}$ to obtain the famous complex Ginzburg–Landau equation (CGLE):

$$\frac{\partial a(x,t)}{\partial t} = a - (1 + ic_3)|a|^2 a + (1 + ic_1)\nabla^2 a, \qquad (11.15)$$

which describes weakly nonlinear oscillations in a continuous medium. Its terms have the following physical meanings: the first term on the r.h.s. describes the linear growth of oscillations; the second term describes nonlinear saturation (real part of the coefficient) and nonlinear shift of frequency (imaginary part); the last term describes the spatial interaction (diffusion) of dissipative (real part) and reactive (imaginary part) types. The purely conservative version of the CGLE (i.e., with only purely imaginary coefficients left on the r.h.s.; formally this corresponds to the limit $c_{1,3} \to \infty$) is the nonlinear Schrödinger equation, a fully integrable Hamiltonian system. In the context of self-sustained oscillations the dissipative terms are essential; moreover, in some situations (isochronous oscillations and purely dissipative coupling) the coefficients $c_1$ and $c_3$ vanish. Not pretending to describe completely the properties of the CGLE (see, e.g., [Shraiman *et al.* 1992; Cross and Hohenberg 1993; Chaté and Manneville 1996; Bohr *et al.* 1998]), we here emphasize only the features important from the synchronization point of view.

The CGLE has plane wave solutions (cf. (11.5))

$$a(x,t) = (1 - \mathcal{K}^2) \exp[i\mathcal{K}x - i(c_3 + (c_1 - c_3)\mathcal{K}^2)t],$$

which can be interpreted as synchronized states in the medium. Not all of these waves are stable, but some stable long-wave solutions exist if

$$1 + c_1 c_3 > 0. \qquad (11.16)$$

To see how criterion (11.16) appears, let us write the phase approximation for the CGLE. This approximation is valid for states that are slowly varying in space, where the diffusive term (proportional to the square of the characteristic wave number) can be considered as a small perturbation. Thus we can apply the general formula (7.14) for perturbations near the spatially homogeneous limit cycle to obtain the equation for the phase. In this formula we insert the phase dependence in the form (cf. Eqs. (7.10) and (7.16))

$$\phi(X, Y) = \tan^{-1}\frac{Y}{X} - \frac{c_3}{2}\ln(X^2 + Y^2)$$

and the perturbation in the form

$$p_X = \nabla^2 X(\phi) - c_1\nabla^2 Y(\phi), \quad p_Y = \nabla^2 Y(\phi) + c_1\nabla^2 X(\phi),$$

with $a = X + iY = \cos\phi + i\sin\phi$ to get

$$\frac{\partial\phi}{\partial t} = -c_3 + (1 + c_3 c_1)\nabla^2\phi + (c_3 - c_1)(\nabla\phi)^2. \qquad (11.17)$$

This equation coincides, of course, with (11.4). The main point that makes the dynamics of the CGLE nontrivial is the possible instability of the phase: the phase diffusion coefficient in Eq. (11.17) is $1+c_3c_1$, and when it is negative the spatially homogeneous synchronous state is unstable. The criterion (11.16) was derived by Newell [1974], but the instability is often called the Benjamin–Feir instability after analogous treatment of the instability of nonlinear water waves [Benjamin and Feir 1967].

The physical mechanism of the instability will become clear if we compare the Newell criterion (11.16) to Eq. (8.17), describing the interaction of two oscillators. As discussed in Section 8.2, a combination of the nonisochrony and reactive coupling leads to repulsion of the phases of two coupled oscillators; exactly the same combination leads to instability in a continuous medium.

Numerical experiments show that near the threshold of phase instability (11.16) the amplitude $|a|$ remains close to the steady state value 1, but the phase changes irregularly in space and time. This regime is called phase turbulence. It can be described by an extension of Eq. (11.17):

$$\frac{\partial \phi}{\partial t} = -c_3 + (1 + c_3c_1)\nabla^2\phi + (c_3 - c_1)(\nabla\phi)^2 - \tfrac{1}{2}c_1^2(1 + c_3^2)\nabla^4\phi, \tag{11.18}$$

where the stabilizing term proportional to the fourth spatial derivative is included. This is the Kuramoto–Sivashinsky equation [Nepomnyashchy 1974; Kuramoto and Tsuzuki 1976; Sivashinsky 1978], describing the nonlinear stage of the phase instability. It demonstrates turbulent solutions if the system size is sufficiently large. It is interesting to note that, due to turbulence of the phase dynamics, the large-scale properties of the phase profile in the Kuramoto–Sivashinsky equation (11.18) are the same as in the Kardar–Parisi–Zhang equation (11.11) [Yakhot 1981; Bohr *et al.* 1998]: the turbulence plays the role of effective noise for large-scale phase variations. In particular, in a large system the phase profile becomes rough, which means a loss of coherence. Far away from the border of instability one observes a regime of amplitude, or defect, turbulence. A characteristic feature of this regime is the appearance of defects – points on the space–time diagram where the amplitude is exactly zero. At the defect the phase difference between neighboring points changes by $\pm 2\pi$ and both synchronization and phase coherence disappear. We illustrate the main regimes in the CGLE in Fig. 11.3.

In two- and three-dimensional oscillatory media new stable objects appear: spirals. A spiral rotates around a topological defect and is stable in a large range of parameters. In the spiral the oscillations are synchronized. Other objects often observed in two-dimensional oscillatory media are targets: concentric oscillations propagating from an oscillatory centrum. In a nonhomogeneous medium, where targets with different periods are possible, usually the wave with the smallest period wins the competition. For more details, as well as properties of spirals and targets in a medium of relaxation oscillators, see works by Cross and Hohenberg [1993], Bohr *et al.* [1998], Mikhailov [1994] and Walgraef [1997] and references therein.

## 11.3.2  Forcing oscillatory media

An interesting but mainly unresolved question concerns synchronization of oscillatory media by external periodic forcing. In particular, recently Petrov *et al.* [1997] performed experiments with a two-dimensional oscillatory chemical reaction (Belousov–Zhabotinsky reaction). With a flashing light they produced a periodic forcing, and studied different phase locked and desynchronized states (see also Section 4.2.4). We summarize here the numerical results obtained for the weakly nonlinear one-dimensional case. The oscillatory medium is described by the CGLE (11.15); an additional periodic sinusoidal force with a frequency close to the natural one leads to the equation (cf. the corresponding equation for a single forced oscillator, Eq. (7.43))

$$\frac{\partial a(x,t)}{\partial t} = (1 - i\nu)a - (1 + ic_3)|a|^2 a + (1 + ic_1)\nabla^2 a - ie. \qquad (11.19)$$

Here $e$ is the amplitude of forcing and $\nu$ is the frequency mismatch. For a spatially homogeneous state, the problem of existence of the synchronized solution $a = $ constant reduces to the analysis of Eq. (7.43). The difference is in the stability of this solution: in a medium, spatially inhomogeneous perturbations can grow even in the region of

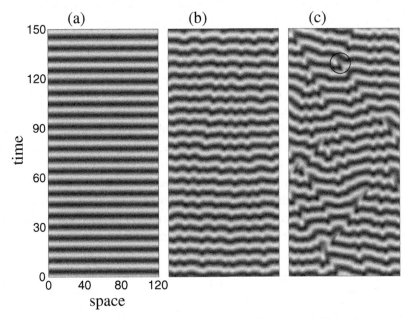

**Figure 11.3.** Regimes in the complex Ginzburg–Landau equation. The real part $Re(a)$ is shown in a gray-scale code, so that the lines of constant color correspond to the lines of constant phase. (a) Stable spatially homogeneous periodic in time oscillations for $c_3 = 1$, $c_1 = 0$. (b) Regime of phase turbulence for $c_3 = 1$, $c_1 = -1.7$; the lines of constant phase fluctuate but remain unbroken. (c) Defect turbulence for $c_3 = 1$, $c_1 = -2$; one defect, where the amplitude vanishes and the lines of constant phase break, is encircled.

stable synchronous solution of the oscillator (Eq. (7.43)). Thus, synchronization can
be spoiled by spatially inhomogeneous regimes. Another interesting point is that even
when the homogeneous synchronous state is stable, it is not necessarily the global
attractor. Indeed, suppose we apply an external force to a state with large deviations
of the phase, or to a plane wave solution with nonzero wave number. The force
tends to bring the phase to a stable phase position $\phi_0$, but all the phases $\phi_0 + 2\pi m$
are equivalently stable. Thus the phase profile will form a sequence of regions with
constant phase, separated with $\pm 2\pi$-kinks, see Fig. 11.4.

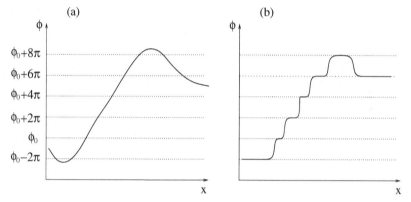

**Figure 11.4.** Formation of kinks from an initial phase profile. The stable phase
positions are marked with dotted lines. (a) An initial profile. (b) The kinks are
formed under external forcing.

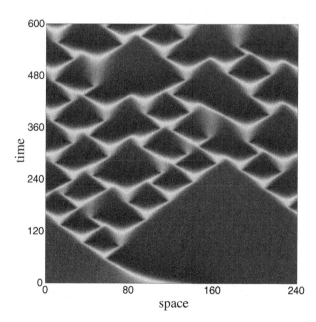

**Figure 11.5.** The process of
kink-breeding in the forced
CGLE (11.19) for $c_3 = 1$,
$c_1 = 0, \nu = 1.1, e = 0.073$.
The field $\mathrm{Re}(a)$ is shown in
gray-scale coding. The
regime of perfect
synchronization (i.e.,
$\mathrm{Re}(a) = \mathrm{constant}$) is stable
towards small perturbations,
but one initially imposed
kink spoils this, resulting in
an intermittent state of
appearing and disappearing
kinks.

Depending on the parameters of the system, these kinks can exist as stable objects (generally moving with some constant velocity) and the oscillatory medium is never completely synchronized. Otherwise, the kink dies through the appearance of a defect, as in the regime of amplitude turbulence. In the latter case there are again two possibilities. The kink can disappear completely, so that eventually the medium will be completely synchronized: the phase at all points is the same and locked to the external force. But for some parameters of the system the perturbations remaining after the disappearance of the kink lead to the creation of two new kinks (see [Chaté *et al.* 1999] for details). Thus a process of kink-breeding appears, typically leading (in a large system) to turbulence. The state is intermittent: on a given site one observes long completely synchronized epochs, but sometime a kink nearby induces a $2\pi$ phase slip. We illustrate the space–time dynamics in Fig. 11.5.

## 11.4   **Bibliographic notes**

One-dimensional lattices of oscillators of different nature have been extensively studied numerically: phase oscillators [Ermentrout and Kopell 1984; Kopell and Ermentrout 1986; Sakaguchi *et al.* 1988a; Rogers and Wille 1996; Zheng *et al.* 1998; Ren and Ermentrout 2000]; weakly nonlinear oscillators [Ermentrout 1985]; phase locked loops [Afraimovich *et al.* 1994; de Sousa Vieira *et al.* 1994]; Josephson junctions [Braiman *et al.* 1995]; relaxation oscillators [Corral *et al.* 1995b,a; Herz and Hopfield 1995; Hopfield and Herz 1995; Drossel 1996; Mousseau 1996; Díaz-Guilera *et al.* 1998]; chaotic [Osipov *et al.* 1997] and noise-excited [Neiman *et al.* 1999b] oscillators. For studies of two-dimensional lattices see [Sakaguchi *et al.* 1988a; Aoyagi and Kuramoto 1991; Blasius *et al.* 1999].

Oscillatory media have attracted much interest, but only a few papers dwell on synchronization effects. Synchronization by a periodic external force was studied experimentally by Petrov *et al.* [1997]; the corresponding forced complex Ginzburg–Landau equation was considered by Coullet and Emilsson [1992b], Coullet and Emilsson [1992a], Schrader *et al.* [1995], Chaté *et al.* [1999] and Elphick *et al.* [1999]. Junge and Parlitz [2000] described phase synchronization in coupled Ginzburg–Landau equations.

# Chapter 12

# Populations of globally coupled oscillators

The effect of mutual synchronization of two coupled oscillators, described in Chapter 8, can be generalized to more complex situations. One way to do this was described in Chapter 11, where we considered lattices of oscillators with nearest-neighbor coupling. However, often oscillators do not form a regular lattice, and, moreover, interact not only with neighbors but also with many other oscillators. Studies of three and four oscillators with all-to-all coupling give a rather complex and inexhaustible picture (see, e.g., [Tass and Haken 1996; Tass 1997]). The situation becomes simpler if the interaction is homogeneous, i.e., all the pairs of oscillators are equally coupled. Moreover, if the number of interacting oscillators is very large, one can consider the thermodynamic limit where the number of elements in the ensemble tends to infinity. Now, when the oscillators are not ordered in space, one usually speaks of an ensemble or a population of coupled oscillators. An analogy to statistical mechanics is extensively exploited, so it is not surprising that the synchronization transition appears as a nonequilibrium phase transition in the ensemble.

We start with the description of the Kuramoto transition in a population of phase oscillators. An ensemble of noisy oscillators is then discussed. Different generalizations of these basic models (e.g., populations of chaotic oscillators) are described in Section 12.3.

## 12.1 The Kuramoto transition

In this section we consider, following Kuramoto [1984], a simple model of $N$ mutually coupled oscillators having different natural frequencies $\omega_k$. The dynamics are governed by equations similar to Eqs. (8.5):

$$\frac{d\phi_k}{dt} = \omega_k + \frac{\varepsilon}{N} \sum_{j=1}^{N} \sin(\phi_j - \phi_k). \qquad (12.1)$$

Here the parameter $\varepsilon$ determines the coupling strength. The coupling between each pair is chosen to be proportional to $N^{-1}$: only in this case we get $N$-independent results in the thermodynamic limit $N \to \infty$. Indeed, if the coupling in each pair is $N$-independent, then the force acting on each oscillator grows with the size of the population, and in the thermodynamic limit this force tends to infinity, obviously leading to synchronization of the whole population. Natural frequencies of the oscillators $\omega_k$ are supposed to be distributed in some range, and for $N \to \infty$ we can describe this distribution with a density $g(\omega)$.[1] We will suppose the density to be symmetric with respect to the single maximum at frequency $\bar{\omega}$. In Eq. (12.1) the coupling is assumed to have the simplest sine form; some generalizations will be discussed later in Section 12.3.

Before discussing the possible onset of mutual synchronization, we rewrite system (12.1) in a way that is more convenient for the analysis. We introduce the complex *mean field* of the population according to

$$Z = X + iY = Ke^{i\Theta} = \frac{1}{N} \sum_{k=1}^{N} e^{i\phi_k}. \qquad (12.2)$$

The mean field has amplitude $K$ and phase $\Theta$:

$$K \cos \Theta = \frac{1}{N} \sum_{k=1}^{N} \cos \phi_k, \qquad K \sin \Theta = \frac{1}{N} \sum_{k=1}^{N} \sin \phi_k. \qquad (12.3)$$

The mean field is an indicator of the onset of coherence due to synchronization in the population. Indeed, if all the frequencies are different, then for any moment of time the phases $\phi_k$ are uniformly distributed in the internal $[0, 2\pi)$, and the mean field vanishes. Conversely, if some oscillators in the population lock to the same frequency, then their fields sum coherently and the mean field is nonzero. An analogy with the ferromagnetic phase transition, where the magnetic mean field appears due to correlations in the directions of elementary magnetic moments (spins), is evident; thus the amplitude of the mean field (12.2) can be taken as a natural order parameter of the synchronization transition.

The basic model (12.1) can be rewritten as a system of oscillators forced by the mean field

$$\frac{d\phi_k}{dt} = \omega_k + \varepsilon K \sin(\Theta - \phi_k). \qquad (12.4)$$

Obviously, the zero mean field implies that the forcing vanishes as well; hence, this noncoherent state is always a solution of the system (12.1). In this case each element of the ensemble oscillates with its own natural frequency $\omega_k$. These frequencies are

[1] We omit the subscript of $\omega$ in arguments of distributions.

generally different, thus the phases are distributed uniformly in the interval $[0, 2\pi)$ and the average according to Eqs. (12.2) and (12.3) gives a zero mean field. The state with a nonzero mean field is less trivial. If it is periodic, $K = \text{constant}$, $\Theta = \omega t$, then each Eq. (12.4) is equivalent to the phase equation of the periodically driven oscillator (Eq. (7.20)). We see that for each oscillator the mean field is acting as an external force that, depending on the parameters, can or cannot synchronize this oscillator.

The appearance of a nonzero mean field can be explained self-consistently: a nonzero mean field synchronizes at least some oscillators so that they become coherent, and this coherent group generates a finite contribution to the mean field. We now make these arguments quantitative.

It is natural to suppose that, due to the symmetry of the distribution $g(\omega)$, the mean field will oscillate with the central frequency $\bar{\omega}$ (in general, we could start with an arbitrary frequency and at the end obtain $\bar{\omega}$ from the self-consistency condition; here, using symmetry, we just assume $\bar{\omega}$ to be the frequency of the mean field, and we will see below that this indeed solves the problem). The main idea is to derive self-consistency conditions, in analogy to the mean field theory of second-order phase transitions. So, we set

$$\Theta = \bar{\omega}t, \quad K = \text{constant}, \quad \psi_k = \phi_k - \bar{\omega}t,$$

to obtain

$$\frac{d\psi_k}{dt} = \omega_k - \bar{\omega} - \varepsilon K \sin \psi_k. \tag{12.5}$$

This equation coincides with (7.24) and has synchronous and asynchronous solutions as follows.

(i) The synchronous solution

$$\psi_k = \sin^{-1} \frac{\omega_k - \bar{\omega}}{\varepsilon K} \tag{12.6}$$

exists if the natural frequency of the $k$th oscillator is close to $\bar{\omega}$: $|\omega_k - \bar{\omega}| \leq \varepsilon K$. The corresponding oscillators are entrained by the mean field.

(ii) The asynchronous solution, where the phase $\psi_k$ rotates obeying equation (12.5), exists if $|\omega_k - \bar{\omega}| > \varepsilon K$. In this asynchronous state the phases are not uniformly distributed; this will be taken into consideration below.

The next step is to find the contributions of the sub-populations of synchronous and asynchronous oscillators to the mean field. To calculate Eqs. (12.3) in the limit $N \to \infty$ we have to know the distribution of the phase differences $n(\psi)$. According to the two types of solution, we decompose this distribution into synchronous and asynchronous parts, $n_{\mathrm{s}}(\psi)$ and $n_{\mathrm{as}}(\psi)$.

For the oscillators entrained by the mean field, the phase difference $\psi$ is time-independent and is determined by the natural frequency according to (12.6), so that the distribution $n_s(\psi)$ can be obtained directly from the distribution of natural frequencies

$$n_s(\psi) = g(\omega)\left|\frac{d\omega}{d\psi}\right| = \varepsilon K g(\bar{\omega} + \varepsilon K \sin\psi)\cos\psi, \quad -\frac{\pi}{2} \le \psi \le \frac{\pi}{2}. \quad (12.7)$$

For nonentrained oscillators the phase is rotating, but for each $\omega_k$ we can obtain the distribution of the phases directly from Eq. (12.5). As the phase rotates nonuniformly in time, the probability of observing a value $\psi$ is inversely proportional to the rotation velocity at this point $|\dot{\psi}|$. Thus, for given $\omega_k$ the distribution of phases is

$$P(\psi, \omega) \sim \frac{1}{|\dot{\psi}|}.$$

Substituting (12.5) into the above and normalizing yields

$$P(\psi, \omega) = \left(|\omega - \bar{\omega} - \varepsilon K \sin\psi| \int_0^{2\pi} \frac{d\psi}{|\omega - \bar{\omega} - \varepsilon K \sin\psi|}\right)^{-1}$$

$$= \frac{\sqrt{(\omega - \bar{\omega})^2 - \varepsilon^2 K^2}}{2\pi|\omega - \bar{\omega} - \varepsilon K \sin\psi|}. \quad (12.8)$$

Now we have to average this distribution over $g(\omega)$ to obtain the distribution of the phases of asynchronous oscillators:

$$n_{as}(\psi) = \int_{|\omega - \bar{\omega}| > \varepsilon K} g(\omega) P(\psi, \omega)\, d\omega$$

$$= \int_{\bar{\omega} + \varepsilon K}^{\infty} \frac{g(\omega)\sqrt{(\omega - \bar{\omega})^2 - \varepsilon^2 K^2}}{2\pi(\omega - \bar{\omega} - \varepsilon K \sin\psi)}\, d\omega$$

$$+ \int_{-\infty}^{\bar{\omega} - \varepsilon K} \frac{g(\omega)\sqrt{(\omega - \bar{\omega})^2 - \varepsilon^2 K^2}}{2\pi(-\omega + \bar{\omega} + \varepsilon K \sin\psi)}\, d\omega.$$

Denoting $\omega - \bar{\omega} = x$ and using the symmetry of the frequency distribution $g(\bar{\omega} + x) = g(\bar{\omega} - x)$, we can rewrite the last formula in the more compact form

$$n_{as}(\psi) = \int_{\varepsilon K}^{\infty} \frac{g(\bar{\omega} + x) x \sqrt{x^2 - \varepsilon^2 K^2}}{\pi[x^2 - \varepsilon^2 K^2 \sin^2\psi]}\, dx. \quad (12.9)$$

Now with the help of the distributions $n_s$ and $n_{as}$ we obtain the self-consistent equation for the mean field

$$K e^{i\bar{\omega}t} = \int_{-\pi}^{\pi} e^{i\psi + i\bar{\omega}t}\left[n_s(\psi) + n_{as}(\psi)\right] d\psi. \quad (12.10)$$

We note first that, according to (12.9), the asynchronous part of the distribution $n_{as}(\psi)$ has period $\pi$ in $\psi$, so it does not contribute to the integral (12.10). We thus obtain two real equations (the real and the imaginary part of (12.10)):

$$K = \varepsilon K \int_{-\pi/2}^{\pi/2} \cos^2\psi \cdot g(\bar{\omega} + \varepsilon K \sin\psi)\, d\psi, \quad (12.11)$$

$$0 = \varepsilon K \int_{-\pi/2}^{\pi/2} \cos \psi \sin \psi \cdot g(\bar{\omega} + \varepsilon K \sin \psi) \, d\psi. \tag{12.12}$$

Equation (12.12) determines the frequency, and we see that $\bar{\omega}$ was the correct choice: this equation is fulfilled due to the symmetry of the frequency distribution. We are left with Eq. (12.11) which determines the amplitude of the mean field $K$. A closed analytical solution can be found only for some special distributions $g(\omega)$.

Consider, as an exactly solvable example, the Lorentzian distribution

$$g(\omega) = \frac{\gamma}{\pi[(\omega - \bar{\omega})^2 + \gamma^2]}. \tag{12.13}$$

For this distribution the integral in (12.11) can then be calculated exactly. After some manipulations one obtains the amplitude of the coherent solution

$$K = \sqrt{1 - \frac{2\gamma}{\varepsilon}}. \tag{12.14}$$

This nontrivial mean field exists if the coupling exceeds the critical value $\varepsilon_c = 2\gamma$. The transition to synchronization resembles a second-order phase transition and can be characterized by the critical exponent $1/2$: $K \sim (\varepsilon - \varepsilon_c)^{1/2}$. This also holds for general unimodal distributions $g(\omega)$. Because for small $K$ only the oscillators with $\omega \cong \bar{\omega}$ are synchronized, only local properties of the function $g$ in the vicinity of the maximum are important near the synchronization threshold. One can see this from Eq. (12.11), where for small $K$ only a neighborhood of $\bar{\omega}$ contributes to the mean field. Thus, supposing $K$ to be small and expanding $g(\bar{\omega} + \varepsilon K \sin \psi)$ in (12.11) in a Taylor series

$$g(\bar{\omega} + \varepsilon K \sin \psi) \approx g(\bar{\omega}) + \frac{g''}{2} \varepsilon^2 K^2 \sin^2 \psi,$$

we get, after substitution in (12.11),

$$\varepsilon_c = \frac{2}{\pi g(\bar{\omega})}, \qquad K^2 \approx \frac{8 g(\bar{\omega})}{|g''| \varepsilon^3} (\varepsilon - \varepsilon_c). \tag{12.15}$$

We illustrate the synchronization transition in a population of oscillators with the results of a numerical simulation in Fig. 12.1.

## 12.2    Noisy oscillators

Here we consider a population of identical oscillators in the presence of external noise. Usually one considers statistically independent equally distributed noisy forces for each oscillator. In such a population the observed frequencies of all elements are obviously equal, but, due to the noise, the oscillators can have completely independent phases. Synchronization appears as a coherence in the ensemble, which manifests itself by a nonzero mean field. A basic model can be written as a system of coupled Langevin equations with noisy forces $\xi_i(t)$:

$$\frac{d\phi_k}{dt} = \omega_0 + \frac{\varepsilon}{N} \sum_{j=1}^{N} \sin(\phi_j - \phi_k) + \xi_k(t). \tag{12.16}$$

The frequencies of all oscillators are equal, so it is convenient to introduce the phases in the rotating frame

$$\psi_k = \phi_k - \omega_0 t$$

to obtain

$$\frac{d\psi_k}{dt} = \frac{\varepsilon}{N} \sum_{j=1}^{N} \sin(\psi_j - \psi_k) + \xi_k(t). \tag{12.17}$$

Below, the noisy terms are supposed to be Gaussian with zero mean, $\delta$-correlated in time and independent for different oscillators:

$$\langle \xi_n \rangle = 0, \qquad \langle \xi_m(t)\xi_n(t') \rangle = 2\sigma^2 \delta(t - t')\delta_{mn}.$$

Qualitatively, possible synchronization phenomena can be described as follows. There are two factors affecting the phase: noise tends to make the distribution of phases

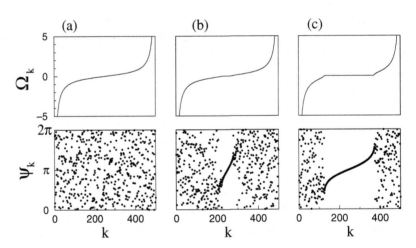

**Figure 12.1.** Dynamics of a population of 500 phase oscillators governed by Eq. (12.1). The distribution of natural frequencies is the Lorentzian one (12.13) with $\gamma = 0.5$ and $\bar{\omega} = 0$; the critical value of coupling is $\varepsilon_c = 1$. (a) Subcritical coupling $\varepsilon = 0.7$. The oscillators are not synchronized, the mean field fluctuates (due to finite-size effects) around zero. (b) Nearly critical situation $\varepsilon = 1.01$. A very small part of the population near the central frequency is synchronized. The observed frequencies $\Omega_k = \langle \dot{\phi}_k \rangle$ are the same for these entrained oscillators. (c) $\varepsilon = 1.2$, a large part of the population is synchronized, the mean field is large. The amplitude of the mean field is $K \approx 0.1$ for (b) and $K \approx 0.41$ for (c), in good agreement with the formula (12.14).

in the population uniform, so it reduces the mean field. Interaction means attraction of the phases which tend to form a cluster and to produce a nonzero mean field. For $\varepsilon/\sigma^2 \to 0$ the noise is stronger and the tendency to noncoherence wins, while for $\varepsilon/\sigma^2 \to \infty$ the interaction wins and the phases adjust their values. We expect to obtain a synchronization transition at some critical coupling strength $\varepsilon_c$.

Similarly to (12.2) we introduce the mean field according to

$$Z = X + iY = \frac{1}{N} \sum_{k=1}^{N} e^{i\psi_k} \tag{12.18}$$

and rewrite system (12.17) as

$$\frac{d\psi_k}{dt} = \varepsilon(-X \sin \psi_k + Y \cos \psi_k) + \xi_k(t). \tag{12.19}$$

The goal of the theory is to write a self-consistent equation for the distribution of the phases. Suppose that the mean field $Z$ is a slow (compared to the noise) function of time, and thus can be considered as a deterministic term in the Langevin equation. Then (12.19) are the Langevin equations for an individual noisy oscillator, similar to Eq. (9.7). The corresponding Fokker–Planck equation for the probability density of the phase is

$$\frac{\partial P(\psi, t)}{\partial t} = \varepsilon \frac{\partial}{\partial \psi} [(X \sin \psi - Y \cos \psi)P] + \sigma^2 \frac{\partial^2 P}{\partial \psi^2}. \tag{12.20}$$

Now we explore the thermodynamic limit $N \to \infty$. In this limit the population average (12.18) can be replaced with the average over the distribution $P(\psi, t)$:

$$Z = X + iY = \int_0^{2\pi} d\psi \, P(\psi, t) \, e^{i\psi}. \tag{12.21}$$

Equations (12.20) and (12.21) represent the final self-consistent system of equations for the unknown distribution function and the mean field. Note that this system is nonlinear as the factors $X$ and $Y$ in (12.20) depend on the $P(\psi, t)$ according to Eq. (12.21).

To analyze the system, we expand the density $P$ into a Fourier series

$$P(\psi, t) = \frac{1}{2\pi} \sum_l P_l(t) e^{il\psi}. \tag{12.22}$$

Note that according to Eq. (12.21) the mean field is exactly the complex amplitude of the first mode $Z = X + iY = P_1^* = P_{-1}$ and, because of normalization, the amplitude of the zero's mode is one, $P_0 = 1$. Substituting (12.22) in (12.20) and separating the Fourier harmonics yields an infinite system of ordinary differential equations

$$\frac{dP_l}{dt} = -\sigma^2 l^2 P_l + \frac{l\varepsilon}{2}(P_{l-1}P_1 - P_{l+1}P_1^*). \tag{12.23}$$

Let us write down the first three equations

$$\dot{P}_1 = \frac{\varepsilon}{2}(P_1 - P_2 P_1^*) - \sigma^2 P_1, \tag{12.24}$$

$$\dot{P}_2 = \varepsilon(P_1^2 - P_3 P_1^*) - 4\sigma^2 P_2, \tag{12.25}$$

$$\dot{P}_3 = \frac{3\varepsilon}{2}(P_2 P_1 - P_4 P_1^*) - 9\sigma^2 P_3. \tag{12.26}$$

First we note that the homogeneous distribution of phases, where all Fourier modes (except for $P_0$) vanish, is a solution of this system. Linearizing around this state we see that the only potentially unstable mode is the first one: it is stable if $\varepsilon < 2\sigma^2$ and unstable if $\varepsilon > 2\sigma^2$. This is exactly the critical value of the coupling, and the unstable mode is the mean field $P_1 = Z^*$. To obtain a steady state beyond the instability threshold, we have to take nonlinear terms into consideration. It is helpful to note that near the threshold $\varepsilon \approx 2\sigma^2$ the second mode decays rather quickly compared to the characteristic time scale of instability (i.e., $|\varepsilon - 2\sigma^2| \ll \sigma^2$). Moreover, we can estimate $|P_2| \sim |P_1|^2$ (from (12.25)) and $|P_3| \sim |P_1|^3$ (from (12.26)). Thus, setting $\dot{P}_2 \approx 0$, $\dot{P}_3 \approx 0$ is a good approximation that allows us to express $P_2$ through $P_1$ algebraically and to get

$$\dot{Z} = \left(\frac{\varepsilon}{2} - \sigma^2\right) Z - \frac{\varepsilon^2}{8\sigma^2}|Z|^2 Z. \tag{12.27}$$

This is the normal form equation for the Hopf bifurcation (in the theory of hydrodynamic instability it is also called the Landau–Stuart equation) describing the appearance of a macroscopic mean field in a population of noisy coupled oscillators. Its stationary solution is

$$|Z|^2 = (\varepsilon - 2\sigma^2)\frac{4\sigma^2}{\varepsilon^2}. \tag{12.28}$$

The mean field grows at the transition as a square root of criticality. This property of the population of noisy oscillators again illustrates analogy with the mean field theory of phase transitions. The phase of the mean field can be arbitrary; in the thermodynamic limit it is constant (in the rotating reference frame).

## 12.3    Generalizations

We have described two basic reasons for noncoherence in a large population of oscillators: distribution of natural frequencies and external noise. A number of related problems, where, e.g., both these factors are present, have attracted much interest recently. In this section we review some of the findings.

### 12.3.1    Models based on phase approximation

#### Noise and distribution of frequencies

A natural generalization of the models described in Sections 12.1 and 12.2 is a combination of the two main sources of disorder: external noise and distribution of natural frequencies. The model is:

$$\frac{d\phi_k}{dt} = \omega_k + \xi_k(t) + \frac{\varepsilon}{N} \sum_{j=1}^{N} \sin(\phi_j - \phi_k). \qquad (12.29)$$

The "one-oscillator" probability density now depends additionally on the frequency $\omega$: $P(\phi, \omega, t)$. The mean field can be defined by averaging over the distributions of phases and frequencies:

$$K e^{i\Theta} = \int_0^{2\pi} d\phi \int_{-\infty}^{\infty} d\omega \, e^{i\phi} P(\phi, \omega, t) g(\omega). \qquad (12.30)$$

This mean field governs the dynamics of an oscillator, so that the distribution function obeys the Fokker–Planck equation

$$\frac{\partial P}{\partial t} = -\frac{\partial}{\partial \phi}[(\omega + \varepsilon K \sin(\Theta - \phi))P] + \sigma^2 \frac{\partial^2 P}{\partial \phi^2}. \qquad (12.31)$$

The system of equations (12.30) and (12.31) gives a self-consistent description of the problem. In general, for small coupling $\varepsilon$ the noncoherent steady state $P(\phi, \omega) = 1/2\pi$, $K = 0$ is stable. As the coupling increases, a transition to a synchronized state with nonzero mean field occurs. The nature of the transition depends on a particular form of the distribution function $g(\omega)$ – it can be a soft (supercritical) bifurcation as described in Section 12.2 or a hard (subcritical) one (see Bonilla et al. [1992]; Acebrón et al. [1998]).

### Hysteretic synchronization transition

Another possible generalization is to consider phase dynamics with inertia, writing instead of (12.29) a system of second-order Langevin equations for coupled rotators

$$m\frac{d^2\phi_k}{dt^2} + \frac{d\phi_k}{dt} = \omega_k + \xi_k(t) + \frac{\varepsilon}{N} \sum_{j=1}^{N} \sin(\phi_j - \phi_k). \qquad (12.32)$$

The transition between nonsynchronous and synchronous states now demonstrates hysteresis [Tanaka et al. 1997a,b; Hong et al. 1999c]. This is closely related to the fact that a single driven rotator governed by equations

$$m\frac{d^2\Psi}{dt^2} + \frac{d\Psi}{dt} + a \sin \Psi = I$$

demonstrates bistability: both a stable rotating solution and a stable steady state coexist in a certain range of the driving torque $I$.

### General coupling function: clusters

In Sections 12.1 and 12.2 only the simplest attractive coupling proportional to the sine of phase difference has been considered.[2] Here we will demonstrate that more complex

---

[2] The sine coupling can be also generalized to the case of a preferred phase shift between the oscillators; we will discuss this case when considering coupled Josephson junctions, see Eq. (12.43) later.

coupling functions can lead to a further complication of the collective dynamics. Okuda [1993] has shown that a general coupling function $q(\phi)$ between identical (i.e., having the same natural frequency) phase oscillators can result in the formation of several clusters. All oscillators in a cluster have the same phase, and there is a constant phase shift between different clusters. The model reads

$$\frac{d\phi_k}{dt} = \omega_0 + \frac{\varepsilon}{N} \sum_{j=1}^{N} q(\phi_j - \phi_k). \qquad (12.33)$$

If the periodic (i.e., $q(\phi) = q(\phi+2\pi)$) coupling function $q$ contains higher harmonics, formation of clusters may be observed for some initial conditions. One such example is demonstrated in Fig. 12.2.

### General coupling function: order function and noise

Daido [1992a, 1993a, 1995, 1996] introduced the concept of "order function" to describe oscillators of type (12.33) with a distribution of natural frequencies:

$$\frac{d\phi_k}{dt} = \omega_k + \frac{\varepsilon}{N} \sum_{j=1}^{N} q(\phi_j - \phi_k). \qquad (12.34)$$

The coupling function $q$ can be, in general, represented as a Fourier series

$$q(\phi) = \sum_{l} q_l e^{i2\pi l\phi}.$$

Supposing that the phases of all synchronous oscillators rotate with a frequency $\bar{\omega}$, we can introduce generalized order parameters as

$$Z_l = \frac{1}{N} \sum_{k=1}^{N} e^{i2\pi l(\phi_k - \bar{\omega}t)},$$

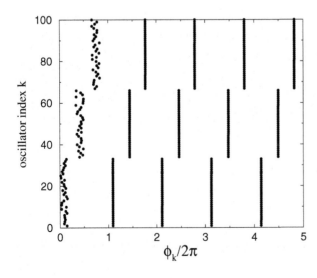

**Figure 12.2.** Dynamics of population of 100 coupled oscillators described by Eq. (12.33) with $q(\phi) = -c^{-1}\tan^{-1}[(c\sin\phi)/(1 - c\cos\phi)]$ for $c = -0.7$ and $\varepsilon = 1$. The population evolves into a three-cluster state, shown stroboscopically at times $n2\pi/\omega_0$. To reach this state we have to choose appropriate initial conditions; the system is multistable and can demonstrate different patterns of synchronization.

and rewrite the equations of motion as

$$\frac{d\phi_k}{dt} = \omega_k - \varepsilon H(\phi_k - \bar{\omega}t),$$

where

$$H(\psi) = -\sum_l q_l Z_l e^{-i2\pi l\psi}.$$

The function $H$ is the mean force that acts on each oscillator, and it is called the order function. It is a generalization of the mean field used by Kuramoto in his analysis of model (12.1). A nonzero order function is an indication of synchronization in the population. Daido [1992a, 1993a, 1995, 1996] has shown analytically that near the synchronization threshold, the norm of the order function is proportional to the bifurcation parameter (and not to the square root of it as in (12.15)):

$$\|H\| \sim \varepsilon - \varepsilon_c.$$

This result demonstrates that the square-root law (12.15) derived by Kuramoto for his model (12.1) is not valid for a general coupling function. Crawford [1995], who included external noise in the model (12.34), came to the same conclusion. His main result is that the amplitude $P_l$ of the $l$th Fourier mode of the distribution, that arises at the synchronization threshold, obeys

$$|P_l| \sim \sqrt{(\varepsilon - \varepsilon_c)(\varepsilon - \varepsilon_c + l^2\sigma^2)}.$$

Thus, in the presence of noise, the amplitudes $P_l$ (which play the role of the order parameter here) grow proportionally with $\sqrt{\varepsilon - \varepsilon_c}$, but for vanishing noise this growth is slower, proportionally with $(\varepsilon - \varepsilon_c)$, in accordance with Daido's results.

## 12.3.2 Globally coupled weakly nonlinear oscillators

A population of globally coupled weakly nonlinear oscillators is modeled with a system (cf. Eqs. (8.12) and (11.14))

$$\frac{dA_k}{dt} = (\mu + i\omega_k)A_k - (\gamma + i\alpha)|A_k|^2 A_k + \frac{\beta + i\delta}{N}\sum_{j=1}^{N}(A_j - A_k). \quad (12.35)$$

The simplest case is that of isochronous oscillators ($\alpha = 0$) with dissipative coupling ($\delta = 0$); it corresponds to the attraction of phases and the synchronization transition in such an ensemble is similar to that in a population of the phase oscillators (see [Matthews and Strogatz 1990; Matthews *et al.* 1991] for details). Therefore we mention here only the features that do not appear in the phase approximation.

### Oscillation death (quenching)

If the coupling constant $\beta$ and the width of the distribution of natural frequencies $\omega_k$ are large, the zero-amplitude state $A_k = 0$ is stable. Qualitatively, this can be explained as follows. In the case of a wide distribution all oscillators have significantly different frequencies and thus their influence on other oscillators is relatively small. On the other hand, the diffusive-type coupling in (12.35) brings additional damping $\sim\beta A_k$ which compensates the growth term $\mu A_k$ and makes the state $A_k = 0$ stable (see [Ermentrout 1990; Mirollo and Strogatz 1990a] for details).

### Collective chaos

In some range of parameters, the mean field defined as $Z = N^{-1} \sum_1^N A_k$ demonstrates behavior that is irregular in time. Matthews and Strogatz [1990] observed this for dissipatively coupled isochronous oscillators with a distribution of natural frequencies; later Hakim and Rappel [1992] and Nakagawa and Kuramoto [1993, 1995] found and studied chaotic dynamics of the mean field in a population of identical nonisochronous oscillators with dissipative and reactive coupling.

## 12.3.3  Coupled relaxation oscillators

Populations of coupled relaxation oscillators are often used for modeling neuronal ensembles (see, e.g., [Hoppensteadt and Izhikevich 1997; Tass 1999]). An individual neuron can be considered an integrate-and-fire oscillator of the type described in Section 8.3, and typically one neuron affects many others through its synapses. One popular model of globally coupled relaxation oscillators was proposed by Mirollo and Strogatz [1990b]; it simply generalizes the model of two interacting integrate-and-fire systems discussed in Section 8.3.

Each of the identical oscillators is described by a variable $x_i$ obeying, in the integrating regime, the equation

$$\frac{dx_k}{dt} = S - \gamma x_k.$$

When the oscillator reaches the threshold $x_k = 1$, it fires, i.e., the variable $x_k$ is reset to zero, and all other variables $x_j$, $j \neq k$ are pulled by an amount $\varepsilon/N$.[3] As the other oscillators are pulled, they can reach the threshold as well, and fire at the same instant of time. We assume, however, that the oscillator at the state $x = 0$ (i.e., just after firing) cannot be affected by the others, so that the state $x = 0$ is absorbing. This latter property ensures the possibility of perfect synchronization: if two oscillators fire at one moment in time, their future dynamics are identical. In general, one cannot exclude the existence of nonsynchronous configurations, but Mirollo and Strogatz [1990b] proved rigorously that the set of all initial conditions that remain nonsynchronous has zero

---

[3] Here again we normalize coupling by the number of oscillators in order to get reasonable behavior in the thermodynamic limit $N \to \infty$.

measure. Thus, with probability one, perfect phase locking, where all oscillators fire simultaneously and periodically, starts in the population. These results are valid for any $N \geq 2$. We illustrate a transition from initially random phases to perfect locking in the Mirollo–Strogatz model in Fig. 12.3.

### 12.3.4  Coupled Josephson junctions

Here we demonstrate that a series array of identical Josephson junctions can be considered as a system of globally coupled rotators. The coupling is ensured by a parallel load, as is shown in Fig. 12.4.

To write down the equations of the system we recall the main properties of the Josephson junction (see Section 7.4 and [Barone and Paterno 1982; Likharev 1991]). Each junction is characterized by the angle $\Psi_k$; the superconducting current is $I_c \sin \Psi_k$, and the junction voltage is $V_k = \dot{\Psi}\hbar/2e$. The current through all the

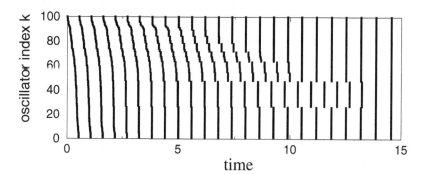

**Figure 12.3.** Dynamics of a population of 100 coupled integrate-and-fire oscillators. The model is described in the text, the values of parameters are $S = 2$, $\gamma = 1$, $\varepsilon = 0.2$. The firing events of each oscillator are shown with dots. For presentation we have sorted the array of variables, so that a set of dots appears as a (broken) line. One can see how the clusters are formed from oscillators with closed phases.

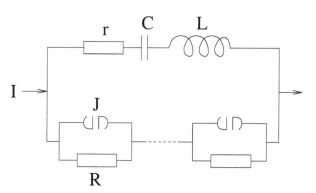

**Figure 12.4.** A series array of many Josephson junctions $J$, coupled by virtue of a parallel $RLC$-load. Capacitances of the junctions are neglected, only the resistance $R$ parallel to a junction is taken into account; this corresponds to a model of a resistively shunted Josephson junction.

junctions is the same, so that we can write

$$\frac{\hbar}{2eR}\frac{d\Psi_k}{dt} + I_c \sin\Psi_k = I - \frac{dQ}{dt}, \tag{12.36}$$

where $\dot{Q}$ is the current through the parallel $RLC$-load. The equation for the load

$$L\frac{d^2Q}{dt^2} + r\frac{dQ}{dt} + \frac{Q}{C} = \frac{\hbar}{2e}\sum_{k=1}^{N}\frac{d\Psi_k}{dt} \tag{12.37}$$

completes the equations of motion. One Josephson junction is equivalent to a rotator, and the system (12.36) and (12.37) is a system of globally coupled rotators. The coupling does not appear directly in the equations of motion of each rotator, because the "global variable" $Q$ has inertia and is governed by a separate equation.

We now demonstrate, following Wiesenfeld and Swift [1995], that for small coupling this system is equivalent to the Kuramoto model (Section 12.1). Let us consider the case of a large external current $I$. The mean voltages across all junctions are then nonzero (all rotators rotate), and we can define the phases of all junctions according to our definition of the phase as a variable that moves on a limit cycle with a constant velocity (see Section 7.1). Uncoupled junctions are governed by Eq. (12.36) with $\dot{Q} = 0$, and the transformation to the uniformly rotating phase $\phi$ can be written explicitly:

$$\phi_k = 2\tan^{-1}\left[\sqrt{\frac{I - I_c}{I + I_c}}\tan\left(\frac{\Psi_k}{2} + \frac{\pi}{4}\right)\right]. \tag{12.38}$$

Calculating the time derivative of $\phi_k$ by virtue of Eq. (12.36) and using the identity

$$I - I_c \sin\Psi = (I^2 - I_c^2)/(I - I_c \cos\phi) \tag{12.39}$$

that follows from (12.38), we obtain equations for the phases

$$\frac{d\phi_k}{dt} = \omega_0 - \dot{Q}\frac{\omega_0(I - I_c\cos\phi_k)}{I^2 - I_c^2}. \tag{12.40}$$

Here we denote the frequency of uncoupled rotations $2eR\sqrt{I^2 - I_c^2}/\hbar$ with $\omega_0$.

So far no approximations have been made, and Eq. (12.40) is exact. We now use the method of averaging similar to that described in Sections 7.1 and 8.1. In the zero approximation all phases $\phi_k$ rotate with the same frequency $\omega_0$: $\phi_k = \omega_0 t + \phi_k^0$. The $RLC$-load is linear, so that we can consider contributions $Q_k$ from different junctions in Eq. (12.37) separately. Each junction drives the load periodically; this periodic force is, however, not sinusoidal but more complex, because the angle variable $\Psi$ is obtained from the linearly rotating phases $\phi$ via the nonlinear transformation (12.38). Thus the force in Eq. (12.37) has many harmonics, and the response of the linear load can be calculated from Eq. (12.37) for each of them:

$$Q_k = \sum_{n=0}^{\infty} Q_{kn} \cos(n\phi_k - \beta_n).$$ (12.41)

Let us consider the main component $n = 1$. The equation of motion for $Q_{k1}$ is obtained after substitution of $d\Psi/dt$ in (12.37) according to (12.36), and expressing $\sin \Psi$ in terms of $\phi$ with the help of Eq. (12.39):

$$L\ddot{Q}_{k1} + (r + NR)\dot{Q}_{k1} + \frac{Q_{k1}}{C} = \left\langle\!\left\langle R\frac{I^2 - I_c^2}{I - I_c \cos(\omega_0 t + \phi_k^0)}\right\rangle\!\right\rangle$$

$$= RI_c^{-1}(I^2 - I_c^2 - I\sqrt{I^2 - I_c^2})\cos(\omega_0 t + \phi_k^0).$$ (12.42)

Here $\langle\!\langle \cdot \rangle\!\rangle$ denotes the first harmonic of the periodic function. The solution of the linear Eq. (12.42) is

$$Q_{k1} = R\frac{I^2 - I_c^2 - I\sqrt{I^2 - I_c^2}}{I_c\sqrt{(1/C - L\omega_0^2)^2 + (r + NR)^2\omega_0^2}} \cos(\omega_0 t + \alpha + \phi_k^0),$$

where

$$\cos \alpha = \frac{L\omega_0^2 - 1/C}{\sqrt{(1/C - L\omega_0^2)^2 + (r + NR)^2\omega_0^2}}.$$

Now we can substitute this solution in (12.40), and average over the period of the fast rotations $2\pi/\omega_0$. Also, we assume now that the phases $\phi_k^0$ are slow functions of time. It is easy to see that higher harmonics $n > 1$ give no contribution to the averaged equations, and, as a result, we obtain

$$\frac{d\phi_k}{dt} = \omega_0 + \frac{\varepsilon}{N}\sum_{j=1}^{N}\sin(\phi_j - \phi_k - \alpha),$$ (12.43)

where

$$\varepsilon = N\frac{2eR^2I\omega_0/\hbar - R\omega_0^2}{\sqrt{(1/C - L\omega_0^2)^2 + (r + NR)^2\omega_0^2}}.$$

The resulting equations coincide with the phase model of Kuramoto (12.1), the only difference being that the interaction term has a slightly more general form. The angle $\alpha$ in the interaction term depends on the properties of the load. If $\alpha = 0$, the interaction between the junctions is attractive, while for $\alpha \neq 0$ each pair of junctions prefers to have a finite phase shift. Nevertheless, even for $\alpha \neq 0$ we can look for an in-phase solution $\phi_1 = \phi_2 = \cdots = \phi_N$. This solution has a frequency that differs from $\omega_0$, and is stable if $\varepsilon \cos \alpha > 0$ (linearization of (12.43) gives a simple matrix with one zero eigenvalue and $N - 1$ eigenvalues $\varepsilon \cos \alpha$). In the case of instability of the phase-synchronized state, another regime with uniformly distributed phases (i.e., $\phi_k^0 = 2\pi k/N$) appears. This so-called splay state is neutrally stable (see [Strogatz and Mirollo 1993; Watanabe and Strogatz 1993, 1994] for details).

If one takes into account small disorder of the Josephson junctions (e.g., due to distribution of critical currents $I_c$), then one obtains a population with different natural frequencies. The synchronization of such Josephson junctions can be thus described as a particular example of the Kuramoto transition at a finite coupling constant $\varepsilon$ [Wiesenfeld *et al.* 1996].

### 12.3.5  Finite-size effects

The synchronization transition in a population of oscillators is expected to be sharp in the thermodynamic limit $N \to \infty$. For finite ensembles one observes finite-size effects, similar to those in statistical mechanics [Cardy 1988]. The main idea is that the finiteness of the population results in fluctuations of the mean field of the order $\sim N^{-1/2}$. For example, Pikovsky and Ruffo [1999] argued that finite ensembles of noise-driven phase oscillators can be described by Eq. (12.27) with an additional fluctuating term:

$$\dot{Z} = \left(\frac{\varepsilon}{2} - \sigma^2\right)Z - \frac{\varepsilon^2}{8\sigma^2}|Z|^2 Z + \eta_1(t) + i\eta_2(t),$$

$$\langle \eta_i(t)\eta_j(t') \rangle = \frac{2d^2}{N}\delta_{ij}\delta(t - t').$$

(12.44)

The noise term is proportional to $1/\sqrt{N}$ and disappears in the thermodynamic limit. Its influence on the dynamics of the mean field can be easily understood if one interprets Eq. (12.44) as the equation for a weakly nonlinear noisy self-sustained oscillator (see, e.g., [Stratonovich 1963]). On the phase plane $X = \text{Re}(Z)$, $Y = \text{Im}(Z)$ we get a smeared limit cycle, and both the amplitude and the phase of the mean field fluctuate, see Fig. 12.5.

### 12.3.6  Ensemble of chaotic oscillators

The phase dynamics of a chaotic oscillator can be similar to those of a noisy periodic oscillator (see Chapter 10). Correspondingly, synchronization in a population of globally coupled chaotic oscillators is similar to the appearance of coherence in an ensemble of noisy oscillators, as described in Sections 12.2 and 12.3.

As an example we present here the results of numerical simulations of a population of globally coupled *identical* Rössler oscillators [Pikovsky *et al.* 1996]:

$$\dot{x}_k = -y_k - z_k + \varepsilon X,$$

$$\dot{y}_k = x_k + a y_k,$$ (12.45)

$$\dot{z}_k = 0.4 + z_k(x_k - 8.5).$$

Here the coupling is realized via the mean field

$$X = \frac{1}{N}\sum_{k=1}^{N} x_k, \qquad Y = \frac{1}{N}\sum_{k=1}^{N} y_k.$$ (12.46)

The mean field vanishes in the nonsynchronous regime, and demonstrates rather regular oscillations beyond the synchronization transition, which in this system occurs at $\varepsilon \approx 0.025$. Remarkably, in the synchronous state each oscillator in the population remains chaotic; coherence appears only due to phase synchronization. We illustrate this in Fig. 12.6, where the phase portraits of one oscillator from the ensemble and of the mean field are presented. The amplitude of the mean field is relatively small, but definitely larger than the fluctuations due to the finite size (these fluctuations are presumably responsible for the amplitude modulation of the mean field).

*Nonidentical* chaotic oscillators can also synchronize. Different natural frequencies can be included in the model (12.45) by introducing an additional parameter governing the time scale of rotations:

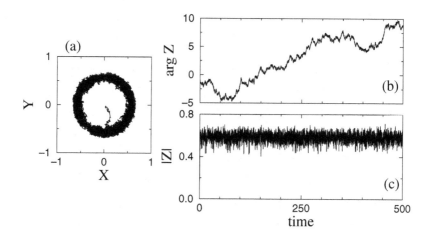

**Figure 12.5.** Evolution of the mean field $Z = X + iY$ in a system of 500 noisy phase oscillators (see Eq. (12.16)). (a) The phase portrait in coordinates $(X, Y)$: after initial transients the trajectory fills a ring of a width proportional to $N^{-1/2}$. (b,c) The time dependence of the phase and the amplitude of the mean field $Z(t)$. From Pikovsky and Ruffo, *Physical Review* E, Vol. 59, 1999, pp. 1633–1636. Copyright 1999 by the American Physical Society.

$$\dot{x}_k = -\omega_k y_k - z_k + \varepsilon X,$$

$$\dot{y}_k = \omega_k x_k + a y_k, \qquad\qquad\qquad (12.47)$$

$$\dot{z}_k = 0.4 + z_k(x_k - 8.5).$$

This model is similar to Eq. (12.29) because here both the distribution of natural frequencies $\omega_k$ and noise (due to internal chaotic dynamics) is present. The synchronization transition in system (12.47) is illustrated in Fig. 12.7. The calculation of the observed frequencies $\Omega_k$ demonstrates that for $\varepsilon = 0.1$ most of the oscillators are mutually synchronized. Additionally, we depict the successive maxima of the variable $x_k$ for each oscillator. These maxima have a broad distribution both below and above the synchronization threshold, which means that the oscillators remain chaotic although they are phase synchronized.

## 12.4  **Bibliographic notes**

Studies of large populations of oscillators have a relatively short history: they became popular only with the appearance of sufficiently powerful computers. Among early works we mention that of Winfree [1967], who also gave a review of underlying biological observations, and of Pavlidis [1969]. Starting from the works of Kuramoto [1975, 1984], who introduced and solved the phase model described in Section 12.1, the problem attracted great interest. Different mathematical approaches have been developed by Strogatz *et al.* [1992], van Hemmen and Wreszinski [1993], Watanabe and Strogatz [1993], Watanabe and Strogatz [1994] and Acebrón *et al.* [1998]. Okuda [1993], Daido [1992a, 1993b,a, 1995, 1996], Crawford [1995], Crawford and Davies [1999], Strogatz [2000] and Balmforth and Sassi [2000] considered general coupling

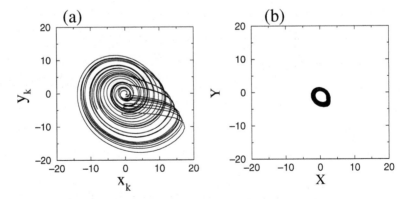

**Figure 12.6.** The projections of the phase portraits of one oscillator (a) and of the mean field (12.46) (b) in the ensemble (12.45) with $a = 0.25$. We note that for this parameter value the Rössler system has a so-called funnel attractor, topologically different from that shown in Fig. 10.1a. The fluctuations of the mean field are presumably due to finite size of the ensemble, $N = 10\,000$.

functions. Ensembles of noisy phase oscillators were studied by Strogatz and Mirollo [1991], Bonilla *et al.* [1992], Bonilla *et al.* [1998], Hansel *et al.* [1993], Crawford [1994], Stange *et al.* [1998, 1999], Hong *et al.* [1999a] and Reimann *et al.* [1999]. A phase shift in coupling function, or, almost equivalently, a delay can change the dynamics, as was discussed by Sakaguchi and Kuramoto [1986], Christiansen *et al.* [1992], Yeung and Strogatz [1999], S. H. Park *et al.* [1999a], Reddy *et al.* [1999] and Choi *et al.* [2000]. Random phase shifts in the coupling can produce glassy states (i.e., states with very many stable configurations) [Daido 1992b; Bonilla *et al.* 1993; Park *et al.* 1998]. Hoppensteadt and Izhikevich [1999] demonstrated that a Kuramoto system with a quasiperiodic multiplicative forcing may operate like a neural network. Phase oscillators with inertia demonstrate a first-order transition with hysteresis [Tanaka *et al.* 1997a,b; Hong *et al.* 1999c,b]. Finite-size effects were described by Dawson and Gärtner [1987], Daido [1990] and Pikovsky and Ruffo [1999].

Globally coupled Josephson junctions (or, equivalently, rotators), with and without noise were considered by Shinomoto and Kuramoto [1986], Sakaguchi *et al.* [1988b],

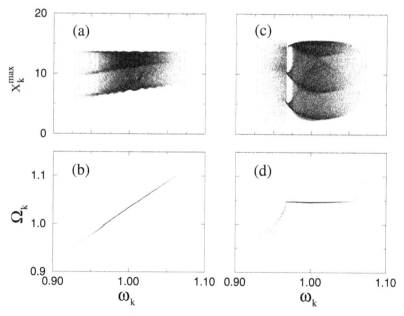

**Figure 12.7.** Successive maxima $x_k^{\max}$ and observed frequencies $\Omega_k$ vs. natural frequencies $\omega_k$ in an ensemble of 5000 coupled Rössler systems (12.47) with $a = 0.15$. The natural frequencies are distributed according to the Gaussian law with mean square value $\Delta\omega = 0.02$. (a,b) The coupling $\varepsilon = 0.05$ is slightly below the transition threshold. The mean field is nearly zero and the observed frequencies coincide with the natural ones. (c,d) Above the threshold ($\varepsilon = 0.1$) most of the oscillators form a coherent state (plateau in (d)), while the amplitudes remain chaotic (with the exception of the period-3 window for $\omega \approx 0.97$). From [Pikovsky *et al.* 1996].

Strogatz *et al.* [1989], Golomb *et al.* [1992], Wiesenfeld *et al.* [1996], Tsang *et al.* [1991b] and Wiesenfeld [1992]. In particular, splay states in this system have attracted much interest [Tsang *et al.* 1991a; Nichols and Wiesenfeld 1992; Swift *et al.* 1992; Strogatz and Mirollo 1993].

Ensembles of weakly nonlinear oscillators were studied by Yamaguchi and Shimizu [1984], Bonilla *et al.* [1987], Mirollo and Strogatz [1990a], Matthews and Strogatz [1990] and Matthews *et al.* [1991]. Some effects here, e.g., collective chaos [Hakim and Rappel 1992; Nakagawa and Kuramoto 1993, 1994, 1995; Banaji and Glendinning 1994] and oscillation death [Ermentrout 1990] do not occur for phase oscillators. Close to these works are studies of coupled laser modes [Winful and Rahman 1990].

Relaxation oscillators demonstrate a large variety of effects [Kuramoto *et al.* 1992; Abbott and van Vreeswijk 1993; Tsodyks *et al.* 1993; Wang *et al.* 1993; Chen 1994; Bottani 1995, 1996; Ernst *et al.* 1995, 1998; Gerstner 1995; Rappel and Karma 1996; van Vreeswijk 1996; Kirk and Stone 1997; Bressloff and Coombes 1999; Wang *et al.* 2000b]. Finally, we mention that Rogers and Wille [1996] studied lattices of oscillators with long-range coupling; here one can tune between local and global coupling.

# Part III

Synchronization of chaotic systems

# Chapter 13

# Complete synchronization I: basic concepts

The goal of this chapter is to describe the basic properties of the complete synchronization of chaotic systems. Our approach is the following: we take a system as simple as possible and describe it in as detailed a way as possible. The simplest chaotic system is a one-dimensional map, it will be our example here. We start with the construction of the coupled map model, and describe phenomenologically, what complete synchronization looks like. The most interesting and nontrivial phenomenon here is the synchronization transition. We will treat it as a transition inside chaos, and follow a twofold approach. On one hand, we exploit the irregularity of chaos and describe the transition statistically. On the other hand, we explore deterministic regular properties of the dynamics and describe the transition topologically, as a bifurcation. We hope to convince the reader that these two approaches complement each other, giving the full picture of the phenomenon. In the next chapter, where we discuss many generalizations of the simplest model, we will see that the basic features of complete synchronization are generally valid for a broad class of chaotic systems.

The prerequisite for this chapter is basic knowledge of the theory of chaos, in particular of Lyapunov exponents. For the analytical statistical description we use the thermodynamic formalism, while for the topological considerations a knowledge of bifurcation theory is helpful. One can find these topics in many textbooks on nonlinear dynamics and chaos [Schuster 1988; Ott 1992; Kaplan and Glass 1995; Alligood *et al.* 1997; Guckenheimer and Holmes 1986] as well as in monographs [Badii and Politi 1997; Beck and Schlögl 1997].

## 13.1    The simplest model: two coupled maps

In this introductory section we demonstrate the effect of complete synchronization, taking the simple model of coupled maps as an example. A single one-dimensional map

$$x(t+1) = f(x(t)) \tag{13.1}$$

represents a dynamical system with discrete time $t = 0, 1, 2, \ldots$ and continuous state variable $x$. Well-known examples exhibiting chaotic behavior are the *logistic map* $f(x) = 4x(1-x)$ and the *tent map* $f(x) = 1 - 2|x|$.

Let us consider two such maps, described by the variables $x$, $y$, respectively. As the dynamics of each variable are chaotic, in the case of independent (noninteracting) systems one observes two independent random-like processes without any mutual correlation. Now let us introduce an interaction between the two systems. Of course, there is no unique way to couple two mappings: any additional term containing both $x$ and $y$ on the right-hand side of the equation will provide some coupling. We want, however, the coupling to have some particular physically relevant properties:

(i) the coupling is contractive, i.e., it tends to make the states $x$ and $y$ closer to each other;

(ii) the coupling does not affect the symmetric synchronous state $x = y$.

The first condition can be also called the dissipative one: it corresponds, e.g., to coupling of electronic elements through resistors.[1] A general form of the linear coupling operator, satisfying the above properties, is

$$\hat{L} = \begin{bmatrix} 1 - \alpha & \alpha \\ \beta & 1 - \beta, \end{bmatrix} \tag{13.2}$$

where $0 < \alpha < 1$, $0 < \beta < 1$. In the simplest case of symmetric coupling we set $\alpha = \beta = \varepsilon$ to get

$$\hat{L} = \begin{bmatrix} 1 - \varepsilon & \varepsilon \\ \varepsilon & 1 - \varepsilon. \end{bmatrix} \tag{13.3}$$

Now the linear coupling (13.3) should be combined with the nonlinear mapping (13.1). The proper way to do this is to alternate the application of the linear and nonlinear mappings, i.e., to multiply[2] the corresponding operators:

---

[1] See also the discussion of dissipative and reactive coupling in Section 8.2.

[2] Contrary to continuous-time dynamics, one should not *add* contributions of different factors, but *multiply* them. To clarify this point, which is important in an understanding of the physical meaning of discrete models, let us consider two ways of introducing additional dissipation to the mapping (13.1). Dissipative terms should decrease the variable $x$. An additive term like $x(t+1) = f(x(t)) - \gamma x(t)$ may or may not reduce $x$, depending on the signs and values of $f(x)$ and $\gamma$. Contrary to this, multiplication with a factor $|\gamma| < 1$ as in $x(t+1) = \gamma f(x(t))$ always reduces the absolute value of $x$.

$$\left[\begin{array}{c} x(t+1) \\ y(t+1) \end{array}\right] = \left[\begin{array}{cc} 1-\varepsilon & \varepsilon \\ \varepsilon & 1-\varepsilon \end{array}\right]\left[\begin{array}{c} f(x(t)) \\ f(y(t)) \end{array}\right]$$

$$= \left[\begin{array}{c} (1-\varepsilon)f(x(t)) + \varepsilon f(y(t)) \\ \varepsilon f(x(t)) + (1-\varepsilon)f(y(t)) \end{array}\right]. \tag{13.4}$$

Note that system (13.4) is completely symmetric with respect to exchange of the variables $x \leftrightarrow y$, due to our choice of symmetrically coupled identical subsystems.

We now describe qualitatively what is observed in the basic model (13.4) when the positive coupling parameter $\varepsilon$ changes. The limiting cases are clear: if $\varepsilon = 0$, the two variables $x$ and $y$ are completely independent and uncorrelated; if $\varepsilon = 1/2$, then even after one iteration the two variables $x$ and $y$ become identical and one immediately observes the state where $x(t) = y(t)$ for all $t$ (the value $\varepsilon = 1/2$ corresponds to the maximally strong coupling). As the coupling does not affect this state, the dynamics of $x$ and $y$ are the same as in the uncoupled systems, i.e., chaotic. Exactly such a regime, where each of the systems demonstrates chaos and their states are identical at each moment in time, is called **complete synchronization** (sometimes the terms "identical", "full", and "chaotic" are also used). Now, if the coupling parameter $\varepsilon$ is considered as a bifurcation parameter and increases gradually from zero, a complex bifurcation structure is generally observed (possibly including nonchaotic states), but one clearly sees a tendency to closer correlation between $x$ and $y$. One can find a critical coupling $\varepsilon_c < 1/2$, such that for $\varepsilon > \varepsilon_c$ the synchronous state $x = y$ is established. The easiest way to see this transition is to represent the dynamics in the plane $(x, y)$ (see Fig. 13.2 later). The points outside the diagonal $x = y$ represent the nonsynchronous state. With increasing of $\varepsilon$ the distribution of the points tends towards the diagonal, and beyond the critical coupling $\varepsilon_c$ all points satisfy $x = y$.

### Coupled skew tent maps

Here we illustrate the effect of complete synchronization using the skew tent map

$$f(x) = \begin{cases} x/a & \text{if } 0 \leq x \leq a, \\ (1-x)/(1-a) & \text{if } a \leq x \leq 1, \end{cases} \tag{13.5}$$

as an example (for $a = 0.5$ one obtains the usual tent map; this symmetric case is, however, degenerate as we will discuss below). A typical nonsymmetric case of this map is shown in Fig. 13.1; it stretches the unit interval and folds it twice. The sequence of stretchings and foldings leads to pure chaos, contrary to the other popular model, the logistic (parabolic) map, where chaotic and periodic states intermingle (as a function of some parameter). Moreover, the invariant measure for the single skew tent map is uniform, allowing us to obtain some analytical results in Section 13.3.

The synchronous and nonsynchronous states are illustrated in Fig. 13.2. The critical value of coupling here is $\varepsilon_c \approx 0.228$ for $a = 0.7$ (we will calculate this quantity below). The complete synchronization in Fig. 13.2a exists for strong couplings $1/2 > \varepsilon > \varepsilon_c$, while for weaker couplings asynchronous states are observed (Fig. 13.2b,c).

In Fig. 13.3 we illustrate the dynamics near the synchronization threshold, which will be the main focus of our attention in this chapter. To characterize the synchronization transition at $\varepsilon = \varepsilon_c$, it is convenient to introduce new variables

$$U = \frac{x+y}{2}, \qquad V = \frac{x-y}{2}. \tag{13.6}$$

Note that in the completely synchronous state the variable $V$ vanishes; for slightly asynchronous states it is small. Geometrically, the variable $U$ is directed along the diagonal $x = y$, while the variable $V$ corresponds to the direction transverse to this diagonal. A characteristic feature of the slightly asynchronous state near $\varepsilon_c$ is the intermittent behavior of the variable $V$, as is illustrated in Fig. 13.3. The rare but large bursts of $V$ are a characteristic feature of this **modulational intermittency** (sometimes called on–off intermittency). In the theoretical description below we will see that the logarithm of $|V|$ is a quantity for which the theory can be constructed, therefore the dynamics of $\ln|V|$ are also shown in Fig. 13.3.

## 13.2    Stability of the synchronous state

First, we would like to note that because the system (13.4) is symmetric with respect to the change of the variables (i.e., it remains invariant under the transformation $x \leftrightarrow y$), the synchronous state $x(t) = y(t)$ is a solution of (13.4) for all values of the coupling constant $\varepsilon$. This means that if the initial conditions are symmetric (i.e., $x(0) = y(0)$ or $V = 0$), symmetry is preserved in time. If we want the synchronous state to be observed not only for specific, but also for general initial states, we must impose the stability condition: the completely synchronized state $V = 0$ should be an attractor, i.e., synchronization should establish even from nonsymmetric initial

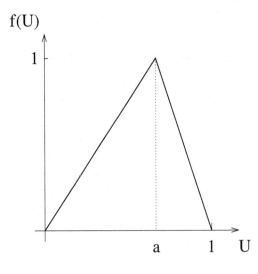

**Figure 13.1.** The skew tent map: a simple solvable model of one-dimensional chaotic dynamics.

states. This stability condition will give us the critical coupling $\varepsilon_c$ for the onset of synchronization. Because $V$ is a transverse variable, the stability of the synchronous state can be described as the *transverse* stability of the symmetric attractor.

Rewriting system (13.4) in the variables $U$ and $V$ (13.6) yields

$$U(t+1) = \tfrac{1}{2}\big[f(U(t)+V(t)) + f(U(t)-V(t))\big], \tag{13.7}$$

$$V(t+1) = \frac{1-2\varepsilon}{2}\big[f(U(t)+V(t)) - f(U(t)-V(t))\big]. \tag{13.8}$$

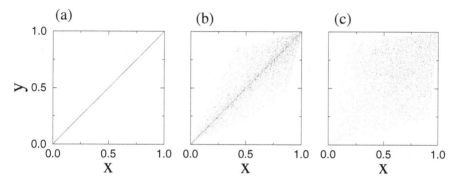

**Figure 13.2.** Attractors in the system of two coupled skew tent maps with $a = 0.7$. The synchronization threshold is $\varepsilon_c \approx 0.228$. (a) The synchronized state ($\varepsilon = 0.3$) fills the diagonal $x = y$. (b) The slightly asynchronous state ($\varepsilon = 0.2$) lies near the diagonal. (c) In the case of small coupling ($\varepsilon = 0.1$) the instant values of the variables $x$ and $y$ are almost uncorrelated. The transition from (a) to (b) is called symmetry-breaking, or blowout bifurcation.

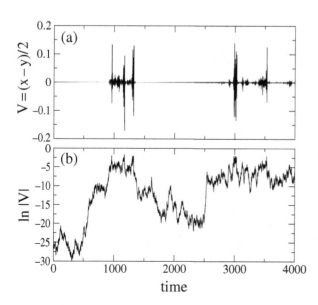

**Figure 13.3.** Modulational intermittency in coupled skew tent maps with $a = 0.7$. The coupling is slightly less than the critical one, $\varepsilon = \varepsilon_c - 0.001$. (a) The difference $V$ between the systems' states demonstrates typical bursts. (b) The logarithm of the difference exhibits a random walk pattern.

Next, we linearize this system near the completely synchronous chaotic state $U(t)$, $V = 0$ and obtain linear mappings for small perturbations $u$ and $v$

$$u(t + 1) = f'(U(t))u(t),\qquad (13.9)$$

$$v(t + 1) = (1 - 2\varepsilon) f'(U(t))v(t).\qquad (13.10)$$

This linear system should be solved together with the nonlinear mapping governing the synchronous chaotic dynamics

$$U(t + 1) = f(U(t)).\qquad (13.11)$$

Since in the linear approximation the perturbations $u$ and $v$ do not interact, the longitudinal (symmetric, $u$) and transverse (asymmetric, $v$) perturbations can be treated separately. The key idea to this treatment comes from the observation that the linearized equations (13.9) and (13.10) govern the growth and decay of the perturbations of a chaotic state, and this growth, respectively decay, is quantitatively measured by the Lyapunov exponents of the system. Indeed, a two-dimensional mapping has two Lyapunov exponents, and because in our cases the dynamics of $u$ and $v$ are separated, these exponents can be simply defined as the average logarithmic growth rates of $u$ and $v$:

$$\lambda_u = \lim_{t \to \infty} \frac{\ln|u(t)| - \ln|u(0)|}{t} = \langle \ln|f'(U)|\rangle,\qquad (13.12)$$

$$\lambda_v = \lim_{t \to \infty} \frac{\ln|v(t)| - \ln|v(0)|}{t} = \ln|1 - 2\varepsilon| + \langle \ln|f'(U)|\rangle.\qquad (13.13)$$

By invoking the ergodicity of the chaotic dynamics, we can replace the time average with the statistical average with respect to the invariant measure, denoting it with the brackets $\langle\rangle$. From Eq. (13.9) one can easily see that the symmetric perturbation $u$ is just the perturbation of the chaotic mapping (13.11), thus the longitudinal (not violating synchronization) Lyapunov exponent $\lambda_u$ is nothing else than the Lyapunov exponent $\lambda$ of the uncoupled chaotic system. The transverse Lyapunov exponent $\lambda_\perp \equiv \lambda_v$ is related to $\lambda$ as

$$\lambda_\perp = \ln|1 - 2\varepsilon| + \lambda.\qquad (13.14)$$

Thus, the average growth or decay of the transverse perturbation $v$ is governed by the transverse Lyapunov exponent $\lambda_\perp$: $\ln|v(t)| \propto \lambda_\perp t$, and the criteria for stability of the synchronous state can be formulated as follows:

$\lambda_\perp > 0$:   synchronous state is unstable,

$\lambda_\perp < 0$:   synchronous state is stable.

The stability threshold is then defined from the condition $\lambda_\perp = 0$:

$$\ln|1 - 2\varepsilon_c| = -\lambda,\qquad \varepsilon_c = \frac{1 - e^{-\lambda}}{2}.\qquad (13.15)$$

For example, for the logistic map, the Lyapunov exponent is $\lambda = \ln 2$, so $\varepsilon_c = 1/4$. We have considered stability on average, through the mean growth rate of perturbations. Relation of this notion to other stability characteristics (e.g., asymptotic stability) is highly nontrivial and is the subject of the presentation below.

## 13.3    Onset of synchronization: statistical theory

In this section we describe what happens at the synchronization threshold in statistical terms. The main idea is to consider the transverse perturbation $v$ as a noise-driven process, considering the synchronous chaos $U(t)$, $V(t) = 0$ as a driving random force. In doing this we neglect almost all dynamical (deterministic) features of chaos; not surprisingly the theory works better in the case of strong chaos. After presenting the theory we illustrate its predictions with the coupled skew tent maps.

### 13.3.1    Perturbation is a random walk process

The linear stability analysis performed in Section 13.2 suggests that the proper variable for the description of the dynamics of transverse perturbation near the synchronization threshold is the logarithm of this perturbation. Thus we introduce new (not independent, but useful) variables

$$w = |v|, \qquad z = \ln w.$$

Moreover, because the transverse Lyapunov exponent $\lambda_\perp$ fully determines the linear stability and completely describes the dependence of the dynamics on the coupling strength $\varepsilon$, we will use it as a bifurcation parameter and write the equations governing the perturbation dynamics as

$$w(t+1) = w(t)e^{\lambda_\perp}e^{g(U(t))}. \tag{13.16}$$

Here we have denoted[3]

$$g(U(t)) = \ln|f'(U)| - \lambda. \tag{13.17}$$

Note that the average of $g$ is zero, but instantaneous values of $g$ fluctuate. We can interpret Eqs. (13.17) and (13.16) as follows: the chaotic process $U(t)$ drives the variable $w$ in (13.16). The forcing here appears as the multiplicative term $\exp(g + \lambda_\perp)$. One can also say that the term $\exp(g + \lambda_\perp)$ modulates the growth rate of $w$: the variable $w$ grows if this term is larger than one and decreases otherwise.

---

[3] This will allow us to write the equations in the most general form, applicable not only for a description of the synchronization transition, but also for other cases of modulational intermittency.

The equation for the variable $z$ follows immediately from (13.16)

$$z(t + 1) = z(t) + g(U(t)) + \lambda_\perp. \tag{13.18}$$

Here the chaotic driving appears as the purely additive term $g + \lambda_\perp$.

We note also that from the mathematical point of view, the system of Eqs. (13.11) and (13.16) and (13.18) is the so-called *skew system*: the variable $U$ affects $w$ and $z$, but there is no influence of $w$ and $z$ on $U$. For our coupled chaotic mappings (13.4) we get a skew system only as an approximation, because we have linearized the equations near the completely synchronous state. However, skew systems naturally appear in physical situations with unidirectional coupling (see Section 14.1). From the physical point of view, the system (13.11), (13.16) and (13.18) is a drive–response system: it contains the driving component $U$ and the driven components $w$, $z$. The peculiarity of (13.11), (13.16) and (13.18) is that the driven subsystem (13.16) and (13.18) is linear.

The main idea in the statistical description of the synchronization onset is to consider the Eqs. (13.16) and (13.18) as *noise-driven* ones. So we consider the chaotic signal $U(t)$ as a random process, and interpret Eq. (13.16) as an equation with multiplicative noise, and Eq. (13.18) as an equation with additive noise. With this interpretation in mind, it becomes evident that Eq. (13.18) describes a *one-dimensional random walk* with a step $g + \lambda_\perp$. The mean value of the fluctuating term $g$ is zero, hence the transverse Lyapunov exponent $\lambda_\perp$ determines whether the walk is biased in the negative direction ($\lambda_\perp < 0$), so that the variables $z$ and $w$ decrease with time (on average), or the walk is biased in the positive direction ($\lambda_\perp > 0$) and the variables $z$ and $w$ grow with time (on average). At the synchronization threshold (13.15) the random walk is unbiased. Thus the value of $\lambda_\perp$ characterizes the directed motion of the random walk. Another important characteristic is the diffusion constant. If the values of $g(U(t))$ were independent random numbers, one could relate, using the law of large numbers, the diffusion coefficient of the random walk to the variance of $g$. In our case, when the process $U(t)$ is generated by the dynamical system (13.1), the values of $g(U(t))$ may be correlated, so one needs a careful treatment of the random walk dynamics.

### 13.3.2 The statistics of finite-time Lyapunov exponents determine diffusion

The solution of Eq. (13.18) can be formally written as

$$z(T) - z(0) - \lambda_\perp T = \sum_{t=0}^{T-1} g(U(t)). \tag{13.19}$$

On the r.h.s. we have a sum of chaotic quantities; to be able to apply the central limit theorem arguments we divide this sum by $T$ and denote

$$\Lambda_T = \frac{1}{T} \sum_{t=0}^{T-1} g(U(t)) = \frac{1}{T} \sum_{t=0}^{T-1} \ln |f'(U)| - \lambda. \tag{13.20}$$

The quantity $\Lambda_T$ is called the *finite-time* (or *local*) Lyapunov exponent (in our formulation (13.20) this quantity is shifted by $\lambda$ and thus tends to zero as $T \to \infty$; one can also define the unshifted finite-time exponent which tends to $\lambda$ as $T \to \infty$). For large $T$ the finite-time Lyapunov exponent $\Lambda_T$ converges to zero, but for us the fluctuations are mostly important. According to the central limit theorem, the probability distribution density of $\Lambda_T$ (for simplicity, below we omit the subscript $T$) should scale as

$$p(\Lambda; T) \propto \exp[T s(\Lambda)] \tag{13.21}$$

with a scaling function $s(\Lambda)$ (see, e.g., [Paladin and Vulpiani 1987; Ott 1992; Crisanti *et al.* 1993]). This function is a one-hump concave function with a single maximum at zero. The first term in the expansion near the maximum $s(\Lambda) \approx -\Lambda^2/(2D)$ gives the Gaussian distribution of $\Lambda$. The coefficient $D$ determines the width of the distribution of the finite-time exponents $\Lambda$; the variance of $\Lambda$ decreases with time $T$ according to the law of large numbers:

$$\langle \Lambda^2 \rangle \approx \frac{D}{T}.$$

Because the transverse variable $z$ in (13.19) is simply related to the finite-time exponent

$$z(T) - z(0) - \lambda_\perp T = \Lambda T,$$

we get for $z$ the diffusive growth with the diffusion constant $D$:

$$\langle (z(T) - \langle z(T) \rangle)^2 \rangle \propto T D.$$

The advantage of the more general form (13.21) is that it allows us to describe properly the large deviations of finite-time exponents as well. In particular, in the Gaussian approximation arbitrary large finite-time exponents are possible, while a more correct description gives lower and upper bounds for the support of the function $s$ (see the example of the skew tent map below). In nonlinear dynamics such an approach to a description of the statistical properties of deviations with the help of a scaling function is known as the *thermodynamic formalism* [Ott 1992; Crisanti *et al.* 1993; Badii and Politi 1997; Beck and Schlögl 1997]; in mathematical statistics one speaks of the *theory of large deviations* [Varadhan 1984].

We now return to the dynamics of the transverse perturbation (13.18). Near the synchronization threshold the mean drift of the random walk of the variable $z$ is small; hence the dynamics are mainly determined by fluctuations. The distribution of $z$ spreads in time, so that rather small as well as rather large values of $z$ can be observed. If we go back from the logarithm of the perturbation $z$ to the perturbation field $w$, we see that $w = \exp(z)$ can reach extremely large and extremely low values. Thus, the random walk of $z$ corresponds to an extremely bursty time series of the transverse perturbation $w$. This regime is called **modulational intermittency** [Fujisaka and

Yamada 1985, 1986; Yamada and Fujisaka 1986] (the term "on–off intermittency" is also used [Platt *et al.* 1989]) and is illustrated in Fig. 13.3.

It is important to state that the main source of this intermittency is fluctuations of the finite-time transverse Lyapunov exponent. In exceptional cases, where the exponent does not fluctuate, the regime differs from the modulational intermittency. In particular, in the symmetric tent map $f(U) = 1 - 2|U|$ the multipliers are equal to two for all values of $U$. This also holds for the logistic map $f(U) = 4U(1 - U)$: here the fluctuations of finite-time Lyapunov exponents vanish for large $T$. The synchronization transition in these systems has specific statistical properties, as discussed by Kuznetsov and Pikovsky [1989] and Pikovsky and Grassberger [1991].

In order to proceed with our statistical analysis of the modulational intermittency at the onset of complete synchronization, we have to go beyond the linear approximation. We will not consider the special cases of the symmetric tent map and the logistic map, but discuss only a general case of fluctuating finite-time Lyapunov exponents.

### 13.3.3   Modulational intermittency: power-law distributions

By virtue of the finite-time Lyapunov exponents, we can rewrite the evolution of the perturbation $z$ (13.18) in the form

$$z(t + T) = z(t) + T\lambda_\perp + T\Lambda.$$

For large $T$ we can neglect the correlations of subsequent $\Lambda_T$ and consider these quantities as independent random variables. This allows us to write an equation for the probability distribution density $W(z; t)$. This density at time $t + T$ is a convolution of two probabilities:

$$W(z; t + T) = \int d\Lambda \, p(\Lambda; T) W(z - T\lambda_\perp - T\Lambda; t).$$

Next, we look for a statistically stationary (time-independent) solution, trying the ansatz

$$W(z) \propto \exp(\kappa z). \tag{13.22}$$

Substituting this and using (13.21) we obtain

$$1 = \int d\Lambda \, p(\Lambda; T) e^{-T\kappa(\lambda_\perp + \Lambda)} \propto \int d\Lambda \, e^{Ts(\Lambda) - T\kappa\lambda_\perp - T\kappa\Lambda}. \tag{13.23}$$

The latter integral can be evaluated for large $T$ if we take the maximum of the expression in the exponent:

$$\int d\Lambda \, e^{Ts(\Lambda) - T\kappa\lambda_\perp - T\kappa\Lambda} \propto \exp[T(s(\Lambda^*) - \kappa\Lambda^* - \kappa\lambda_\perp)],$$

where the value of $\Lambda^*$ is determined by the maximum condition

$$\frac{ds(\Lambda^*)}{d\Lambda} = \kappa.$$

Substituting these two expressions in (13.23), we obtain the equation for $\Lambda^*$

$$s(\Lambda^*) - (\Lambda^* + \lambda_\perp)\frac{ds(\Lambda^*)}{d\Lambda} = 0, \tag{13.24}$$

from which we can find the exponent of the probability distribution $\kappa$. Those familiar with the thermodynamic formalism can easily recognize here the usual formulae for the Legendre transform (see textbooks on statistical mechanics for a general introduction and, e.g., Ott [1992] for applications to chaotic systems).

Note that in (13.24) the scaling function $s(\Lambda)$ characterizes the properties of fluctuations of local multipliers of the symmetric chaotic state, and it is independent of the coupling parameter $\varepsilon$. The $\varepsilon$-dependence comes into (13.24) only through the transverse Lyapunov exponent $\lambda_\perp$. Because $s(0) = s'(0) = 0$, at criticality, where the transverse Lyapunov exponent $\lambda_\perp$ changes sign, the exponent $\kappa$ changes sign as well. For small $\lambda_\perp$, the exponent $\kappa$ linearly depends on $\lambda_\perp$. As follows from (13.14), the dependence of $\kappa$ on $\varepsilon - \varepsilon_c$ is also linear.

In terms of the perturbation $w$, the stationary distribution (13.22) is the power law

$$W(w) \propto w^{\kappa-1}. \tag{13.25}$$

As with any perfect power law, this distribution is not normalizable. As has already been mentioned, at the threshold (13.15) the exponent $\kappa$ changes sign. Thus the distribution (13.25) diverges for $w \to 0$ in the regime of complete synchronization (where $\kappa < 0$), and diverges for $w \to \infty$ for small couplings, when the synchronous regime is unstable and $\kappa > 0$. To obtain a normalizable distribution, we have to go beyond the linear approximation, taking into account terms additional to the linear mapping (13.16).

### Below the synchronization threshold $\varepsilon < \varepsilon_c$

The divergence for $w \to \infty$ appears because we have neglected the saturation effects in the system (13.9) and (13.10). It is clear that the difference $x - y$ cannot grow without limit, because the attractor of the coupled maps occupies a finite region in the phase space. In general, the saturation should be described with the help of nonlinear terms in both Eqs. (13.9) and (13.10). The corresponding theory, however, has not yet been developed. As a simplified model we still use the linear equation (13.18), and simulate the saturation with an artificial upper boundary at $z = z_{max} = \ln w_{max}$: the random walk is "reflected" by this boundary so that the distribution (13.25) has a cutoff at $w_{max}$.

With this cutoff we can normalize the distribution density (13.25):

$$W(w) = \begin{cases} \kappa w_{max}^{-\kappa} w^{\kappa-1} & \text{for } w \leq w_{max}, \\ 0 & \text{otherwise.} \end{cases}$$

From this relation one easily obtains the moments of the perturbation $w$:

$$\langle w^q \rangle = \frac{\kappa}{q + \kappa} w_{\mathrm{max}}^q.$$

Exactly at the synchronization threshold all the moments vanish, and for small deviations from criticality they grow linearly with $\varepsilon_c - \varepsilon$ (because $\kappa$ linearly depends on the transverse Lyapunov exponent $\lambda_\perp$). This is a rather unusual behavior at a bifurcation point; it is directly related to the power-law character of the distribution.

### A numerical trap for identical systems

An interesting effect can be observed in the nonsynchronized state close to the synchronization threshold. If one simulates the dynamics of the coupled identical system (13.4) on a computer, then complete synchronization can be observed even in the region where the transverse Lyapunov exponent is positive. This "unstable" synchronization appears due to the finite precision of numerical calculations. Indeed, on a computer, if the states of two symmetric systems are equal up to the last digit, the evolution of these states is identical and one observes complete synchronization. For example, the accuracy of computer calculations performed with double precision is usually $10^{-15}$. If, in the course of the dynamics, the perturbation $w$ is less than this quantity at time $t_0$, then for all $t > t_0$ one has $w \equiv 0$ and this looks like complete synchronization. In terms of the random walk of the logarithm of the perturbation $z$, one can interpret this effect as the existence of an absorbing boundary at $z_{\mathrm{min}} = \ln(10^{-15})$: as soon as the random walk hits this boundary, it sticks to it. Near the synchronization transition the probability for $z$ to reach $z_{\mathrm{min}}$ is not small, and presumably "unstable" synchronization has been reported in several publications. This numerical artifact can be avoided by introducing a small asymmetry in the system (e.g., via small parameter mismatch). It also helps to add a small noise which should be, of course, different for both systems.

### Beyond the synchronization threshold $\varepsilon > \varepsilon_c$

The divergence for $w \to 0$ means complete synchronization (so that in fact the distribution is a singular $\delta$-function) and in the perfectly symmetric system one cannot observe the power-law distribution (13.25) for large couplings. However, the distribution (13.25) will be observed if the perfect symmetry of interacting systems is broken, due to one of the following factors.

(i) **Nonidentity.** If the interacting systems are slightly different, perfect synchronization is impossible. Suppose that two slightly different maps are interacting, so that instead of (13.4) we write

$$\begin{bmatrix} x(t+1) \\ y(t+1) \end{bmatrix} = \begin{bmatrix} 1-\varepsilon & \varepsilon \\ \varepsilon & 1-\varepsilon \end{bmatrix} \begin{bmatrix} f_1(x(t)) \\ f_2(y(t)) \end{bmatrix}$$

$$= \begin{bmatrix} (1-\varepsilon)f_1(x(t)) + \varepsilon f_2(y(t)) \\ \varepsilon f_1(x(t)) + (1-\varepsilon) f_2(y(t)) \end{bmatrix}. \tag{13.26}$$

Now for the variables (13.6) we obtain

$$U(t+1) = \tfrac{1}{2}[f_1(U(t) + V(t)) + f_2(U(t) - V(t))], \tag{13.27}$$

$$V(t+1) = \frac{1-2\varepsilon}{2}[f_1(U(t) + V(t)) - f_2(U(t) - V(t))]. \tag{13.28}$$

The symmetric state $V = 0$ is no longer the solution of these equations, but for a small mismatch we can expect $V$ to be small. Moreover, we will neglect the effect of small $v$ on the dynamics of the variable $U$ and consider only the second equation (Eq. (13.28)), which for small $v$ can be rewritten as

$$v(t+1) = \frac{(1-2\varepsilon)}{2}[(f_1'(U) + f_2'(U))v + (f_1(U) - f_2(U))]. \tag{13.29}$$

The main difference with the pure symmetric case Eq. (13.10) is the last inhomogeneous term on the r.h.s. It is proportional to the mismatch, and we will assume that this chaotic term is of the order $\delta \ll 1$. Then, the dynamics can be qualitatively represented as follows. If the difference between the systems' states $v$ is larger than $\delta$, the inhomogeneous term is not important and we have pure random walk dynamics described by Eq. (13.18). If the difference is of order $\delta$, then the inhomogeneous term acts as a random force and prevents $v$ from becoming smaller than $\delta$. For the random walk of the variable $z$ this can be described as the existence of a reflecting boundary at $z \approx \ln \delta$. With such a boundary, the random walk takes place in the region $z > \ln \delta$ even for negative biases $\lambda_\perp < 0$. This means that bursts of $v$ are also observed in the synchronized state with negative transverse Lyapunov exponent. In other words, the synchronized chaos appears to be extremely sensitive to perturbations: even small mismatch may result in large bursts.

(ii) **Noise.** The same effect as that produced by nonidentity is caused by a small noise acting on the coupled mappings: it produces small transverse perturbations even if the fully synchronized state is stable. Thus, the same estimates as above are valid, and $\delta$ is the noise level.

These factors lead to a lower cutoff of the probability density at $w_{\min}$, where $w_{\min}$ roughly corresponds to the inhomogeneity and/or noise level. The existence of a power-law tail for the stationary distribution means that the probability of observing large deviations from the fully synchronous state is relatively large (compared, say, to the Gaussian distribution where large deviations have exponentially small probability). This is another manifestation of sensitivity to perturbations; this sensitivity results in a rather large sporadic response. We will see in Section 13.4 that this sensitivity is related to a nontrivial topological structure of the attractor and its basin in the phase space.

### An example: coupled skew tent maps

We see that the statistics of the variable $z$ are related to the statistics of fluctuations of local Lyapunov exponents, the latter being described by the scaling function $s(\Lambda)$. We illustrate the theory above with the skew tent map (13.5). Because the invariant probability density for the skew tent map (13.5) is uniform in the interval $(0, 1)$, and the visits of the regions $(0, a)$ and $(a, 1)$ are noncorrelated, we obtain

$$|\ln f'| = \begin{cases} -\ln a & \text{with probability } a, \\ -\ln(1-a) & \text{with probability } (1-a). \end{cases}$$

Hence, the Lyapunov exponent is

$$\lambda = -a \ln a - (1-a) \ln(1-a). \tag{13.30}$$

Thus we obtain a random walk with two types of steps:

$$-\ln a - \lambda \qquad \text{with probability } a,$$
$$-\ln(1-a) - \lambda \qquad \text{with probability } (1-a).$$

The distribution $p(\Lambda, T)$ is then the binomial distribution. If we denote the number of iterations with $U < a$ as $n$, and set $h = n/T$, then from the binomial distribution

$$p(n, T) = \frac{T!}{n!(T-n)!} a^n (1-a)^{T-n}$$

and from the relation

$$\Lambda = \frac{n}{T}(-\ln a) + \frac{T-n}{T}(-\ln(1-a)) - \lambda,$$

by virtue of the Stirling formula, we obtain the scaling function $s(\Lambda)$ in the parametric form:

$$s(h) = h \ln \frac{a}{h} + (1-h) \ln \frac{1-a}{1-h},$$

$$\Lambda(h) = -h \ln a - (1-h) \ln(1-a) - \lambda.$$

It is easy to check that this scaling function (Fig. 13.4a) has its single maximum at zero.

Using the basic formula (13.24) we can represent the exponent $\kappa$ and the transverse Lyapunov exponent $\lambda_\perp$ in a parametric form as functions of $h$:

$$\kappa = \frac{ds}{d\Lambda} = \frac{ds/dh}{d\Lambda/dh} = \ln\left(\frac{1-h}{h}\right) / \ln\left(\frac{1-a}{a}\right) - 1,$$

$$\lambda_\perp = -\Lambda(h) + s(h)/\kappa.$$

This dependence of the exponent $\kappa$ on the transverse Lyapunov exponent is shown in Fig. 13.4b. Before discussing this relation, we demonstrate in Fig. 13.5 the validity of the power-law ansatz (13.22).

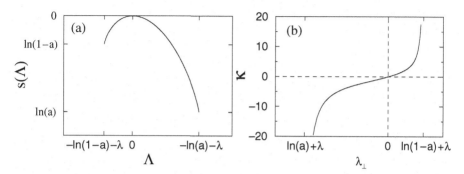

**Figure 13.4.** The scaling function describing the fluctuations of the finite-time Lyapunov exponents (a) and the exponent $\kappa$ vs. the transverse Lyapunov exponent (b), for the skew tent map. We note that, according to (13.14), $\lambda_\perp = \ln|1 - 2\varepsilon| + \lambda$.

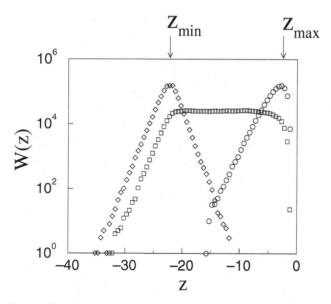

**Figure 13.5.** The exponential distributions of the variable $z$ (corresponding to the power-law distributions of $w$) for coupled skew tent maps as in Fig. 13.3, for three values of coupling: critical coupling $\varepsilon = \varepsilon_c$ (squares); coupling larger than critical $\varepsilon = \varepsilon_c + 0.025$ (diamonds); coupling smaller than critical $\varepsilon = \varepsilon_c - 0.025$ (circles). An additional parameter mismatch $a \pm 10^{-10}$ has been introduced to provide the lower cutoff $z_{min}$. In the region $z_{min} < z < z_{max}$ the distributions are, to a good accuracy, exponential, in accordance with the ansatz (13.22).

From Fig. 13.4 one can see that the nontrivial power-law solution exists only in some neighborhood of the synchronization transition point. This is due to the fact that the scaling function $s(\Lambda)$ has a finite support

$$-\ln a - \lambda \leq \Lambda \leq \ln(1-a) - \lambda. \tag{13.31}$$

This finiteness is a general property of dynamical systems, and it is violated in the parabolic approximation for $s$. Indeed, the parabolic approximation of $s$ means the Gaussian approximation for the distribution of the finite-time Lyapunov exponents. The Gaussian distribution has indefinitely extended tails, which correspond to arbitrary large local expansion/contraction rates. Conversely, linear perturbation analysis for trajectories of a dynamical system leads to equations with finite coefficients, so that in real dynamical systems the finite-time Lyapunov exponents cannot be arbitrarily large.

Accordingly to the finiteness of the finite-time Lyapunov exponents, the dependence $\kappa$ on $\lambda_\perp$ is defined in the interval (13.31) only. As one can easily see, the minimum and the maximum values of the finite-time exponent $\Lambda$ correspond to the minimum and the maximum of local expansion rates: for the tent map these are $-\ln(1-a)$ and $-\ln a$. In the region (13.31) the transverse dynamics are nontrivial as some trajectories are transversally stable and some are unstable. Outside this interval all trajectories are either stable or unstable, and the power-law tails of the distribution do not exist (in terms of the random walk, all the steps have the same sign). We will discuss this situation once more in Section 13.4.

### 13.3.4  Modulational intermittency: correlation properties

The process near the threshold of complete synchronization is very intermittent: it looks like a sequence of bursts separated by long silent epochs (Fig. 13.3). We have already given a qualitative explanation of this modulational intermittency: while the logarithm of the perturbation performs a random walk, only maxima of the logarithm appear as relatively sharp peaks.

The analogy to the random walk can also be used for a quantitative description of time correlation properties of the modulational intermittency. To this end we model the discrete time random walk of the variable $z$ with a continuous-time diffusion process with mean drift $\lambda_\perp$ and diffusion constant $D$. Note that the mean drift depends on the coupling constant while the diffusion is determined by the properties of the symmetric chaotic mapping only. Such a process can be described with the Fokker–Planck equation (see, e.g., [Feller 1974]):

$$\frac{\partial W(z,t)}{\partial t} = -\lambda_\perp \frac{\partial W(z,t)}{\partial z} + \frac{D}{2}\frac{\partial^2 W(z,t)}{\partial z^2}. \tag{13.32}$$

This equation should be supplemented with boundary conditions. As we have discussed above, both nonlinear saturation and noise/inhomogeneity can be modeled with reflecting boundary conditions at $z_{max}$ and $z_{min}$, respectively.

Let us now determine the statistics of the time intervals between the bursts in the modulational intermittency slightly below the synchronization threshold, $\varepsilon < \varepsilon_c$. Here it is sufficient to introduce the reflecting boundary at $z_{max}$ only. A burst is observed if the variable $z$ is close to $z_{max}$. To estimate the time till the next burst occurs, we can take some value $z_0 < z_{max}$ and consider the time until the random walk reaches $z_{max}$. In the language of the theory of random processes, this is the first passage time. For the diffusion processes described by the Fokker–Planck equation (13.32) the statistics of the first passage time are well known (see, e.g., Feller [1974]). The probability density for the time intervals $\tau$ between the bursts is

$$W(\tau) = \frac{|z_{max} - z_0|}{\sqrt{2\pi D \tau^3}} \exp\left(-\frac{(z_{max} - z_0 - \lambda_\perp \tau)^2}{2D\tau}\right). \tag{13.33}$$

Near the threshold, where the characteristic time interval between bursts is large, we can write for large $\tau$

$$W(\tau) \propto \tau^{-3/2} \exp\left(-\frac{\lambda_\perp^2}{2D}\tau\right).$$

The distribution density is a power-law with the exponential cutoff at $\tau^* \sim D\lambda_\perp^{-2}$. Very large intervals between bursts are possible, and the mean duration of the "laminar" phase[4]

$$\langle \tau \rangle = \frac{z_{max} - z_0}{\lambda_\perp}$$

diverges at the threshold $\lambda_\perp = 0$.

Assuming statistical independence of the intervals between the bursts, one can obtain (from the statistics of these intervals) an estimate of the power spectrum $S(\zeta)$ of the process. If we denote the Fourier transform of the probability density (13.33) by $\theta(\zeta)$, then (see, e.g., [Rytov et al. 1988])

$$S(\zeta) \propto \mathrm{Re}\,\frac{\theta(\zeta)}{1 - \theta(\zeta)}.$$

Fortunately, the corresponding integrals can be computed analytically, and in the limit $z_{max} - z_0 \to 0$ one obtains the power-law spectrum for the modulational intermittency:

$$S(\zeta) \propto \zeta^{-1/2} \qquad \text{for} \quad D \gg \zeta \gg D\lambda_\perp^{-2}.$$

Concluding this discussion of the statistical properties of modulational intermittency, we would like to mention that they are fully applicable to noise-driven systems, where the noise acts multiplicatively. Usually such systems are considered in continuous time, and the transition from a steady state to macroscopic oscillations is called the noise-induced nonequilibrium phase transition (see [Horsthemke and Lefever 1989] and references therein).

---

[4] Other moments of the distribution (13.33) can be also calculated, see [Fujisaka et al. 1997].

## 13.4    Onset of synchronization: topological aspects

We now discuss the transition to complete synchronization from the topological viewpoint, looking at structures and bifurcations in the phase space. The transition in Fig. 13.2 can be seen as a transition from a completely synchronized symmetric attractor lying on the diagonal $x = y$ to an asymmetric one occupying a neighborhood of the diagonal (asymmetry here means that $x(t) \neq y(t)$; the probability distribution on the plane $(x, y)$ may remain symmetric, see Fig. 13.2). This can be considered as a bifurcation of the strange attractor, and we want to establish a relation with bifurcation properties of individual trajectories. It is convenient to look at *periodic* orbits inside chaos, because unstable periodic trajectories represent the skeleton of a chaotic set. They are dense in the chaotic attractor, and many quantities characterizing chaotic motion (such as the invariant measure, the largest Lyapunov exponent) can be expressed in terms of periodic orbits.[5] The advantage here is that we can use the results of elementary bifurcation theory (see, e.g., [Iooss and Joseph 1980; Guckenheimer and Holmes 1986; Hale and Koçak 1991]) as they are directly applicable to periodic orbits.

### 13.4.1    Transverse bifurcations of periodic orbits

We start from the fully synchronous state (i.e., $\varepsilon \approx 1/2$) and follow the symmetry-breaking transition as the coupling constant $\varepsilon$ decreases. Let us first look at the simplest periodic orbit – a fixed point. If the mapping $x \rightarrow f(x)$ has a fixed point $x^*$, then for all $\varepsilon$ there exists a synchronous fixed point solution $x(t) = y(t) = x^*$. The stability of this fixed point can be determined from the linear Eqs. (13.9) and (13.10). There are two multipliers:

$$\mu_u = f'(x^*), \qquad \mu_v = (1 - 2\varepsilon) f'(x^*), \qquad (13.34)$$

corresponding to the two perturbation modes $u$ and $v$. Because the fixed point belongs to the chaotic mapping $f$, the multiplier $\mu_u$ is larger than one in absolute value, so that the direction $u$ is always unstable. The transverse direction $v$ is stable if $|(1 - 2\varepsilon) f'(x^*)| < 1$ and unstable otherwise. Thus, a bifurcation happens at $\varepsilon_c(x^*)$ determined from the condition

$$\varepsilon_c(x^*) = \frac{1 - |f'(x^*)|^{-1}}{2}. \qquad (13.35)$$

The type of bifurcation is determined by the sign of the multiplier at criticality: if $\mu_v = 1$ a pitchfork bifurcation is observed; if $\mu_v = -1$ a period-doubling bifurcation occurs. If these bifurcations are supercritical (which cannot be determined from linear theory and depends on the nonlinearity of the mapping), two symmetric fixed points in the case of the pitchfork bifurcation or a period-two orbit in the case of period-doubling appear. These solutions are stable in the $V$-direction, but they inherit instability in the $U$-direction from the symmetric fixed point. The bifurcation is sketched in Fig. 13.6.

---

[5] For the characterization of chaos via periodic orbits see, e.g., [Artuso *et al.* 1990a,b; Ott 1992].

This picture is valid for all fixed points and periodic orbits of the mapping $f(x)$, so that a symmetric period-$T$ trajectory $x^p(t) = y^p(t)$ bifurcates either to a pair of symmetric orbits through a pitchfork bifurcation, or to a period-$2T$ orbit through a period-doubling bifurcation. The bifurcation point is determined from the generalization of (13.34): for a period-$T$ orbit the multiplier is the product of local multipliers

$$\mu_v = (1 - 2\varepsilon)^T \prod_{t=1}^{T} f'(x^p(t)),$$

therefore, similar to (13.35),

$$\varepsilon_c(x^p) = \frac{1 - \left[\prod_{t=1}^{T} |f'(x^p)|\right]^{-1/T}}{2}. \tag{13.36}$$

It is worth noting the analogy of this expression with the statistical criterion of the synchronization onset (13.15): instead of the average on the whole chaotic attractor multiplier $e^\lambda$, we have in (13.36) the mean multiplier of the particular periodic orbit.

## 13.4.2  Weak vs. strong synchronization

One very important observation is that the bifurcation points (13.36) in general do not coincide with the critical value (13.15). Typically, the multipliers for different periodic orbits are different, so we have a whole bifurcation region $(\varepsilon_{c,\min}, \varepsilon_{c,\max})$, where different periodic orbits become transversally unstable. Thus, contrary to bifurcation of a single periodic orbit, the transition for a chaotic set occupies an interval of parameters. We first describe the transition for the case of supercritical transverse bifurcation.

We can distinguish the following stages (see Fig. 13.7).

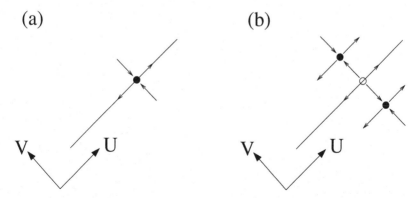

**Figure 13.6.** A sketch of the supercritical transverse bifurcation of an unstable fixed point. (a) For $\varepsilon > \varepsilon_c(x^*)$, the fixed point is unstable in the longitudinal direction and stable in the transverse direction. (b) For $\varepsilon < \varepsilon_c(x^*)$, the transverse instability results in the birth of a cycle (or of a pair of fixed points).

*Strong synchronization, $\varepsilon > \varepsilon_{c,max}$*
All periodic orbits are transversally stable. Hence, any point in the vicinity of the diagonal is attracted to the synchronous state $x = y$ and remains there.

*Weak synchronization, $\varepsilon_c < \varepsilon < \varepsilon_{c,max}$*
Some[6] periodic orbits are transversally unstable, but the synchronous state is stable on average. Now, in the near vicinity of the diagonal, almost all (in the sense of the Lebesgue measure) initial points are attracted to the completely synchronous state, but there are also initial conditions that escape from this locality.

*Weakly asynchronous state, $\varepsilon_{c,min} < \varepsilon < \varepsilon_c$*
The synchronous state is unstable on average, but some synchronous periodic orbits are still transversally stable.

---

[6]  In fact, an infinite number of.

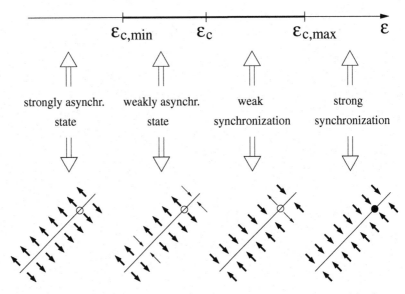

**Figure 13.7.** A synchronization transition bifurcation diagram. The transition is "smeared" over the region $(\varepsilon_{c,min}, \varepsilon_{c,max})$. In the bottom panels the bold arrows show the overwhelming dynamics near the symmetric state (attraction for $\varepsilon > \varepsilon_c$ and repulsion for $\varepsilon < \varepsilon_c$). The thin arrows show exceptional dynamics. For $\varepsilon < \varepsilon_{c,min}$, all symmetric trajectories are transversally unstable; this state is strongly asynchronous. For $\varepsilon_{c,min} < \varepsilon < \varepsilon_c$, almost all symmetric trajectories are transversally unstable (weakly asynchronous state). For $\varepsilon_c < \varepsilon < \varepsilon_{c,max}$, almost all orbits are attracted to the symmetric attractor, although some trajectories on it are transversally unstable (weak synchronization). For $\varepsilon > \varepsilon_{c,max}$, all orbits on the symmetric attractor are stable (strong synchronization). One fixed point that undergoes bifurcation at $\varepsilon = \varepsilon_{max}$ is shown by the circle (cf. Fig. 13.6).

*Strongly asynchronous state, $\varepsilon < \varepsilon_{c,min}$*
All periodic orbits are transversally unstable.

The stability of periodic orbits as described here directly corresponds to the statistical consideration of Section 13.3: the largest (smallest) multiplier corresponds to the largest (respectively, the smallest) finite-time Lyapunov exponent; the region where a power-law distribution exists corresponds to the weakly synchronized and weakly desynchronized stages described above.

The least trivial regime is the stage of weak synchronization, where the synchronous symmetric state is stable on average, but some periodic orbits are transversally unstable. Note that even from the existence of a single transversally unstable periodic orbit, it follows that transversally unstable trajectories can be found everywhere: indeed, every periodic orbit has a dense set of points that converge to it, a trajectory starting at one of these points will eventually follow the periodic orbit and will be transversally unstable. So we have an apparent paradoxical situation: an attractor has a dense set of nonattracting trajectories. This is an example of where rather subtle differences between mathematical definitions of attractor become important. We thus remind the reader here of two popular definitions of an attractor of a dynamical system.

*Topological definition*
Here an attractor is defined (see, e.g., Katok and Hasselblatt [1995]) as a compact set $\mathcal{A}$ which has a neighborhood $\mathcal{U}$ such that $\bigcap_{n>0} f^n(\mathcal{U}) = \mathcal{A}$ and $f^k(\mathcal{U}) \in \mathcal{U}$ for some $k > 0$. This definition means that there exists an open neighborhood, from which *all* the points are attracted to $\mathcal{A}$.

*Probabilistic (Milnor) definition*
Milnor [1985] defined an attractor as a closed set for which the realm of attraction $\rho(\mathcal{A})$ has strictly positive measure, and there is no strictly smaller set $\mathcal{A}' \in \mathcal{A}$ whose realm of attraction coincides with $\rho(\mathcal{A})$ up to a set of Lebesgue measure zero.

The difference between these definitions is clear: in the Milnor definition it is allowable for some points to escape from the attractor; this is forbidden in the topological definition. From the consideration above it follows that the strongly synchronous state for large couplings is a topological attractor, while the weakly synchronous regime corresponds to an attractor in the Milnor sense and not to a topological attractor. The topological attractor for $\varepsilon_c < \varepsilon < \varepsilon_{c,max}$ is larger than the Milnor one – it includes trajectories bifurcating from the transversally unstable orbits and their unstable manifolds. As we have seen in Section 13.3, in the state of weak synchronization the probability density can have a power-law tail provided a lower cutoff is present due to a small inhomogeneity or noise. Geometrically, this can be interpreted as an expansion of the Milnor attractor to the topological attractor, due to small noise or/and inhomogeneity.

### 13.4.3  Local and global riddling

We have discussed the case of supercritical transverse instability of periodic orbits in the symmetric attractor, as shown in Fig. 13.6. In this case, new transversally stable periodic points appear in the vicinity of the symmetric state $x = y$. Correspondingly, the topological attractor, which can be considered as an "envelope" of these newly born asymmetric periodic orbits, appears. The Milnor attractor $x = y$ attracts almost all points from its neighborhood, but there are exceptional transverse perturbations that grow. Nevertheless, these growing perturbations cannot go very far from the symmetric state, especially if the coupling is close to the critical coupling of the first instability $\varepsilon \lesssim \varepsilon_{c,max}$ because the growth is bounded by the unstable manifolds of the newly born asymmetric periodic orbits. In fact, almost all growing perturbations come back to the symmetric state (except for those that lie on the stable manifolds of asymmetric fixed points). This situation is called *local riddling*. It exists near the symmetric state for $\varepsilon_c < \varepsilon < \varepsilon_{c,max}$, and it becomes visible at the synchronization transition $\varepsilon = \varepsilon_c$.

Another situation is observed in the case of a subcritical transverse pitchfork bifurcation of the periodic orbits belonging to a symmetric attractor. Here at $\varepsilon = \varepsilon_{c,max}$ a pair of symmetric fixed points (periodic orbits) collides with the symmetric fixed point (periodic orbit) and the latter becomes transversally unstable. In contrast to the case of local riddling, transverse perturbations now do not stay in the vicinity of the symmetric orbit, but move to some attractor far away from the diagonal (or even escape to infinity). The same happens for the subcritical period-doubling bifurcation; we illustrate both cases in Fig. 13.8.

The regime of weak synchronization is now even more subtle, as there are points in any vicinity of the diagonal that leave this vicinity and are attracted to some other

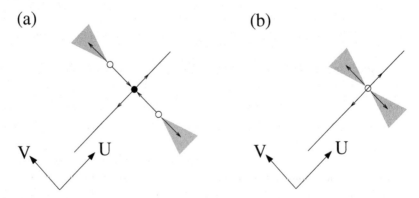

**Figure 13.8.** Subcritical transverse bifurcation of a fixed point. (a) For $\varepsilon > \varepsilon_{c,max}$, the point is stable. The shadowed regions are attracted to a remote attractor. (b) At the subcritical bifurcation, $\varepsilon = \varepsilon_{c,max}$, the basin of the remote attractor touches the symmetric one. The shadowed regions also have pre-images (not shown here) that are dense in the vicinity of the diagonal.

remote attractor. Moreover, these points are dense (due to the same argument used above in discussing the structure of weak synchronization), albeit their measure tends to zero as we approach the symmetric attractor $x = y$. This structure of the attractor's basin is called *globally riddled*; this yields extreme sensitivity to noise. Indeed, noise kicks the system out of the diagonal, and therefore there is a finite probability for points to start escaping at every iteration. Thus, such a synchronization can only be transient – eventually all trajectories leave the vicinity of the synchronous state.

Summarizing the properties of the synchronization transition, we emphasize that here we have a bifurcation inside chaos, and this bifurcation is smeared. Indeed, the whole parameter range $(\varepsilon_{c,min}, \varepsilon_{c,max})$ demonstrates a nontrivial behavior that can be described both topologically and statistically. This situation is typical for chaotic systems with fluctuating finite-time Lyapunov exponents. The unusual properties of the transition also manifest themselves in more complex situations, to be discussed in the next chapter.

## 13.5  Bibliographic notes

In presenting complete synchronization we followed the early papers [Fujisaka and Yamada 1983; Yamada and Fujisaka 1983; Pikovsky 1984a]. Statistical properties at the onset of synchronization were discussed by Pikovsky [1984a], Yamada and Fujisaka [1986], Fujisaka *et al.* [1986], Fujisaka and Yamada [1987], Yamada and Fujisaka [1987] and Pikovsky and Grassberger [1991]. For studies of modulational intermittency see also [Hammer *et al.* 1994; Heagy *et al.* 1994c; Platt *et al.* 1994; Venkataramani *et al.* 1995; Xie *et al.* 1995; Yu *et al.* 1995; Čenys *et al.* 1996; Lai 1996a; Venkataramani *et al.* 1996; Yang and Ding 1996; Čenys *et al.* 1997a,b; Ding and Yang 1997; Fujisaka *et al.* 1997; Kim 1997; Fujisaka *et al.* 1998; Miyazaki and Hata 1998; Nakao 1998].

The topological properties of the synchronization transition were discussed in [Pikovsky and Grassberger 1991; Ashwin *et al.* 1994, 1996, 1998; Aston and Dellnitz 1995; Heagy *et al.* 1995; Ashwin and Aston 1998; Maistrenko *et al.* 1998]. For general properties of riddling and blowout bifurcation see [Heagy *et al.* 1994a; Lai *et al.* 1996; Lai and Grebogi 1996; Maistrenko and Kapitaniak 1996; Billings *et al.* 1997; Lai 1997; Maistrenko *et al.* 1997, 1998, 1999a; Nagai and Lai 1997a,b; Kapitaniak *et al.* 1998; Kapitaniak and Maistrenko 1998; Manscher *et al.* 1998; Astakhov *et al.* 1999].

Experiments on complete synchronization have been performed with electronic circuits [Schuster *et al.* 1986; Heagy *et al.* 1994b, 1995; Yu *et al.* 1995; Čenys *et al.* 1996; Lorenzo *et al.* 1996; Rulkov 1996] and lasers [Roy and Thornburg 1994; Terry *et al.* 1999]. See also the special issue [Pecora (ed.) 1997] and references therein.

# Chapter 14

## Complete synchronization II: generalizations and complex systems

In the previous chapter we considered the simplest possible case of complete synchronization: the symmetric interaction of two identical chaotic one-dimensional mappings. Below we describe more general situations: many coupled maps, continuous-time dynamical systems, distributed systems. Moreover, we consider synchronization in a general context as a symmetric state inside chaos. We mainly restrict our description to the linear stability theory that provides the synchronization threshold; we describe the nonlinear effects observed at the synchronization transition only in a few cases.

## 14.1   Identical maps, general coupling operator

The simplest generalization of the theory of Chapter 13 is to study complete synchronization in a large ensemble of coupled chaotic systems. We consider $N$ identical coupled chaotic mappings (13.1) subject to linear coupling. We represent coupling by means of a general linear operator $\hat{L}$ given by an $N \times N$ matrix:

$$x_k(t+1) = \sum_{j=1}^{N} L_{kj} f(x_j(t)). \tag{14.1}$$

The dissipative conditions on the coupling that we discussed in Section 13.1 can be now formulated as follows.

(i) The system (14.1) possesses a symmetric completely synchronous solution where the states of all subsystems are identical

$$x_1(t) = x_2(t) = \cdots = x_N(t) = U(t).$$

This is the case if the constant vector $e_1 = (1, \ldots, 1)$ is the eigenvector of the matrix $\hat{L}$ corresponding to the eigenvalue $\sigma_1 = 1$.

(ii) All other eigenvalues of $\hat{L}$ are in absolute value less than 1. This causes damping of inhomogeneous perturbations due to coupling.

The stability of the synchronous state can be determined from linearization of Eq. (14.1). Contrary to the simplest case of Chapter 13, there are many transverse modes now, and the largest transverse Lyapunov exponent gives the stability condition. Let us look at the evolution of an inhomogeneous perturbation $\delta x_k(0)$ of a chaotic solution $U(t)$. After $T$ iterations we have

$$\delta x_k(T) = \hat{L}^T \prod_{t=1}^{T} f'(U(t)) \delta x_k(0).$$

Because the factors $f'(U)$ are $k$-independent, the evolution of the perturbation can be decomposed into the eigenvectors of the matrix $L_{kj}$. For large $T$ the largest inhomogeneous perturbation is dominated by the second eigenvector $e_2$ because it has the eigenvalue $\sigma_2$ closest to 1; hence we obtain the growth rate (similar to (13.14))

$$|\delta x_k(T)| \propto c_2 |\sigma_2|^T e^{T\lambda} - e_2 e^{T\lambda_\perp}, \qquad \lambda_\perp = \lambda + \ln |\sigma_2|. \qquad (14.2)$$

With this notation the linear stability criterion for the synchronous state coincides with that of Section 13.2: $\lambda_\perp < 0$.

This criterion can be applied both for small and large ensembles of interacting chaotic systems. In the latter case, the natural question arises of whether complete synchronization of a chaotic ensemble is possible for a large number of interacting elements $N$, or even in the thermodynamic limit $N \to \infty$. As can be seen from (14.2), the answer depends on the behavior of the spectrum of the operator $\hat{L}$. It always has the eigenvalue $\sigma_1 = 1$, so that we can write the stability criterion of the synchronous state in the form

$$\ln |\sigma_1| - \ln |\sigma_2| > \lambda.$$

This means that there must be a gap (of the size at least $\lambda$) in the spectrum of the linear coupling operator $\hat{L}$ between the first and the second eigenvalues. In other words, the dynamics of the nonsymmetric modes must be fast enough: the damping due to coupling should be faster than the instability due to chaos. It is clear that not all types of couplings possess such a gap. We consider here, as examples, some cases of physical importance.

## 14.1.1  Unidirectional coupling

Physically, unidirectional coupling means that the signal from one chaotic oscillator forces another one. Electronically, it is easy to implement such a scheme by coupling chaotic circuits through an amplifier. Usually, the unidirectional coupling is considered for a regular lattice, but it can occur also in complex networks, see Fig. 14.1.

For two interacting systems the situation of unidirectional coupling is described by the interaction matrix

$$\hat{L} = \begin{bmatrix} 1 & 0 \\ \varepsilon & 1 - \varepsilon \end{bmatrix}. \tag{14.3}$$

The eigenvalues can be readily calculated

$$\sigma_1 = 1, \qquad \sigma_2 = 1 - \varepsilon,$$

and the synchronous chaotic state is linearly stable if

$$\lambda + \ln|1 - \varepsilon| < 0. \tag{14.4}$$

We can easily generalize this result to the case of a lattice of $N$ unidirectionally coupled systems (Fig. 14.1a), where the matrix has the form

$$\hat{L} = \begin{bmatrix} 1 & 0 & 0 & 0 & \cdots & 0 \\ \varepsilon & 1-\varepsilon & 0 & 0 & \cdots & 0 \\ 0 & \varepsilon & 1-\varepsilon & 0 & \cdots & 0 \\ & & \cdots & & & \\ 0 & 0 & 0 & \cdots & \varepsilon & 1-\varepsilon \end{bmatrix}. \tag{14.5}$$

Here the second eigenvalue (which is in fact $(N-1)$-times degenerate) is also $\sigma_2 = 1 - \varepsilon$ and for the stability of the synchronous state we obtain (14.4), independently of the number $N$ of interacting systems: the operator describing the unidirectional coupling possesses a gap in the spectrum.

It is worth mentioning that the existence of such a gap depends crucially on the boundary conditions used in a lattice. In (14.5) we have assumed no interaction between the first and the last elements in the lattice. If we take the one-way coupled

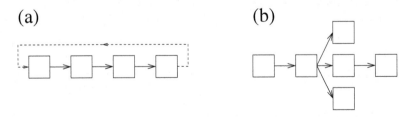

**Figure 14.1.** Schematic view of unidirectional coupling in a lattice (a) and in a network (b). The lattice can form a ring if the last element drives the first one (shown with a dashed line in (a)).

lattice with *periodic* boundary condition (see Fig. 14.1a), where the first element is coupled to the last one, the interaction matrix reads

$$\hat{L} = \begin{bmatrix} 1-\varepsilon & 0 & 0 & 0 & \dots & \varepsilon \\ \varepsilon & 1-\varepsilon & 0 & 0 & \dots & 0 \\ 0 & \varepsilon & 1-\varepsilon & 0 & \dots & 0 \\ & & \dots & & & \\ 0 & 0 & 0 & \dots & \varepsilon & 1-\varepsilon \end{bmatrix}, \qquad (14.6)$$

and the spectrum changes drastically. As the lattice is homogeneous, the eigenfunctions are Fourier modes, and the spectrum can be represented as a function of the wave number $\mathcal{K}$

$$|\sigma(\mathcal{K})|^2 = (1-\varepsilon)^2 + 2\varepsilon(1-\varepsilon)\cos\mathcal{K} + \varepsilon^2, \qquad -\pi \le \mathcal{K} < \pi.$$

This spectrum has no gap (because $|\sigma(\mathcal{K})| \to 1$ as $\mathcal{K} \to 0$), thus no synchronization can be observed for large enough $N$.

This enormous effect of boundary conditions on the dynamics in the lattice has a clear physical explanation. In the one-way coupled lattice, a local inhomogeneous perturbation decays at each site; however, it does not disappear but propagates to the neighboring site. For the case of boundary condition (14.5) it disappears eventually at the last site, while for the system (14.6) it comes back to the first site. In the language of instabilities in distributed dynamical systems this corresponds to *convective instability*, i.e., the perturbation disappears at the place where it was imposed, but propagates and grows downstream. The other case, when the perturbation grows at the same place where it was imposed, is called *absolute instability* (see, e.g., [Lifshitz and Pitaevskii 1981]).

## 14.1.2 Asymmetric local coupling

The case of asymmetric local coupling unites the features of diffusive coupling (13.3) and one-way coupling (14.5):

$$\hat{L} = \begin{bmatrix} 1-\gamma & \gamma & 0 & 0 & \dots & 0 \\ \varepsilon & 1-\varepsilon-\gamma & \gamma & 0 & \dots & 0 \\ 0 & \varepsilon & 1-\varepsilon-\gamma & \gamma & \dots & 0 \\ & & \dots & & & \\ 0 & 0 & 0 & \dots & \varepsilon & 1-\varepsilon \end{bmatrix}. \qquad (14.7)$$

In this model, called the coupled map lattice, chaotic subsystems form a one-dimensional lattice with a nearest neighbor coupling. In the thermodynamic limit (the number of interacting maps tends to infinity) the spectrum of eigenvalues has the form

$$\sigma(\mathcal{K}) = 1 - \varepsilon - \gamma + 2\sqrt{\varepsilon\gamma}\cos\mathcal{K}, \qquad -\pi < \mathcal{K} < \pi.$$

For $\varepsilon \ne \gamma$ this spectrum possesses a gap at $\mathcal{K} = 0$:

$$\sigma(\mathcal{K} \to 0) = 1 - (\sqrt{\varepsilon} - \sqrt{\gamma})^2 < 1.$$

This gap ensures the possibility of complete synchronization in the lattice in the same way as for the one-way coupling described above. For purely symmetric diffusive coupling $\varepsilon = \gamma$, the gap disappears: a long diffusively coupled lattice of chaotic elements cannot be synchronized because the long-wave modes with small $\mathcal{K}$ become unstable.

### 14.1.3  Global (mean field) coupling

In global coupling, each element interacts with all others, and the interaction between two elements does not depend on the "distance" between them (see Section 4.3, Chapter 12, and Fig. 4.24). In a typical situation one considers $N$ identical chaotic maps (13.1) interacting with each other through a dissipative coupling of type (13.3). As the number of interactions for one element is $N - 1$, it is convenient to normalize the coupling constant by $N$. The resulting equations are

$$x_k(t+1) = (1 - \varepsilon) f(x_k(t)) + \frac{\varepsilon}{N} \sum_{j=1}^{N} f(x_j(t)), \qquad k = 1, \ldots, N. \quad (14.8)$$

The system (14.8) is often called the system of globally coupled maps, or the system with mean field coupling, because the last term on the r.h.s. is the mean over all elements of the ensemble.

The interaction matrix $\hat{L}$ can be represented in the form

$$\hat{L} = (1 - \varepsilon)\hat{I} + \frac{\varepsilon}{N}\hat{J},$$

where $\hat{I}$ is the unity matrix and all the elements of the matrix $\hat{J}$ are equal to one. Complete synchronization is a regime when all states are equal $x_1 = x_2 = \cdots = x_N$ and obey the mapping (13.1), this is evidently a solution of system (14.8). The eigenvalues of the coupling operator $\hat{L}$ can be straightforwardly calculated: one eigenvalue is 1 and all other $N - 1$ eigenvalues are $1 - \varepsilon$. Thus, the stability consideration leads to the synchronization condition

$$\lambda + \ln|1 - \varepsilon| < 0,$$

where $\lambda$ is the Lyapunov exponent of the local mapping. As all perturbation modes have the same stability properties (the second eigenvalue of the coupling matrix is $(N-1)$-times degenerate), according to the linear theory all modes grow and the states of all maps become different. However, numerical simulations of the model (14.8) show a much more ordered picture. After some transients one observes clustering: large groups of systems having exactly the same state. We illustrate this effect in Fig. 14.2. The number of clusters and the distribution of chaotic systems among them strongly depends on initial conditions. For large $N$ one observes situations both with a few clusters, and those where the number of clusters is of the order of $N$.

Clustering can be considered as weak synchronization, where the coupling is strong enough to bring some systems to the same state, but complete synchronization of all systems is nevertheless unstable. Note that, similarly to the case of two coupled maps (see the discussion in Section 13.4), the cluster attractors can have riddled basin boundaries, and are thus very sensitive to small noise. Moreover, the formation of clusters may be caused by finite numerical precision of calculations (similar to the numerical trap for two interacting systems, see Section 13.3): if the states of two systems coincide in a computer representation at some time, all future iterations will also be identical, even if the synchronized state is unstable.

## 14.2 Continuous-time systems

The treatment of complete synchronization in continuous-time chaotic systems can be performed in the same way as for discrete mappings. The simplest model is a linear interaction of two identical chaotic systems. Each $M$-dimensional chaotic system is described by a set of nonlinear ordinary differential equations (ODEs)

$$\frac{dx_k}{dt} = f_k(x_1, \ldots, x_M, t), \qquad k = 1, \ldots, M. \tag{14.9}$$

These equations may be nonautonomous, which means that forced systems are also included in the consideration.

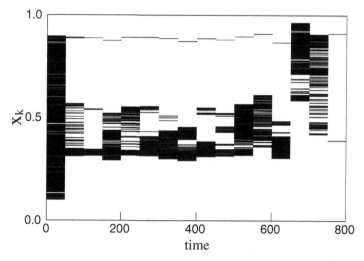

**Figure 14.2.** Evolution of an ensemble of 1000 globally coupled logistic maps $x \to 4x(1 - x)$ with the coupling constant $\varepsilon = 0.28$. The iterations start from random initial conditions, uniformly distributed in the interval $(0.1, 0.9)$. The values of $x_k$ are shown with bars at every 50th iteration. After $\approx 750$ iterations the system arrives at a two-cluster state.

A natural way to introduce dissipative coupling into a pair of identical systems[1] is by adding linear symmetric coupling terms to the r.h.s. (cf. discussion of dissipative coupling in Section 8.2)

$$\frac{dx_k}{dt} = f_k(x_1, \ldots, x_M, t) + \varepsilon_k(y_k - x_k), \tag{14.10}$$

$$\frac{dy_k}{dt} = f_k(y_1, \ldots, y_M, t) + \varepsilon_k(x_k - y_k). \tag{14.11}$$

With this coupling, the system possesses a synchronous chaotic solution $x_k(t) = y_k(t) = U_k(t)$ and one can straightforwardly study its stability.

For the deviations $v_k = y_k(t) - x_k(t)$ we obtain the linearized system

$$\frac{dv_k}{dt} = J_{kj} v_j - 2\varepsilon_k v_k, \tag{14.12}$$

where $J_{kj} = \partial f_k(U(t))/\partial x_j$ is the Jacobian matrix of partial derivatives. Asymptotically, for large $t$ the solutions of this linear system grow exponentially. The system (14.12) is $M$-dimensional; therefore there are $M$ transverse Lyapunov exponents (similarly to the existence of $M$ usual Lyapunov exponents in the $M$-dimensional noncoupled system (14.9)), and the largest of them determines the stability of inhomogeneous perturbations. Since the r.h.s. of (14.12) depends on the coupling constants $\varepsilon_k$, the largest transverse Lyapunov exponent $\lambda_\perp$ is a function of these constants, and the condition

$$\lambda_\perp(\varepsilon_k) < 0$$

defines the synchronization region. The simplest case is when all coupling constants have the same value $\varepsilon_1 = \cdots = \varepsilon_M = \varepsilon$. Then, with the ansatz $v_k = e^{-2\varepsilon t} \tilde{v}_k$ we can reduce (14.12) to the equations

$$\frac{d\tilde{v}_k}{dt} = J_{kj} \tilde{v}_j, \tag{14.13}$$

describing linear perturbations of the single chaotic system; their solutions grow with the largest Lyapunov exponent $\lambda_{\max}$. The stability condition now has the simple form

$$2\varepsilon > \lambda_{\max}.$$

The other type of synchronization for continuous-time systems, phase synchronization, was discussed in Chapter 10. Complete synchronization is more general in the sense that it can happen in any chaotic system independently of whether one can introduce phases and frequencies; it can be observed in self-sustained systems as well as in driven ones and in systems with discrete time maps. However, complete synchronization sets in only for large couplings and only for identical systems, whereas phase synchronization may be observed for small couplings and for different systems.

---

[1] Identity means also that the right-hand sides have an identical explicit time dependence, i.e., the drivings of both systems are identical.

## 14.3    Spatially distributed systems

Chaotic motion in spatially extended dynamical systems is often referred to as *space–time chaos*, with decaying correlations both in space and time. Popular models of space–time chaos include coupled map lattices, partial differential equations, and lattices of continuous-time oscillators. We refer to books [Kaneko 1993; Bohr *et al.* 1998] and review articles [Chaté and Manneville 1992; Cross and Hohenberg 1993] for details.

The notion of complete synchronization, when applied to space–time chaos, is used in two senses. One possible type of synchronization is the identity of chaotic motions at all sites. In this case one observes a spatially homogeneous chaotic (in time) regime in a distributed system. Another possibility is that the dynamics are chaotic both in space and time, but possess some symmetry. For example, in two spatial dimensions a field $u(x, y, t)$ may be $y$-independent while demonstrating one-dimensional space–time chaos in $x$. We discuss these two aspects of complete synchronization of space–time chaos below.

### 14.3.1    Spatially homogeneous chaos

One way of constructing a model with space–time chaos is to build a spatially distributed system as an array or a continuous set of individual chaotic elements. Such a system can be considered as a direct extension of the lattice of coupled periodic oscillators (see Eq. (11.14)). Apart from the coupled map lattices described by (14.7), two models are popular in this context. One is a discrete lattice of coupled chaotic oscillators. Similarly to the system of two oscillators (Eqs. (14.10) and (14.11)), the model can be written as

$$\frac{d\mathbf{x}_k}{dt} = \mathbf{f}(\mathbf{x}_k) + \sum_{\{j\}} \varepsilon(\mathbf{x}_j - \mathbf{x}_k),$$

where the summation in the coupling term is over the neighbors of the element $k$. In the case of a field that is continuous in space $\mathbf{u}(\mathbf{r}, t) = (u_1, \ldots, u_M)$, a model of reaction–diffusion equations is often used as a generalization of Eqs. (14.9)

$$\frac{\partial u_k}{\partial t} = f_k(\mathbf{u}) + D_k \nabla^2 u_k, \qquad k = 1, \ldots, M. \tag{14.14}$$

This system naturally appears in descriptions of chemical turbulence [Kuramoto 1984; Kapral and Showalter 1995], where the variables $u_k$ describe concentrations of different chemical reagents; they evolve in time due to chemical reactions and in space due to diffusion.

The synchronous solution of this system is the spatially homogeneous chaotic (in time) state $\mathbf{U}(t)$ governed by Eqs. (14.9). It exists if the boundary conditions admit such a solution, e.g., they are flux-free $\nabla u_k|_{\mathrm{B}} = 0$ (these boundary conditions are natural for chemical systems). In the chemical interpretation, the synchronous regime means a homogeneous-in-space distribution of the reagents. For a distributed system there are many transverse modes, and the stability condition can be formulated as the condition for all transverse Lyapunov exponents to be negative. For the system (14.14), these modes are eigenmodes of the Laplace operator with the corresponding boundary conditions. Let $v_k^{(l)}(\mathbf{r})$ be an eigenfunction of the Laplacian with the eigenvalue $\sigma$:

$$D_k \nabla^2 v_k^{(l)} = -\sigma^{(l)} v_k^{(l)}.$$

Then for the stability of the perturbation $v^{(l)}$ we obtain a linear system

$$\frac{dv_k^{(l)}}{dt} = J_{kj} v_j^{(l)} - \sigma v_k^{(l)}, \tag{14.15}$$

which is equivalent to (14.12). Thus for each spatially inhomogeneous mode the linearized equations (14.15) yield a spectrum of transverse Lyapunov exponents, and for stability of the spatially homogeneous chaos all of them must be negative. As the coupling in (14.14) is symmetric (purely diffusive), no effects of convective instability, as discussed in Section 14.1.1, can occur. For a domain of a characteristic length $L$, the second eigenvalue is of order $L^{-2}$. Therefore, in the thermodynamic limit $L \to \infty$ the spectrum of the diffusion operator has no gap (long-wave perturbations decay slowly). Complete synchronization in this limit is not possible: only for relatively small systems can the transverse Lyapunov exponents all be negative. Physically, this means that in a large system there are always long-wave modes whose Lyapunov exponents are close to the largest exponent of the chaotic spatially homogeneous solution, and these modes are thus unstable.

One remark is in order here. Above we have always assumed the space–time chaos to be "normal", i.e., with positive Lyapunov exponents. There are several examples of "anomalous" space–time chaos where the Lyapunov exponents are negative, see [Crutchfield and Kaneko 1988; Politi *et al.* 1993; Bonaccini and Politi 1997]. This "stable" chaos appears due to an unstable finite-size perturbation that cannot be described by a linearization [Torcini *et al.* 1995]. Here all our arguments based on the linear stability analysis are not applicable.

## 14.3.2  Transverse synchronization of space–time chaos

A space–time chaos may be inhomogeneous in some spatial directions, but be synchronized in other ones. The simplest example is given by two distributed systems with turbulent behavior that are coupled via dissipative coupling. As a representative example we consider two coupled Kuramoto–Sivashinsky equations

$$\frac{\partial u_1}{\partial t} + \frac{\partial^2 u_1}{\partial x^2} + \frac{\partial^4 u_1}{\partial x^4} + u_1 \frac{\partial u_1}{\partial x} = \varepsilon(u_2 - u_1),$$

$$\frac{\partial u_2}{\partial t} + \frac{\partial^2 u_2}{\partial x^2} + \frac{\partial^4 u_2}{\partial x^4} + u_2 \frac{\partial u_2}{\partial x} = \varepsilon(u_1 - u_2). \tag{14.16}$$

The Kuramoto–Sivashinsky equation is known to demonstrate space–time chaos if the spatial domain is large enough (see, e.g., [Hyman *et al.* 1986; Bohr *et al.* 1998]). The dissipative coupling is proportional to $\varepsilon$ and tends to make the states of two spatio-temporal systems equal at all sites and at every time. In fact, this model can be treated in the same way as the system of coupled ordinary differential equations (14.10) and (14.11): the synchronization threshold is determined by the largest Lyapunov exponent of space–time chaos:

$$2\varepsilon_c = \lambda_{\max}.$$

We illustrate synchronization in coupled Kuramoto–Sivashinsky equations in Fig. 14.3.

An interesting phenomenon is observed near the transition to complete synchronization. As shown in Chapter 13, such a transition in a nondistributed system is accompanied by a strong modulational intermittency, due to fluctuations of local growth

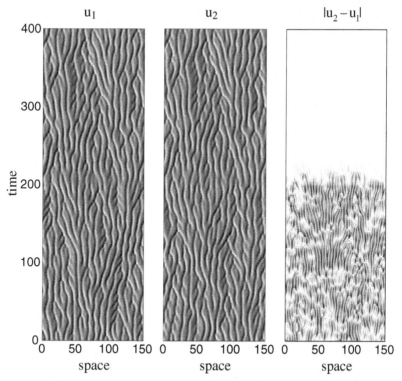

**Figure 14.3.** Synchronization transition in coupled Kuramoto–Sivashinsky equations (14.16). The time evolution of the fields $u_{1,2}$ and $|u_1 - u_2|$ is represented by gray-scale coding. During the time interval $0 < t < 200$ the coupling is zero, $\varepsilon = 0$, and the fields are independent. At time $t = 200$ the coupling with $\varepsilon = 0.1 > \varepsilon_c$ is switched on, resulting in the synchronous space–time chaos, $u_1(x, t) = u_2(x, t)$.

rates. In a distributed system (like the coupled Kuramoto–Sivashinsky equations) the growth rate fluctuates both in space and time. Thus an intermittent pattern is observed, with bursts appearing, wandering, and disappearing rather irregularly [Kurths and Pikovsky 1995].

A transversally synchronized space–time chaos appears naturally not only in artificially constructed models like (14.16), but in two- and three-dimensional (in space) partial differential equations. What is needed is an asymmetry of the diffusion operator with respect to spatial coordinates. Consider, e.g., the reaction–diffusion system (14.14) in a two-dimensional rectangular domain $0 < x < L_x, 0 < y < L_y$, where the diffusion is described by the operator

$$D_x \frac{\partial}{\partial x^2} + D_y \frac{\partial}{\partial y^2}.$$

The eigenmodes $u \propto \cos(\pi n_x x / L_x) \cos(\pi n_y y / L_y)$ have the eigenvalues

$$\sigma = \pi^2 (D_x n_x^2 L_x^{-2} + D_y n_y^2 L_y^{-2}), \qquad n_{x,y} = 0, 1, 2, \ldots .$$

Suppose now that $D_x L_x^{-2} \ll D_y L_y^{-2}$, e.g., the size in the $x$-direction $L_x$ is much larger than that in the $y$-direction $L_y$. Then the homogeneous state may be unstable with respect to many $x$-dependent modes, but spatial homogeneity in $y$ is preserved. Thus we can get a space–time chaos with respect to coordinate $x$, but complete synchronization in $y$-direction.

We illustrate this with the two-dimensional anisotropic complex Ginzburg–Landau equation (cf. (11.15))

$$\frac{\partial a(x, y, t)}{\partial t} = a - (1 + i c_3) |a|^2 a + (1 + i c_1) \frac{\partial^2 a}{\partial x^2} + d_1 \frac{\partial^2 a}{\partial y^2}. \tag{14.17}$$

Here the purely diffusive coupling in the $y$-direction can prevent the transverse instability and the observed field is homogeneous in $y$ (Fig. 14.4b). If the transverse diffusion constant $d_1$ is small, a two-dimensional turbulent field is observed (Fig. 14.4a).

### 14.3.3   Synchronization of coupled cellular automata

Cellular automata (see [Gutowitz 1990] and references therein) demonstrate weaker chaotic properties than coupled map lattices. Indeed, in cellular automata not only are time and space discrete, but also the field itself has discrete values only (usually one considers automata with two states "0" and "1"). Nevertheless, some cellular automata can demonstrate irregular behavior (this is possible in infinite lattices only). Dissipative coupling of cellular automata is not so trivial as for discrete mappings, as the states between "0" and "1" do not exist. So one uses a statistical coupling: the states at some spatial sites (chosen with a probability $p$) become completely identical, while the states of other sites remain different. The probability $p$ plays the role of the coupling constant: for $p = 1$ the synchronization is perfect after one time step; for

$p = 0$ there is no synchronization. Complete synchronization occurs if the coupling probability $p$ exceeds a critical value $p_c$. The synchronization transition is of directed percolation type (see e.g., [Grassberger 1995] and references therein): the sites where the states of two cellular automata are different constitute a spatiotemporal domain[2] that is fractal at the threshold, finite (i.e., ending after a finite time) in the synchronized state, and infinite (with constant density) in the asynchronous state.

## 14.4 Synchronization as a general symmetric state

In the previous consideration we described synchronization in the physical situation of two identical chaotic systems that become synchronized because of a mutual interaction. In a more general context, the synchronized solution is just a symmetric state, and the problem is to determine whether or not this state is possible and under what conditions it is attracting. Moreover, we can face situations where we cannot divide the whole system into interacting subsystems, but the same kind of "synchronization" can still occur.

Suppose we have two sets of variables

$$x_1, \ldots, x_M \quad \text{and} \quad y_1, \ldots, y_M$$

that are governed by a system of ordinary differential equations

$$\frac{dx_k}{dt} = F_k^x(x, y, t), \qquad \frac{dy_k}{dt} = F_k^y(x, y, t), \tag{14.18}$$

where the only assumption is the existence of a symmetric chaotic solution $x_k(t) = y_k(t) = U_k(t)$ for all $k$. Note that we do not assume that the whole system is symmetric

---

[2] This domain is called a cluster in the context of cellular automata.

(a)

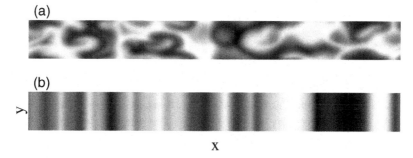

(b)

y

x

**Figure 14.4.** Snapshots of the real part of the field $a(x, y, t)$ that is governed by the complex Ginzburg–Landau equation (14.17). Panels (a) and (b) differ by the value of the diffusion constant $d_1$: in the asynchronous state (a) $d_1 = 0.5$, while in the synchronous (in the $y$-direction) state (b) $d_1 = 2$. Other parameters $c_1 = -1$ and $c_3 = 1.5$ are the same for both cases.

(see, e.g., the example of unidirectional coupling in Section 14.1.1 above, where the coupling is asymmetric but nevertheless a symmetric chaotic solution exists), and we do not even assume that the diagonal $x_k = y_k$ is invariant. The symmetric chaotic state can be interpreted as a synchronized one, and its stability towards transverse pertur-bations requires the study of the linearization of (14.18). Some linear perturbations do not violate the symmetry, but those that are asymmetric yield transverse Lyapunov exponents. The largest transverse exponent defines the stability of the symmetric state; depending on the parameters of interest this state can either be attracting or not.

As a simple example of a nonsymmetric system let us consider a popular model of two coupled one-dimensional maps

$$x(t+1) = f(x(t)) + \varepsilon_1(y(t) - x(t)),$$

$$y(t+1) = f(y(t)) + \varepsilon_2(x(t) - y(t)).$$

Note that although the coupling looks linear, in fact it is of complex nature and does not necessarily lead to synchronization. Stability of a symmetric state $x = y = U$ with respect to transverse perturbations leads to the linear problem for $v = \delta x - \delta y$

$$v(t+1) = v(t)[f'(U(t)) - \varepsilon_1 - \varepsilon_2].$$

Thus the transverse Lyapunov exponent is

$$\lambda_\perp = \langle \ln|f'(U(t)) - \varepsilon_1 - \varepsilon_2| \rangle.$$

Its dependence on the parameters $\varepsilon_{1,2}$ may be nontrivial; the regions of negative $\lambda_\perp$ are the regions of linearly stable symmetric dynamics. A simple expression for the transverse Lyapunov exponent can be written out for the coupled skew tent maps (13.5) (cf. (13.30))

$$\lambda_\perp = a \ln \left| \frac{1}{a} - \varepsilon_1 - \varepsilon_2 \right| + (1-a) \ln \left| \frac{1}{a-1} - \varepsilon_1 - \varepsilon_2 \right|.$$

This example shows that the transverse exponent is not directly related to the Lya-punov exponent of the symmetric chaos.

## 14.4.1 Replica-symmetric systems

One specific case of complete synchronization in symmetric systems has been sug-gested by Pecora and Carroll [1990]. Suppose we have a system of ODEs (14.9) and make a replica of one (or several) equations. For simplicity of presentation we write this for a three-dimensional autonomous model

$$\dot{x} = f_x(x, y, z),$$

$$\dot{y} = f_y(x, y, z),$$

$$\dot{z} = f_z(x, y, z), \qquad \dot{z}' = f_z(x, y, z').$$

Here the replica of the equation for $z$, with the same function $f_z$, is written for a new variable $z'$. Clearly, the symmetric state $x = x^0(t)$, $y = y^0(t)$, $z = z' = z^0(t)$ is always a solution of the system. In order to check the stability of this state, we write the equation for a small deviation $v = z - z'$:

$$\dot{v} = f_z'(x^0(t), y^0(t), z^0(t))v.$$

The linear perturbation $v$ grows exponentially in time

$$v \propto \exp(\lambda_\perp t),$$

and the transverse exponent (sometimes also called the conditional exponent, or the sub-Lyapunov one) $\lambda_\perp$ is given by

$$\lambda_\perp = \langle f_z' \rangle.$$

If this exponent is negative, the synchronous state $z = z'$ is linearly stable. Note that the transverse Lyapunov exponent is one of the Lyapunov exponents of the symmetric regime in the full four-dimensional system $(x, y, z, z')$ (because the symmetry-breaking perturbation $v$ is independent from other perturbations), but has nothing to do with the Lyapunov exponents of the original three-dimensional system $(x, y, z)$ (for the original system any perturbation of $z$ depends on the perturbations in $x$ and $y$).

The idea of replica synchronization is straightforwardly generalized for an $M$-dimensional chaotic system with $n < M$ equations being replicated. One says that the replicated variables constitute the driven, or slave, subsystem and the not-replicated variables constitute the driving, or master, subsystem. Note that for replica synchronization there is no parameter describing the coupling: one can say that the coupling is always strong, although partial – only $m$ variables are coupled. Thus, one can only check for a given chaotic system and for given replicated variables, if complete synchronization is possible. In some sense replica synchronization represents the property of chaotic systems to possess some stable directions in the phase space; sometimes this stability is associated with particular variables and groups of variables.

As an example we show in Fig. 14.5 replica synchronization for the Lorenz model (10.4). Here the variables $y$ and $z$ are replicated. One can see that the differences $|y' - y|$ and $|z' - z|$ decrease in time, so that eventually a symmetric chaotic state establishes in the full five-dimensional system. Pecora and Carroll [1990] found that when the coordinate $z$ is used as the driver, one of the transverse Lyapunov exponents is positive and synchronization does not occur.

## 14.5    Bibliographic notes

General coupling of chaotic systems has been considered in [de Sousa Vieira *et al.* 1992; Heagy *et al.* 1994b; Gade 1996; Güémez and Matías 1996; Lorenzo *et al.* 1996;

Pecora and Carroll 1998; Femat and Solis-Perales 1999; Pasemann 1999]. A particular case of global coupling was studied in [Kaneko 1990, 1991, 1997, 1998; Crisanti *et al.* 1996; Hasler *et al.* 1998; Zanette and Mikhailov 1998a,b; Balmforth *et al.* 1999; Glendinning 1999; Mendes 1999; Maistrenko *et al.* 2000]. Experiments on globally coupled arrays of chaotic electrochemical oscillators were performed by Wang *et al.* [2000a]. Gade [1996] and Manrubia and Mikhailov [1999] considered random coupling of chaotic elements.

Dolnik and Epstein [1996] considered synchronization of two coupled chaotic chemical oscillators. The effects of complete synchronization in distributed systems have been considered in [Pikovsky 1984a; Yamada and Fujisaka 1984]. Partial synchronization in a lattice of lasers was studied theoretically and experimentally by Terry *et al.* [1999], see also [Vieira 1999].

Synchronization of a pair of spatiotemporal chaotic systems was discussed in [Pikovsky and Kurths 1994; Kurths and Pikovsky 1995; Parekh *et al.* 1996; Sauer and Kaiser 1996; Hu *et al.* 1997; Kocarev *et al.* 1997; Jiang and Parmananda 1998; Boccaletti *et al.* 1999; Grassberger 1999]. Synchronization of cellular automata was considered in [Morelli and Zanette 1998; Urias *et al.* 1998; Bagnoli *et al.* 1999; Bagnoli and Rechtman 1999; Grassberger 1999].

Replica synchronization was discussed in [Pecora and Carroll 1990, 1998; Carroll and Pecora 1993a,b, 1998; Gupte and Amritkar 1993; Heagy and Carroll 1994; Tresser *et al.* 1995; Carroll 1996; Carroll *et al.* 1996; González-Miranda 1996a,b; Konnur 1996; Balmforth *et al.* 1997; Boccaletti *et al.* 1997; de Sousa Vieira and Lichtenberg 1997; Duane 1997; Güémez *et al.* 1997; Kim 1997; Pecora *et al.* 1997b,c; Zonghua and Shigang 1997a,b; Carroll and Johnson 1998; Johnson *et al.* 1998; He and Vaidya 1999; Mainieri and Rehacek 1999; Matsumoto and Nishi 1999; Morgül 1999; Voss 2000].

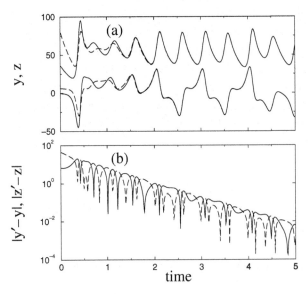

**Figure 14.5.** Replica synchronization in the Lorenz model. (a) Evolution of the variables $z$ and $z'$ (upper curve) and $y$ and $y'$ (lower curve); the replicated variables are shown with dashed lines. (b) Differences $|y - y'|$ (solid line) and $|z - z'|$ (dashed line) decrease exponentially in time.

Asymptotic stability of replica synchronization was studied in [Rong He and Vaidya 1992] by means of Lyapunov functions. An experimental observation of this effect in lasers was reported by Sugawara *et al.* [1994]. Some aspects of replica synchronization appeared to be useful for building communication schemes [Cuomo and Oppenheim 1993; Cuomo *et al.* 1993a,b; Gonzalez-Miranda 1999; Morgül and Feki 1999].

# Chapter 15

# Synchronization of complex dynamics by external forces

In this chapter we describe synchronization by external forces; we shall discuss effects other than those presented in Chapters 7 and 10. The content is not homogeneous: different types of systems and different types of forces are discussed here. Nevertheless, it is possible to state a common property of all situations: synchronization occurs when the driven system loses its own dynamics and follows those of the external force. In other words, the dynamics of the driven system are synchronized if they are stable with respect to internal perturbations. Quantitatively this is measured by the largest Lyapunov exponent: a negative largest exponent results in synchronization (note that here we are speaking not about a transverse or conditional Lyapunov exponent, but about the "canonical" Lyapunov exponent of a dynamical system). This general rule does not depend on the type of forcing or on the type of system; nevertheless, there are some problem-specific features. Therefore, we consider in the following sections the cases of periodic, noisy, and chaotic forcing separately. In passing, it is interesting to note that we can also interpret phase locking of periodic oscillations by periodic forcing (Chapter 7) as stabilization of the dynamics: a nonsynchronized motion (unforced, or outside the synchronization region) has a zero largest Lyapunov exponent, while in the phase locked state it is negative.[1]

Another important concept we present in this chapter is sensitivity to the perturbation of the forcing. In contrast to sensitivity to initial conditions, which is measured via the Lyapunov exponent, sensitivity to forcing has no universal quantitative characteristics. Moreover, one can speak of such sensitivity if one really can make small

---

[1] We remind the reader that there are exceptions to this general approach; one was mentioned in Section 14.3, where we described "stable" (in the sense of negative Lyapunov exponents) space–time chaos.

perturbations in the forcing. This makes sense only for chaotic or quasiperiodic driving, as in this case the force is determined by rather complex dynamics; for periodic forcing we do not have such perturbations. We will see that even synchronized (in the above explained sense) dynamics may be insensitive or sensitive to changes in the force; these two cases correspond to smooth and nonsmooth (fractal) relations between the driving and driven variables.

## 15.1 Synchronization by periodic forcing

In many systems chaos disappears if a periodic external force with sufficiently large amplitude is applied. In this context synchronization means that periodic forced oscillations are observed instead of chaos. General properties of this chaos-destroying synchronization are not universal: typically for very strong forcing regular regimes can be observed, but the dependence on the amplitude and frequency of the forcing, as well as on its form, does not follow any general rule. In any case, in the driven system the attractor is a limit cycle, so that the relation between the driven and driving variables is given by a smooth function.

We present here numerical results for the periodically driven Lorenz system

$$\frac{dx}{dt} = 10(y - x),$$

$$\frac{dy}{dt} = 28x - y - xz, \tag{15.1}$$

$$\frac{dz}{dt} = -\frac{8}{3}z + xy + \varepsilon \sin \omega t.$$

A region of periodic oscillations is shown in Fig. 15.1. The threshold of synchronization is large: no periodic regimes are observed if the amplitude of the forcing is smaller than 20. Two typical stable attractors are shown in Fig. 15.2.

## 15.2 Synchronization by noisy forcing

Synchronization by external noise means that the system forgets its own dynamics and its own initial conditions and follows the driving noise. For a particular system the synchronization transition is not seen, but it can be observed if we consider a replica of the system, i.e., if we compare two *identical* systems driven with the *same* noise, but having different initial conditions. (This also explains what kind of stability is measured by the Lyapunov exponent in a noisy system: it is the sensitivity to perturbations of the initial conditions, and not sensitivity to the driving noise.) In this setup, the difference between positive and negative Lyapunov exponents will be readily seen: for positive exponents, the systems' trajectories will follow their initial

conditions and will remain different; while for negative exponents they will forget their
initial conditions and approach each other, i.e., they will synchronize. This coincidence
of the dynamics of two systems driven by the same noise is trivial if one considers the
simplest linear case:

$$\frac{dx}{dt} = -x + \xi(t), \qquad \frac{dy}{dt} = -y + \xi(t).$$

The behavior of a single system consists of decaying, free oscillations depending on
the initial conditions (the homogeneous part) plus forced oscillations depending on
noise only (the inhomogeneous part). For large times the whole dependence on the
initial conditions disappears, and the states become identical $x = y$ (this can also

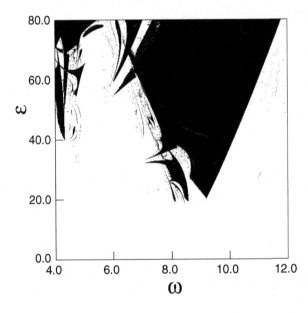

Figure 15.1. The region of
periodic regimes (shown by
black dots) in the forced
Lorenz system (15.1) in the
plane of the parameters of
the forcing. We have started
the simulation from
particular initial conditions
$x = y = 0.001, z = 0$, so
that the possible
multistability (e.g.,
coexistence of periodic and
chaotic solutions) is not
revealed.

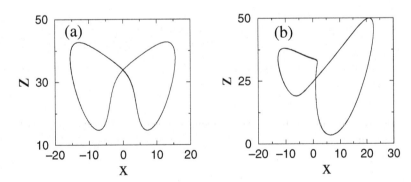

Figure 15.2. Two periodic regimes in the forced Lorenz system. (a) $\varepsilon = 60, \omega = 10$.
(b) $\varepsilon = 80, \omega = 4$.

be seen from the equation for the difference $x - y$). Essentially the same happens in nonlinear systems as well, but the problem of determining the stability is less trivial.

Synchronization by common noise occurs without any direct interaction between the oscillators, and is independent of the number of oscillators. Thus, the same effect will be observed for any large ensemble of identical nonlinear systems driven by the same noise: all systems will synchronize provided the Lyapunov exponent is negative. Below we discuss the cases of noisy forced periodic and chaotic oscillations.

### 15.2.1 Noisy forced periodic oscillations

We have already discussed the effect of noise on periodic oscillations in Chapter 9. There we studied the diffusion properties of the phase; now the main objects of our interest are the stability properties of the dynamics. They are nontrivial because an autonomous periodic oscillator has zero Lyapunov exponent corresponding to the phase shift. With a noisy external force, this exponent is generally nonzero, and the main question is whether it is positive or negative.

As outlined in Section 9.1, the simplest equation for the phase dynamics in the presence of noise reads

$$\frac{d\phi}{dt} = \omega_0 + \xi(t).$$

The r.h.s. does not depend on the value of the phase and therefore the Lyapunov exponent vanishes, $\lambda = \langle d\dot{\phi}/d\phi \rangle = 0$. In other words, the phase remains neutral to the perturbations of initial conditions. This degeneracy disappears if the forcing term depends on the phase (cf. Eq. (9.1)):

$$\frac{d\phi}{dt} = \omega_0 + \varepsilon Q(\phi, \xi(t)),$$

where $\varepsilon$ characterizes the amplitude of the forcing. Thus, in general one can expect either a positive or negative Lyapunov exponent.

As a particular simple example of a noise-driven self-sustained oscillator we take a generalization of the pulse-driven model considered in Section 7.3.3. There we had an oscillator forced with a periodic sequence of $\delta$-pulses (7.64). Now we consider a random sequence of pulses

$$p(t) = \sum_{n=-\infty}^{\infty} \xi_n \delta(t - t_n), \tag{15.2}$$

with random amplitudes $\xi_n$ and applied at random times $t_n$. An obvious modification of the circle map (7.68) gives a noisy map (we take here for simplicity the vanishing parameter $\alpha = 0$)

$$\phi_{n+1} = \phi_n + \omega_0 T_n + \varepsilon \xi_n \cos \phi_n. \tag{15.3}$$

Due to the randomness of the time intervals $T_n$ and the amplitudes of the pulses $\xi_n$, the time evolution of the phase is irregular, so one can hardly speak of synchronization or phase locking in the common sense. However, we will show that the phase may become adjusted to the external force and in this sense some synchronization features do appear.

Let us calculate the sensitivity of the phase $\phi_n$ to variations of the initial phase $\phi_0$. This is nothing but the usual calculation of the Lyapunov exponent:

$$\lambda = \frac{\langle \ln |d\phi_{n+1}/d\phi_n| \rangle}{\langle T \rangle} = \frac{\langle \ln |1 - \varepsilon \xi \sin \phi| \rangle}{\langle T \rangle}. \tag{15.4}$$

Here $\langle T \rangle$ is the mean time interval between consecutive pulses. The dependence on the force amplitude $\varepsilon$ (Fig. 15.3) is nonmonotonic: for small amplitudes the exponent is negative; for large amplitudes it is positive.

To show these properties analytically, for small $\varepsilon$ we can expand (15.4) in a Taylor series in $\varepsilon$ to obtain

$$\lambda \approx \frac{1}{\langle T \rangle} \left( -\varepsilon \langle \xi \sin \phi \rangle - \frac{\varepsilon^2}{2} \langle \xi^2 \sin^2 \phi \rangle \right).$$

In general, in order to perform averaging, one has to know the probability distribution of the phase. For small $\varepsilon$ and large fluctuations of the time intervals $T_n$ this distribution is nearly homogeneous, thus the first term vanishes and the leading term in the Lyapunov exponent is negative:

$$\lambda \propto -\frac{\varepsilon^2}{\langle T \rangle}.$$

The negative Lyapunov exponent means that after a finite time, the oscillator phase "forgets" its initial value and follows the external force (Fig. 15.4). We give the following "hand-waving" explanation of why the Lyapunov exponent is negative. Let us consider the random sequence of pulses as consisting of periodic patches. For each periodic patch, according to the theory of phase locking (Chapter 7), either a synchronized (with negative Lyapunov exponent) or a quasiperiodic (with zero Lyapunov exponent) state can be observed. Randomly "switching" these patches we "mix" these two situations, so that the average exponent is negative.

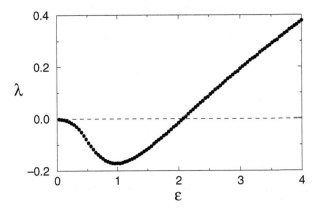

Figure 15.3. The Lyapunov exponent $\lambda$ of the random mapping (15.3) as a function of the noise amplitude $\varepsilon$. The time intervals $\omega_0 T_n$ are independent random numbers distributed exponentially with mean value 1; the pulse amplitudes $\xi_n$ have Gaussian distribution with zero mean and unit variance.

For large amplitude $\varepsilon$ of pulses, we can use another approximation,[2] neglecting the constant term under the logarithm in (15.4):

$$\lambda \approx \frac{\langle \ln |\varepsilon \xi \sin \phi| \rangle}{\langle T \rangle} \propto \frac{\ln \varepsilon + \langle \ln |\xi \sin \phi| \rangle}{\langle T \rangle}.$$

The Lyapunov exponent is thus positive for large $\varepsilon$. For a single forced periodic oscillator this does not have any meaning. But if we prepare two replica systems driven with the same noise and having close (even nearly identical) initial conditions, the difference between these systems will grow exponentially, and after a short time they will demonstrate almost independent oscillations, i.e., desynchronization.

Concluding our considerations we mention that in the mathematical literature, the object that appears during the evolution of an ensemble of identical systems driven by the same noise is called a *random attractor* [Crauel and Flandoli 1994; Arnold 1998].

### 15.2.2  Synchronization of chaotic oscillations by noisy forcing

The largest Lyapunov exponent of a chaotic motion is positive, but an external noise may make it negative. One possible mechanism is the change of stability induced by a coordinate-dependent (modulational) noisy force, as in the case of periodic oscillations described in Section 15.2.1. Another possibility is that noise does not influence stability directly, but changes the distribution in the phase space. Due to this change, some "stable" regions in the phase space may be visited more frequently, thus decreasing the largest Lyapunov exponent. This mechanism even works with additive noise. Let us consider as an example two one-dimensional maps subject to the same noise:

---

[2] Although in the derivation of (15.3) we assumed smallness of the forcing, we can consider the random mapping (15.3) as a model of its own and study it for arbitrary values of the parameters.

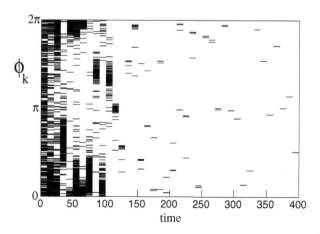

**Figure 15.4.** Evolution of an ensemble of 500 oscillators obeying the random mapping (15.3). The phases are shown at every tenth iteration by a horizontal bar. The statistical distributions of $T_n, \xi_n$ are as in Fig. 15.3. The forcing amplitude $\varepsilon = 0.4$ corresponds to negative Lyapunov exponent: all initially uniformly distributed phases after $\approx 300$ iterations collapse to a single state.

$$x(t+1) = f(x(t)) + \xi(t),$$
$$y(t+1) = f(y(t)) + \xi(t).$$

Clearly, the synchronous state $x(t) = y(t) = U(t)$ is a solution. In order to find its stability, we look at a small difference between the variables $v = x - y$ obeying the linearized equation

$$v(t+1) = f'(U(t))v(t). \tag{15.5}$$

This is exactly the linearized equation for perturbations in the single map whose divergence rate is the Lyapunov exponent

$$\lambda = \langle \ln |f'(U)| \rangle. \tag{15.6}$$

The averaging in (15.6) is performed over the invariant measure in the mapping with noise, and depends on the noise intensity. Thus changing the noise and/or the parameters of the map one can observe a transition from positive to negative Lyapunov exponent, i.e., a transition from asynchronous to synchronous states.

We emphasize that Eq. (15.5) is very similar to Eq. (13.10), describing the linear stage of synchronization of coupled autonomous chaotic mappings. Thus the whole statistical theory developed in Section 13.3 is valid for noisy systems as well. At the transition to synchronization one observes modulational intermittency with the properties described in Section 13.3. In particular, the numerical trap described in Section 13.3 exists for identical noise-driven systems too: even if the Lyapunov exponent is positive, two (or many) systems can appear as synchronous in numerical simulations if their states at some moment in time coincide in the computer representation (i.e., if the distance is smaller than the numerical precision). This effect, of course, disappears if a small mismatch is taken into account.

The difference with the purely deterministic case is that here we cannot consider topological properties of attractors in the phase space, as we did in Section 13.4. Indeed, in the noisy case, the topological structures of a deterministic system (periodic orbits, etc.) are smeared out. Nontrivial topological structures can be observed if the forcing is chaotic, as discussed in Section 15.3.

## 15.3    Synchronization of chaotic oscillations by chaotic forcing

### 15.3.1    Complete synchronization

One possible realization of chaotic forcing has already been discussed in Chapter 14 as unidirectional coupling. We rewrite the system defined by the interaction operator (14.3) as

$$x(t+1) = f(x(t)), \tag{15.7}$$

$$y(t + 1) = (1 - \varepsilon) f(y(t)) + \varepsilon f(x(t)). \tag{15.8}$$

The chaotic force generated by mapping (15.7) acts on system (15.8) in such a way that a completely synchronous state $y = x$ is possible. For the stability of the synchronous regime, we can again use the theory described in Chapters 13 and 14. In an experimental realization, one does not need to construct two identical systems: it is enough to record a signal generated by a chaotic oscillator and to use this signal as the forcing (see [Tsukamoto *et al.* 1996, 1997]). The synchronization can be easily identified as the coincidence of the generated signal with the forcing.

## 15.3.2  Generalized synchronization

Complete synchronization by chaotic forcing is possible only when the system possesses a symmetry, so that a regime where all the variables of the driven and the driving systems are equal is possible. If there is no symmetry, it is still possible that the driven system follows the driving, although in a weaker sense. Let us consider a general situation with unidirectional coupling:

$$\mathbf{x}(t + 1) = \mathbf{f}(\mathbf{x}(t)), \tag{15.9}$$

$$\mathbf{y}(t + 1) = \mathbf{g}(\mathbf{x}(t), \mathbf{y}(t)), \tag{15.10}$$

where $\mathbf{x}$ and $\mathbf{y}$ are vectors. Here the variables $\mathbf{x}$ belong to the driving system, while the variables $\mathbf{y}$ describe the driven one. We do not assume any symmetry between $\mathbf{x}$ and $\mathbf{y}$, moreover, the dimensions of these two parts can be different. When the state of the driven system $\mathbf{y}$ is completely determined by the state of the driving system, one speaks of **generalized synchronization**. In mathematical language, the existence of a one-to-one (injective) function that maps $\mathbf{x}$ to $\mathbf{y}$

$$\mathbf{y} = \mathbf{H}(\mathbf{x}) \tag{15.11}$$

is assumed. Technically, it is sometimes more convenient to establish a relation between $\mathbf{y}(t + 1)$ and $\mathbf{x}(t)$ writing $\mathbf{y}(t + 1) = \tilde{\mathbf{H}}(\mathbf{x}(t))$ (see, e.g., Fig. 15.5 later), because these two variables are directly related through Eq. (15.10). If the map (15.9) is invertible, this is equivalent to the definition (15.11).

Another way to characterize generalized synchronization is to make a replica of the driven system. In this case we enlarge Eqs. (15.9) and (15.10) to

$$\mathbf{x}(t + 1) = \mathbf{f}(\mathbf{x}(t)),$$

$$\mathbf{y}(t + 1) = \mathbf{g}(\mathbf{x}, \mathbf{y}),$$

$$\mathbf{y}'(t + 1) = \mathbf{g}(\mathbf{x}, \mathbf{y}').$$

This can be considered as an example of replica synchronization discussed in Section 14.4: in the case of a stable response one observes a symmetric solution with $\mathbf{y} = \mathbf{y}'$.

### Nonsmooth generalized synchronization

Equation (15.10) describes the dynamics of the forced system, and it is clear that the state of $\mathbf{y}$ follows the forcing $\mathbf{x}$ independently of the initial condition $\mathbf{y}(0)$ only if the dynamics of $\mathbf{y}$ are stable, i.e., the maximal Lyapunov exponent corresponding to $\mathbf{y}$ is negative. This condition is necessary but not sufficient. Indeed, it can happen that there is a multistability in $\mathbf{y}$, i.e., one trajectory $\mathbf{x}(t)$ may cause two (or more) stable responses $\mathbf{y}(t)$. Another problem is that the stability in $\mathbf{y}$ may ensure the existence of the function $\mathbf{H}$, but not its smoothness. Actually, the function $\mathbf{H}$ can be fractal.

To demonstrate this, we consider as an example a one-dimensional[3] driven system with a negative Lyapunov exponent $\lambda_y$. Then, following Paoli et al. [1989b], we compare the Lyapunov dimensions[4] of the attractor in the driving and composed (driving and driven) systems; we denote them as $D_x^{(L)}$ and $D_{xy}^{(L)}$, respectively. For a smooth relation between $y$ and $x$, the dimension of the composed system should be the same as that of the driving; if $D_{xy}^{(L)} > D_x^{(L)}$, then the relation is not smooth. Because the coupling is unidirectional, the Lyapunov exponents $\lambda_j$ of the driving system are not influenced by the driven one, hence $D_{xy}$ cannot be smaller than $D_x$. The Lyapunov dimension is given by the formula of Kaplan and Yorke [1979]

$$D_x^{(L)} = \mathcal{D} + \frac{1}{|\lambda_{\mathcal{D}+1}|} \sum_{j=1}^{\mathcal{D}} \lambda_j, \tag{15.12}$$

where the Lyapunov exponents are sorted in descending order and the integer part $\mathcal{D}$ of the dimension is defined as the largest integer such that

$$\sum_{j=1}^{\mathcal{D}} \lambda_j > 0.$$

For the composed system, the dimension is determined by the exponents $\lambda_j$ and $\lambda_y$. Because only the first $\mathcal{D} + 1$ Lyapunov exponents appear in (15.12), if $\lambda_y < \lambda_{\mathcal{D}+1}$ then $D_{xy}^{(L)} = D_x^{(L)}$. Suppose now that $\lambda_{\mathcal{D}} > \lambda_y > \lambda_{\mathcal{D}+1}$. Then

$$D_{xy}^{(L)} = \mathcal{D} + \frac{1}{|\lambda_y|} \sum_{j=1}^{\mathcal{D}} \lambda_j > D_x^{(L)}.$$

The dimension increases even more if $\lambda_{\mathcal{D}} < \lambda_y$. This increase of the dimension through driving means that the response $\mathbf{y}$ is not a smooth function of $\mathbf{x}$, otherwise the dimension does not increase. Thus, generalized synchronization with a smooth relation (15.11) can be observed only if the stability of the driven system is sufficiently strong.

In particular, from the above arguments it follows that when the driving chaotic system is a one-dimensional map (e.g., the tent or the logistic map), the function $\mathbf{H}$

---

[3] It is easy to see that the same consideration is valid for an arbitrary dimension of the driven system.

[4] Discussion of the Lyapunov dimension can be found, e.g., in [Schuster 1988; Ott 1992].

is always fractal. Indeed, a one-dimensional map can be considered as a limiting case of a two-dimensional mapping, the negative Lyapunov exponent of which tends to $-\infty$. Thus, this map has Lyapunov exponents $\lambda_1, -\infty$. For the composed system, the Kaplan–Yorke formula yields $D_{xy}^{(L)} = \min(2, 1 + \lambda_1/|\lambda_y|)$, where $\lambda_y$ is the maximal exponent of the driven system; and this dimension is always larger than one. Hence, to have a nontrivial transition between smooth and nonsmooth functions $\mathbf{H}$, one has to consider at least a two-dimensional driving chaotic map.

### Smooth and nonsmooth generalized synchronization: an example

The function $\mathbf{H}$ can be a fractal one even in a simple case when the subsystem (15.10) is a one-dimensional linear driven system

$$y(t+1) = \gamma y(t) + q(\mathbf{x}(t)). \tag{15.13}$$

The response is stable (i.e., (15.11) holds) if $|\gamma| < 1$; for definiteness we assume $\gamma > 0$.

A detailed investigation of this problem for a generalized baker's map (see Eq. (15.16) below) as the drive and a linear map of type (15.13) as the response was carried out by Paoli *et al.* [1989a]. They analysed the spectrum of singularities of the response (the so-called $f(\alpha)$-spectrum) and described its metamorphoses when the time constant of the driven system $\gamma$ is changed. The $f(\alpha)$-spectrum is a Legendre transform of generalized dimensions (see [Badii and Politi 1997] for details). From the results of Paoli *et al.* [1989a] it follows that small values of $\gamma$ do not influence the dimensions. As $\gamma$ grows and becomes larger than the smallest contraction rate of the baker's map, some generalized dimensions of the response differ from that of the driving. This is an indication that the function $\mathbf{H}$ is nonsmooth. For large $\gamma$ all the dimensions change.

Below we describe a simplified version of the analysis by Paoli *et al.* [1989a], following Hunt *et al.* [1997]. We start with iterations of (15.13)

$$y(t+1) = q(\mathbf{x}(t)) + \gamma y(t)$$
$$= q[\mathbf{f}^{-1}(\mathbf{x}(t+1))] + \gamma\{\gamma y(t-1) + q[\mathbf{f}^{-2}(\mathbf{x}(t+1))]\} = \cdots$$

to obtain a formal expression for the function $H$ relating $y(t+1)$ and $\mathbf{x}(t+1)$:

$$H(\mathbf{x}) = \sum_{j=1}^{\infty} \gamma^{j-1} q(\mathbf{f}^{-j}(\mathbf{x})). \tag{15.14}$$

Clearly, this function exists if $\gamma < 1$ and the forcing function $q$ is bounded. To check the smoothness, we differentiate (15.14) and obtain

$$\nabla H = \sum_{j=1}^{\infty} \gamma^{j-1} J[\mathbf{f}^{-j}(\mathbf{x})] \nabla q(\mathbf{f}^{-j}(\mathbf{x})), \tag{15.15}$$

where $J$ is the Jacobian. A sufficient condition for the derivatives at the point $\mathbf{x}$ to exist is the convergence of the series (15.15), which will be ensured if the values of $\gamma^j \|J\mathbf{f}^{-j}(\mathbf{x})\|$ decay geometrically. The most dangerous for convergence are large values of the Jacobian $\|J\mathbf{f}^{-j}(\mathbf{x})\|$, which is calculated from *inverse* iterations of the mapping (15.9). This corresponds to small values of the derivative calculated for *forward* iterations, i.e., to the most stable direction of the mapping (15.9). We see again that the smoothness of the function $H$ is determined by the relation between the Lyapunov exponent of the driven system $\lambda_y = \ln \gamma$ and the most negative exponent of the driving system (15.9).

Let us assume that the driving map (15.9) is two-dimensional, with one positive and one negative Lyapunov exponent, the latter to be denoted as $\lambda_x$. Then from (15.15) it follows that the function $H$ is continuous "on average" (i.e., the derivative exists at almost all points) provided that

$$\lambda_y - \lambda_x < 0,$$

i.e., if the (averaged) contraction rate of the driven system is stronger than the (averaged) contraction rate in the driving. One can recognize that this is exactly the condition of nongrowth of the Lyapunov dimension calculated according to (15.12).

A more precise analysis can be performed if we define the $(\mathbf{x}, y)$-dependent Lyapunov exponent, in analogy to the thermodynamic formalism approach used in Chapter 13. We define the so-called past-history Lyapunov exponents: for the driving subsystem we write

$$\|J\mathbf{f}^{-T}(\mathbf{x})\| \propto e^{-T\Lambda_x(\mathbf{x})},$$

and for the driven subsystem

$$\frac{dy(t)}{dy(t-T)} \propto e^{T\Lambda_y(\mathbf{x},y)},$$

(in the particular case of (15.13), $\Lambda_y = \ln \gamma$, but in general this exponent is coordinate-dependent). Then the function $H$ has a derivative at point $(\mathbf{x}, y)$ if

$$\Lambda_y(\mathbf{x}, y) - \Lambda_x(\mathbf{x}) < 0.$$

If the negative Lyapunov exponent $\Lambda_x(\mathbf{x})$ of the mapping (15.9) fluctuates from point to point, the function $H$ may be smooth at some points and fractal at others.

We illustrate the transition between smooth and nonsmooth generalized synchronization in Fig. 15.5. Here the driving system is a two-dimensional generalized baker's map defined in the unit square $0 \le x_1, x_2 < 1$

$$[x_1(t+1), x_2(t+1)] = \begin{cases} [\beta x_1(t), x_2(t)/\alpha] \\ \quad \text{if } x_2(t) < \alpha, \\ [\beta + (1-\beta)x_1(t), (x_2(t)-\alpha)/(1-\alpha)] \\ \quad \text{if } x_2(t) > \alpha. \end{cases} \quad (15.16)$$

The parameters $\alpha$ and $\beta$ are both less than $1/2$. The driven system is linear

$$y(t+1) = \gamma y(t) + \cos(2\pi x_1(t)). \quad (15.17)$$

It is easy to see that the unstable direction in the driving map is $x_2$ and the stable direction is $x_1$. The natural measure is uniform in the $x_2$-direction and varies wildly in the $x_1$-direction, provided that $\alpha \neq \beta$. To calculate the past-history stable Lyapunov exponent for a given point $x_1, x_2$, we have to determine where the pre-images of $x_1, x_2$ were wandering. Denoting

$$a_- = \lim_{T \to \infty} \frac{T_-}{T}, \qquad a_+ = \lim_{T \to \infty} \frac{T_+}{T},$$

where $T_-$ ($T_+$) is the number of pre-images with $x_2 < \alpha$ ($x_2 > \alpha$), we get for the past-history exponent

$$\Lambda(x_1, x_2) = a_- \ln \beta + a_+ \ln(1 - \beta).$$

The values of $a_\pm$ are different for different trajectories; they can be obtained using the methods of symbolic dynamics. The natural symbolic description associates two

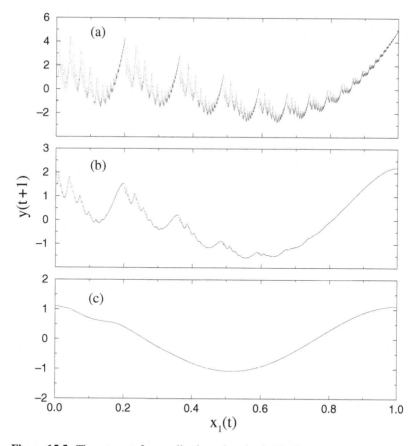

**Figure 15.5.** Three types of generalized synchronization in the system (15.16) and (15.17), with $\alpha = 0.1$ and $\beta = 0.2$. (a) $\gamma = 0.8$; the curve $\tilde{H}$ is fractal. (b) $\gamma = 0.6$; the curve is not differentiable in a dense set of points, but differentiable "on average". (c) $\gamma = 0.1$; the curve is smooth.

symbols with the regions $x < \alpha$ and $x > \alpha$, and all possible sequences of these symbols exist. Thus trajectories with all $0 \le a_\pm \le 1$ exist. This means that

$$\ln \beta \le \Lambda(x_1, x_2) \le \ln(1 - \beta).$$

But the most probable values for $a_\pm$ are $a_- = \alpha$, $a_+ = 1 - \alpha$, hence the average negative Lyapunov exponent of the mapping is

$$\lambda_x = \alpha \ln \beta + (1 - \alpha) \ln(1 - \beta).$$

The Lyapunov exponent of the driven system is a constant $\lambda_y = \ln \gamma$. So, according to the Kaplan–Yorke formula, the Lyapunov dimension of the whole system is larger than that of the driving if $|\lambda_x| > |\ln \gamma|$. But even if this inequality does not hold, in the region

$$|\lambda_x| < |\ln \gamma| < |\ln \beta|$$

there are points $x_1$, $x_2$ for which the past-history Lyapunov exponent $\Lambda$ is larger that $\ln \gamma$. At these points (they are everywhere dense, although their measure is zero) the function $H$ is not differentiable. We illustrate different cases of generalized synchronization in system (15.16) and (15.17) in Fig. 15.5.

### 15.3.3  Generalized synchronization by quasiperiodic driving

It is interesting to note that the transition from smooth to nonsmooth generalized synchronization occurs not only for chaotically driven systems, but also for quasiperiodically driven ones. In the latter case, the driving system is described by a torus in the phase space. If the driven system also behaves quasiperiodically, then the trajectory lies on a torus in the enlarged phase space. But there is also another possibility – when the driven system, being stable, has a *strange nonchaotic attractor*. The strange nonchaotic attractor has a negative largest Lyapunov exponent (it is therefore nonchaotic), but is fractal (therefore strange). Similarly to the case of nonsmooth generalized synchronization of chaos, the relation between the driven and driving systems for strange nonchaotic attractors is highly nontrivial: a functional relation between the quasiperiodic driving and the driven system either does not exist, or is represented by a fractal curve.

We illustrate such a strange nonchaotic attractor with the quasiperiodically driven logistic map

$$\begin{aligned} x(t+1) &= x(t) + \omega, \\ y(t+1) &= a - y^2(t) + b \cos 2\pi x(t). \end{aligned} \tag{15.18}$$

Here $\omega = (\sqrt{5} - 1)/2$ can be interpreted as the irrational forcing frequency. Two cases are shown in Fig. 15.6, corresponding to a smooth and a fractal relation between $y$ and $x$. In both cases the Lyapunov exponent in the driven system is negative, so that the regime in Fig. 15.6b is nonchaotic.

## 15.4  **Bibliographic notes**

The effect of periodic forcing on different chaotic systems has been studied numerically in papers [Aizawa and Uezu 1982; Anishchenko and Astakhov 1983; Kuznetsov *et al.* 1985; Bezaeva *et al.* 1987; Landa and Perminov 1987; Landa *et al.* 1989; Dykman *et al.* 1991; Rosenblum 1993; Franz and Zhang 1995; Tamura *et al.* 1999]. Experiments with a chaotic backward-wave oscillator have been performed by Bezruchko [1980] and Bezruchko *et al.* [1981].

Synchronization of identical nonlinear systems driven by the same noise was described in papers [Pikovsky 1984b,c, 1992; Yu *et al.* 1990]. In our presentation we follow [Pikovsky 1984b,c]. Dependence of the largest Lyapunov exponent on the noise in chaotic systems was studied by Matsumoto and Tsuda [1983]. For examples of identical chaotic systems driven with the same noise see [Maritan and Banavar 1994; Khoury *et al.* 1996, 1998; Ali 1997; Longa *et al.* 1997; Sánchez *et al.* 1997; Minai and Anand 1998, 1999a; Shuai and Wong 1998]. In particular, [Maritan and Banavar 1994; Shuai and Wong 1998] observed fake synchronization due to finite precision of computer simulations, see [Pikovsky 1994; Herzel and Freund 1995] for a discussion of this artifact. Experiments with noise-driven electronic circuits are reported in [Khoury *et al.* 1998].

Unidirectional coupling of chaotic systems have been studied theoretically and numerically in [Pecora and Carroll 1991; Rulkov *et al.* 1995; Abarbanel *et al.* 1996; Kapitaniak *et al.* 1996; Kocarev and Parlitz 1996; Konnur 1996; Pyragas 1996, 1997; Rulkov and Suschik 1996; Ali and Fang 1997; Brown and Rulkov 1997a,b; Hunt *et al.* 1997; Liu and Chen 1997; Parlitz *et al.* 1997; Carroll and Johnson 1998; Johnson *et al.* 1998; Baker *et al.* 1999; Liu *et al.* 1999; Minai and Anand 1999b; Parlitz and Kocarev 1999; Santoboni *et al.* 1999]. In our discussion of smooth and nonsmooth generalized synchronization we follow Paoli *et al.* [1989a]; Hunt *et al.* [1997] and

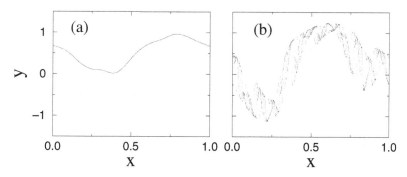

**Figure 15.6.** The attractors in the quasiperiodically forced logistic map (15.18). (a) For $\varepsilon = 0.3$ and $a = 0.9$, the driven variable $y$ is a smooth function of $x$; hence one can speak of smooth generalized synchronization between $x$ and $y$. (b) For $\varepsilon = 0.45$ and $a = 0.8$, the fractal strange nonchaotic attractor is observed: although the driven system follows the driving, there is no smooth relation between the variables $y$ and $x$.

Stark [1997], see also papers [Kaplan *et al.* 1984; Badii *et al.* 1988; Mitschke *et al.* 1988; Paoli *et al.* 1989b; Mitschke 1990; Pecora and Carroll 1996] where the response of linear systems to chaotic signals was considered, as well as [de Sousa Vieira and Lichtenberg 1997]. For experimental observations of generalized synchronization see [Peterman *et al.* 1995; Abarbanel *et al.* 1996; Gauthier and Bienfang 1996; Rulkov and Suschik 1996; Tsukamoto *et al.* 1997; Tang *et al.* 1998b]. Peng *et al.* [1996] and Tamasevicius and Čenys [1997] discuss how synchronization of hyperchaotic systems may be achieved using one scalar chaotic signal.

Strange nonchaotic attractors in quasiperiodically driven systems were introduced in [Grebogi *et al.* 1984] and since then have been studied both theoretically [Romeiras *et al.* 1987; Ding *et al.* 1989; Brindley and Kapitaniak 1991; Heagy and Hammel 1994; Pikovsky and Feudel 1994, 1995; Kuznetsov *et al.* 1995; Keller 1996; Lai 1996b; Nishikawa and Kaneko 1996; Yalcinkaya and Lai 1997; Prasad *et al.* 1998 and references therein] and experimentally [Ditto *et al.* 1990; Zhou *et al.* 1992; Yang and Bilimgut 1997; Zhu and Liu 1997].

# Appendices

# Appendix A1

# Discovery of synchronization by Christiaan Huygens

In this Appendix we present translations of the original texts of Christiaan Huygens where he describes the discovery of synchronization [Huygens 1967a,b].

## A1.1 A letter from Christiaan Huygens to his father, Constantyn Huygens[1]

26 February 1665.

While I was forced to stay in bed for a few days and made observations on my two clocks of the new workshop, I noticed a wonderful effect that nobody could have thought of before. The two clocks, while hanging [on the wall] side by side with a distance of one or two feet between, kept in pace relative to each other with a precision so high that the two pendulums always swung together, and never varied. While I admired this for some time, I finally found that this happened due to a sort of sympathy: when I made the pendulums swing at differing paces, I found that half an hour later, they always returned to synchronism and kept it constantly afterwards, as long as I let them go. Then, I put them further away from one another, hanging one on one side of the room and the other one fifteen feet away. I saw that after one day, there was a difference of five seconds between them and, consequently, their earlier agreement was only due to some sympathy that, in my opinion, cannot be caused by anything other than the imperceptible stirring of the air due to the motion of the pendulums. Yet the clocks are inside closed boxes that weigh, including all the lead, a little less than a hundred pounds each. And the vibrations of the pendulums, when

---

[1] Translation from French by Carsten Henkel.

they have reached synchronism, are not such that one pendulum is parallel to the other, but on the contrary, they approach and recede by opposite motions. When I put the clocks closer together, I saw that the pendulums adopted the same way of swinging. In addition, I took a square table of three feet, one inch thick, and put it between the two [clocks] so that it touched the ground below and was so high that it entirely covered the clocks and in this way, separated one from the other. Nevertheless, the synchronism remained as it had been before, over whole days and nights; even when I perturbed them, it was reestablished in a short time. I plan now to get them well in pace while they are far apart, and I shall try to determine the distance to which the sympathy mentioned above extends. From what I have already seen, I imagine that it will be out to five or six feet. But to obtain a greater certainty of these things, you shall have to wait, please, until I have examined them further and found out their origin more precisely. However, here we have found two clocks that never come to disagree, which seems unbelievable and yet is very true. Never before have other clocks been able to do the same thing as those of this new invention, and one can see from that how precise they are, since something so small is needed to keep them in eternal agreement.

## A1.2   Sea clocks (sympathy of clocks). Part V[2]

22 February 1665.

Within 4 or 5 days I had noticed an amazing agreement between the two new clockworks, in which were small chains (Fig. A1.1), so that not even for the slightest

[2] Translation from Latin by Dorothea Prell.

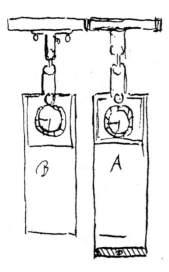

**Figure A1.1.**

deviation was one superior to the other. But the swinging of both pendula stayed constantly reciprocal. Hence, because the clockworks were at a small distance separation, I began to suspect a certain sympathy, as if one were affected by the other. To start an experiment I changed the movement of one pendulum, so that they were not moving together at the same time, but a quarter of an hour or half an hour later I found them to be concurrent again.

Both of the clockworks were hanging down from its own beam of an approximate thickness of 3 inches, the ends of which were laid on two chairs as supports. Every time the beams lay close together, the one clockwork B was not completely at the side of clockwork A, but was projecting forwards. B was also a bit shorter than A and it had no lead at the lower end, which in clockwork A is indicated as $D$. They were both hanging down with weights with lead fixed at the bottom for keeping the balance up to 80 or 90 pounds and a bit more for A because of weight $D$. The length of the pendula was 7 inches. Their swinging was arranged so that they always approached each other simultaneously and moved apart simultaneously; when forced to move in another way they did not stay so, but returned there automatically and afterwards remained unchanged.

On 22 February I turned round both clockworks so that the clockfaces were facing each other; in this position the agreement did not last – clockwork A had some fore-run and this continued until I brought them both back into their former positions. In the evening I moved B 4 feet away from A, and arranged it so that one side of clockwork B was facing the front of clockwork A. In the morning, after exactly $10\frac{1}{2}$ hours, I found that A was ahead of B, but only for two tiny seconds.

23 February. I slightly accelerated the pendulum of clockwork B. I saw, not very much later, the pendula returning to their concordance once again, though not by way of similar motion, but with mixed beats, and so they stayed the whole day. However, at about 9 o'clock in the evening B was ahead at last, that is for half the oscillation. Maybe the colder evening air has interrupted the concordance, as it is not surprising that at such a distance the agreement would be affected by the slightest perturbation. The side of clockwork B faced the front of A.

At half past ten in the evening of this 23 February I put both clockworks back to the former place. They returned immediately to their agreement as on many occasions previously, with the beats corresponding to each other in movements in opposite direction.

24 February. At 9 o'clock in the morning they seemed to be in perfect agreement.

24 February. I slightly turned the front of clockwork A to the side of B. Thereby the agreement did not last, but B was ahead, and this continued until it was repositioned as before.

I then separated the clockworks from each other by a distance of six and a half feet whilst they had previously lain only 2 inches apart. They stayed for a period of an hour or two with mixed movements, but after this time they returned to the agreement of equally swinging beats, as they had done at a lesser separation. At 3 o'clock in the afternoon I put a rectangular bar of 2 feet in length between the clockworks, which should have stopped the stream of air, at least in the lower section. In the upper section the clockworks were projecting approximately half a foot, but there the movement (of the air), if any at all, was minimal. Nonetheless the clockworks continued agreeing as before.

25 February. Although in the previous evening I had put another plank of three feet in length and about one inch thickness between them, completely hiding one clock-work from the other, I nevertheless found this morning that they had been moving in agreement with equally swinging beats through the whole night. So they had remained until 6 o'clock in the evening, when clockwork B stopped because of damage to the larger chain. With the obstacle removed they returned, although I had made them start with mixed beats, after a delay of half an hour, to their agreement and stayed like this until at 11 o'clock I made a further trial by moving them 12 feet apart.

26 February. At half past nine in the morning I found clockwork B 5 seconds ahead, whereby it has been made clear to what extent this sympathy (apparently through communication via air movement) had been reached before.

27 February. I reduced the speed of clockwork B, so the clockworks would move in better agreement between themselves at the same 12 feet separation, but I was not free to stay longer. At 9 o'clock in the evening I arranged them more accurately.

28 February. At 9 o'clock in the morning I found clockwork A hardly one second ahead; they moved with mixed beats, and they stayed so until 6 o'clock in the evening. At half past six they moved in the same time, with B being ahead this half swinging. I doubt whether this consonance lasted 9 hours without the assistance of sympathy, but which could not withstand the change of the air in the evening (which was much less cold, because the cold ceased after it had been with us for three days). But since A was ahead previously, and B later on, they could not have been fitted more accurately between themselves.

1 March. At 10 o'clock in the morning A was three seconds ahead.

Both clockworks had two seats forming a support (see Fig. 1.2 in Section 1.1) whose scarce and totally invisible movement caused by the agitation of the pendula

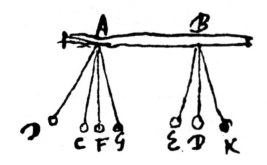

**Figure A1.2.**

was the reason for the predicted sympathy, and forced the pendula to move constantly together with opposing beats. Each singular pendulum draws the support by itself with the highest force at the instant when it goes through a seat. If therefore pendulum B should be in position BD (Fig. A1.2) while A is only in AC, and B should be moving to the left and A to the right, the point of the suspension A is moved to the left whereby the vibration of pendulum A is being accelerated. When B has again passed through BE, when A is in position AF, the suspension of B is moved to the right, and therefore the vibration of pendulum B slows down. B has gone again through the position BD when A is at AG, whereby the suspension A is drawn to the right, and therefore the vibration of pendulum A is being accelerated. B is again in BK, when A has been returned to position AF, whereby the suspension of B is drawn to the left, and therefore the vibration of pendulum B slows down. And so, when the vibration of pendulum B is steadily slowing down, and A is being accelerated, it is necessary that, for a short time they should move together in opposite beats: at the same time A is moving to the right and B to the left, and vice versa. At this point they cannot withdraw from this agreement because for the same reason they are at once brought back in the same way. The supports, of course, are motionless, but if the consonance should initially be even slightly disturbed, then it is restored through the smallest motion of the supports. This motion cannot of course be perceived, and it is therefore not amazing to have given cause for error.

# Appendix A2

## Instantaneous phase and frequency of a signal

## A2.1  Analytic signal and the Hilbert transform

A consistent way to define the phase of an *arbitrary signal* is known in signal processing as the analytic signal concept [Panter 1965; Rabiner and Gold 1975; Boashash
1992; Smith and Mersereau 1992]. This general approach, based on the *Hilbert
transform* (HT) and originally introduced by Gabor [1946], unambiguously gives the
*instantaneous phase and amplitude* for a signal $s(t)$ via construction of the *analytic
signal* $\zeta(t)$, which is a complex function of time defined as

$$\zeta(t) = s(t) + i s_H(t) = A(t) e^{i\phi(t)}. \tag{A2.1}$$

Here the function $s_H(t)$ is the HT of $s(t)$

$$s_H(t) = \pi^{-1} \text{P.V.} \int_{-\infty}^{\infty} \frac{s(\tau)}{t - \tau} d\tau, \tag{A2.2}$$

where P.V. means that the integral is taken in the sense of the Cauchy principal value.
The instantaneous amplitude $A(t)$ and the instantaneous phase $\phi(t)$ of the signal $s(t)$
are thus uniquely defined from (A2.1). We emphasize that the HT is parameter-free.
Note also that computation of the instantaneous characteristics of a signal requires its
knowledge in the whole time domain, i.e., the HT is nonlocal in time. However, the
main contribution to the integral (A2.2) is made in the vicinity of the chosen instant of
time.

As one can see from (A2.2), the HT can be considered as the convolution of the
functions $s(t)$ and $1/\pi t$. Due to the properties of convolution, the Fourier transform
$S_H(\varsigma)$ of $s_H(t)$ is the product of the Fourier transforms of $s(t)$ and $1/\pi t$. For physically
relevant Fourier frequencies $\varsigma > 0$, $S_H(\varsigma) = -i S(\varsigma)$. This means that the HT can be

realized by an ideal filter whose amplitude response is unity and phase response is a
constant $\pi/2$ lag at all Fourier frequencies.

A harmonic oscillation $s(t) = A \cos \omega t$ is often represented in the complex form
as $A \cos \omega t + i A \sin \omega t$. This means that the real oscillation is complemented by the
imaginary part which is delayed in phase by $\pi/2$, that is related to $s(t)$ by the HT. The
analytic signal is the direct and natural extension of this technique, as the HT performs
a $-\pi/2$ phase shift for every spectral component of $s(t)$.

Although formally $A(t)$ and $\phi(t)$ can be computed for an arbitrary $s(t)$, they have a
clear physical meaning only if $s(t)$ is a narrow-band signal (see the detailed discussion
in [Boashash 1992]). In this case, the amplitude $A(t)$ coincides with the envelope
of $s(t)$, and the *instantaneous frequency* $d\phi/dt$ corresponds to the frequency of the
maximum of the power spectrum computed in a running window.

## A2.2    Examples

We illustrate the properties of the Hilbert transform by the following examples.

*Damped oscillations*
Let us model the measured signals by computing the free oscillations of the linear
oscillator

$$\ddot{x} + 0.05\dot{x} + x = 0 \tag{A2.3}$$

and the nonlinear Duffing oscillator

$$\ddot{x} + 0.05\dot{x} + x + x^3 = 0, \tag{A2.4}$$

and calculate from $x(t)$ the instantaneous amplitudes $A(t)$ and the frequencies $d\phi/dt$
(Fig. A2.1). The amplitudes, shown as thick lines, are really envelopes of the decaying
processes. The frequency of the linear oscillator is constant, while the frequency of the
Duffing oscillator is amplitude-dependent, as expected. Note, that although only about
20 periods of oscillations have been used, the nonlinear properties of the system can
be easily revealed from the time series, because the frequency and the amplitude are
estimated not as averaged quantities, but at every point of the signal. This method is
used in mechanical engineering for the identification of elastic and damping properties
of a vibrating system [Feldman 1994]. This example illustrates the important property
of the HT: it can be applied to *nonstationary data*.

*Periodic signal*
For the next example we take the solution of the van der Pol equation (cf. (7.2))

$$\ddot{x} - (1 - x^2)\dot{x} + x = 0. \tag{A2.5}$$

For the chosen parameters, the waveform essentially differs from the sine (Fig. A2.2a),
and the phase portrait of the system (A2.5) is not a circle. Correspondingly, the

instantaneous amplitude $A(T)$ is not constant but oscillates (Fig. A2.2a) with the frequency $2\omega = 2 \cdot 2\pi/T$, where $T$ is the period of the oscillation. The growth of the instantaneous phase is not exactly linear; indeed, $\phi(t) - \omega t$ oscillates with the frequency $2\omega$ (Fig. A2.2b,c). We remind the reader that the analytic signal approach provides only an estimate of the true phase (see Chapter 7) which should increase linearly with time.

*Chaotic signal*
From the viewpoint of nonlinear time series analysis (see, e.g., [Kantz and Schreiber 1997]) the HT can be considered as a two-dimensional embedding in coordinates

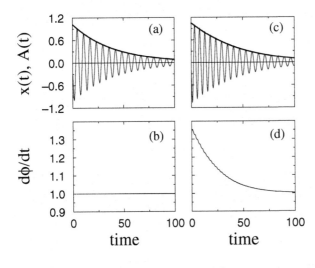

**Figure A2.1.** Free vibrations $x(t)$ of the linear (a) and the nonlinear (Duffing) (c) oscillators. The instantaneous amplitudes $A(t)$ calculated via the Hilbert transform are shown by bold lines. Corresponding instantaneous frequencies $d\phi/dt$ are shown in (b) and (d). From [Rosenblum and Kurths 1998], Fig. 1, Copyright Springer-Verlag.

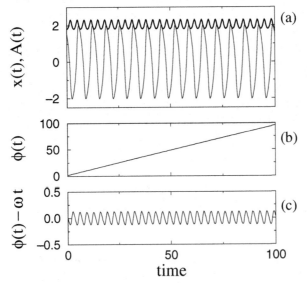

**Figure A2.2.** A solution of the van der Pol equation $x(t)$ and its instantaneous amplitude $A(t)$ (bold line) (a). The instantaneous phase $\phi$ grows practically linearly in (b); nevertheless, small oscillations are seen in an enhanced resolution (c); here $\omega$ is the average frequency.

($s$, $s_H$). Let us choose as an observable the $x$ coordinate of the Rössler system (10.2). The phase portrait of this system in coordinates ($x$, $x_H$) is shown in Fig. A2.3a; one can see that it is very similar to the "true" portrait of the Rössler oscillator in coordinates ($x$, $y$) (Fig. A2.3b).[1] The instantaneous amplitude and phase are shown in Fig. A2.4. Although the phase $\phi$ grows practically linearly, small irregular fluctuations of that growth are also seen. This agrees with the known fact that oscillations of the system are chaotic, but the power spectrum of $x(t)$ contains a very sharp peak.

[1] We note that the HT provides in some sense an optimal two-dimensional embedding. Indeed, if one uses the Takens [1981] technique to reconstruct a phase portrait from a signal with the period $T$, then the time delay $T/4$ ensures that the attractor is not stretched along the diagonal. Performing the HT is equivalent to choosing such a delay for every spectral component of a signal.

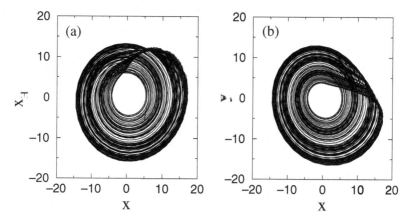

**Figure A2.3.** The phase portrait of the Rössler system in ($x$, $x_H$) coordinates (a) and in the original coordinates ($x$, $y$) (b).

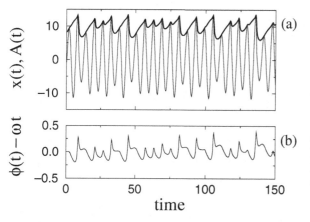

**Figure A2.4.** A solution of the Rössler system $x(t)$ and its instantaneous amplitude $A(t)$ (bold line) (a). The instantaneous phase $\phi$ grows practically linearly, nevertheless, small irregular fluctuations are seen in an enhanced resolution (b). From [Rosenblum and Kurths 1998], Fig. 2, Copyright Springer-Verlag.

*Human electrocardiogram*

As an example of a complex signal, we take a human ECG record (Fig. A2.5). We see that the point in the $(s, s_H)$-plane makes two rotations corresponding to the so-called R- and T-waves, respectively (small loops corresponding to the P-waves are not seen in this magnification). What is important is that the trajectories in $(s, s_H)$ pass through the origin, and therefore the phase is not always defined. We did not encounter this problem in the previous examples because of the simple structure of the signals there. Indeed, the normal procedure before computing the HT is to subtract the mean value from the signal. Often this ensures that trajectories revolve around the origin; we implicitly used this fact before. In order to compute the phase for the cardiogram, we have to translate the origin to some point $s^*, s_H^*$ and compute the phase and the amplitude according to

$$A(t)e^{i\phi(t)} = (s - s^*) + i(s_H - s_H^*). \tag{A2.6}$$

Definitely, in this way we loose the unambiguity in the determination of phase: now it depends on the choice of the origin. Two reasonable choices are shown in Fig. A2.5c by two arrows. Obviously, depending on the new origin in this plane, one cardiocycle (interval between two heartbeats) would correspond to a phase increase of either $2\pi$ or $4\pi$. This reflects the fact that our understanding of what is "one oscillation" depends on the particular problem and our physical intuition.

## A2.3    Numerics: practical hints and know-hows

An important advantage of the analytic signal approach is that it can be easily implemented numerically. Here we point out the main steps of the computation.

*Computing HT in the frequency domain*

The easiest way to compute the HT is to perform a fast Fourier transform (FFT) of the original time series, shift the phase of every frequency component by $-\pi/2$ and to apply the inverse FFT.[2] Zero padding should be used to make the length of the time series suitable for the FFT. To reduce boundary effects, it is advisable to eliminate about ten characteristic periods at the beginning and the end of the signal. Such a computation with double precision allows one to obtain the HT with a precision of about 1%. (The precision was estimated by computing the variance of $s(t) + H^2(s(t))$, where $H^2$ means that the HT was performed twice; theoretically $s(t) + H^2(s(t)) \equiv 0$.)

*Computing the HT in the time domain*

Numerically, this can be done via convolution of the experimental data with a pre-computed characteristic of the filter (Hilbert transformer) [Rabiner and Gold 1975;

---

[2] The phase shift can be conveniently implemented by swapping the imaginary and real parts of the Fourier transform: $\text{Re}(\omega_i) \to \text{tmp}, \text{Im}(\omega_i) \to \text{Re}(\omega_i), -\text{tmp} \to \text{Im}(\omega_i)$, where tmp is some dummy variable.

Little and Shure 1992; Smith and Mersereau 1992]. Such filters are implemented, e.g., in software packages MATLAB [Little and Shure 1992] and RLAB (public domain, URL: http://rlab.sourceforge.net). Although the HT requires computation on an infinite time scale (i.e., the Hilbert transformer is an infinite impulse response filter), an acceptable precision of about 1% can be obtained with the 256-point filter characteristic. The sampling rate must be chosen in order to have at least 20 points per characteristic period of oscillation. In the process of computation of the convolution $L/2$ points are lost at both ends of the time series, where $L$ is the length of the transformer.

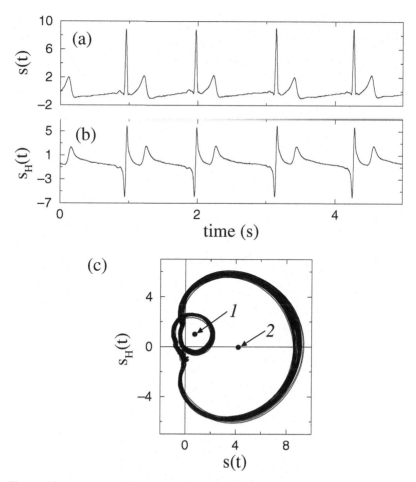

**Figure A2.5.** A human ECG (a), its Hilbert transform (b) and ECG vs. its Hilbert transform (c). Determination of the phase depends on the choice of the origin in the $(s, s_H)$ plane; two reasonable choices are shown in (c) by two arrows.

*Computing and unwrapping the phase*

A convenient way to compute the phase is to use the functions DATAN2($s_H$, $s$) (FORTRAN) or atan2($s_H$, $s$) (C) that give a cyclic phase in the $[-\pi, \pi]$ interval. The phase difference of two signals $s_1(t)$ and $s_2(t)$, can be obtained via the HT as

$$\phi_1(t) - \phi_2(t) = \tan^{-1} \frac{s_{H,1}(t)s_2(t) - s_1(t)s_{H,2}(t)}{s_1(t)s_2(t) + s_{H,1}(t)s_{H,2}(t)}. \qquad (A2.7)$$

For the detection of synchronization, it is usually necessary to use a phase that is defined not on the circle, but on the whole real line (i.e., it varies from $-\infty$ to $\infty$). For this purpose, the phase (or the phase difference) can be unwrapped by tracing the $\approx 2\pi$ jumps in the time course of $\phi(t)$.

*Sensitivity to low-frequency trends*

We have already discussed the fact that the phase is well-defined only if the trajectories in the $(s, s_H)$-plane always go around the origin and $s$ and $s_H$ do not vanish simultaneously. This may be violated if the signal contains low-frequency trends, e.g., due to the drift of the zero level of the measuring equipment. As a result, some loops that do not encircle the origin will not be counted as a cycle, and $2\pi$ will be lost in the overall increase of the phase. To illustrate this, we add an artificial trend to the ECG signal; the embedding of this signal in coordinates $s$, $s_H$ is shown in Fig. A2.6, to be compared with the same presentation for the original data in Fig. A2.5. Obviously, the origin denoted in Fig. A2.5 by the first arrow would be a wrong choice in this case.

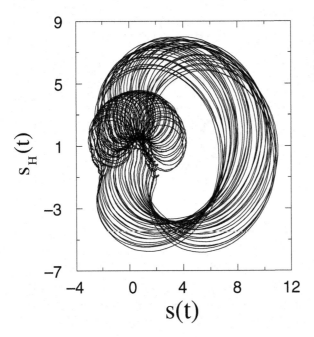

**Figure A2.6.** An illustration of the sensitivity of the Hilbert transform to low-frequency trends.

To avoid these problems, we recommend always plotting the signal vs. its HT and checking whether the origin has been chosen correctly.

## A2.4   Computation of the instantaneous frequency

### Frequency of a continuous signal

Estimation of the instantaneous frequency $\omega(t)$ of a signal is rather cumbersome. The direct approach, i.e., numerical differentiation of $\phi(t)$, naturally results in very large

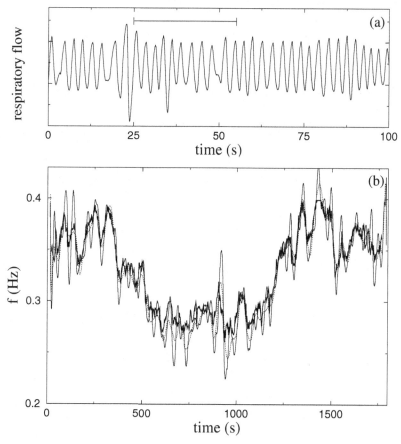

**Figure A2.7.** The instantaneous frequency of human respiration. The signal itself is represented in (a). In (b), the bold line shows the frequency that corresponds to the maximum of the power spectrum computed in a running 30 s length window by means of the autoregression technique (Burg method [Press *et al.* 1992]). The length of the window is indicated by the horizontal bar in (a); it corresponds to about ten characteristic periods of respiration. The solid and dashed lines in (b) present the instantaneous frequency $f(t)$ obtained with the help of the HT and by fitting the polynomial over windows of length 30 and 60 s, respectively.

fluctuations in the estimate of $\omega(t)$. Moreover, one may encounter that $\omega(t) < 0$ for some $t$. This happens not only because of the influence of noise, but also as a result of a complicated form of the signal. For example, some characteristic patterns in the ECG (e.g., the T-wave) result in negative values of the instantaneous frequency. From the physical point of view, we expect the instantaneous frequency to be a positive function of time that varies slowly with respect to the characteristic period of oscillations and to have the meaning of a number of oscillations per time unit. This is especially important for the problem of synchronization, where we are not interested in the behavior of the phase on a time scale smaller than the characteristic oscillation period. There exist several methods to obtain estimates of $\omega(t)$ in accordance with this viewpoint; for a discussion and comparison see [Boashash 1992].

Here we take, for an illustration, a record of human breathing (respiratory flow measured at the nose), see Fig. A2.7a, and use a technique that is called in [Boashash 1992] a "maximum likelihood frequency estimator". Suppose that the instantaneous phase $\phi(t)$ is unwrapped into the infinite domain, so that this function is growing, although not necessarily monotonic. Then we perform for each instant of time a local polynomial fit over an interval essentially larger than the characteristic period of oscillations. The (analytically obtained) derivative of that polynomial function at this instant gives an always positive estimate of the frequency. Practically, we perform this by means of a Savitzky–Golay filter; a fourth-order polynomial and an interval of approximation equal to approximately ten characteristic periods seems to be a reasonable parameter choice. The instantaneous frequency computed in this way practically coincides with the frequency of the maximum of the running autoregression spectrum obtained, e.g., by means of the Burg technique [Press $et\ al.$ 1992], see Fig. A2.7b.

### Frequency of a point process

The phase and the slowly varying frequency of a point process can be easily obtained. Indeed, if the time interval between two events corresponds to one complete cycle of the oscillatory process, then the phase increase during this time interval is exactly $2\pi$. Hence, we can assign to the times $t_i$ the values of the phase $\phi_i = \phi(t_i) = 2\pi i$. It is difficult to handle this time series because it is not equidistantly spaced. Nevertheless, we can make use of the fact that it is a monotonically increasing function of time, and invert it. The resulting process $t(\phi_i)$ is equidistant, as the phase step is $2\pi$. Now we can apply the polynomial fitting technique described above to obtain the instantaneous period $T_i = T(\phi_i)$. Inverting the series once again we obtain the frequency $\omega_i = \omega(t_i) = 2\pi / T_i$.

# References

H. D. I. Abarbanel. *Analysis of Observed Chaotic Data*. Springer, New York, Berlin, Heidelberg, 1996.

H. D. I. Abarbanel, N. F. Rulkov, and M. M. Suschik. Generalized synchronization of chaos: The auxillary system approach. *Phys. Rev. E*, 53(5):4528–4535, 1996.

L. F. Abbott and C. van Vreeswijk. Asynchronous states in networks of pulse-coupled oscillators. *Phys. Rev. E*, 48(2):1483–1490, 1993.

J. A. Acebrón, L. L. Bonilla, S. De Leo, and R. Spigler. Breaking the symmetry in bimodal frequency distributions of globally coupled oscillators. *Phys. Rev. E*, 57(5):5287–5290, 1998.

U. Achenbach and K. E. Wohlfarth-Bottermann. Synchronization and signal transmission in protoplasmatic strands of *Physarum*. Reaction to varying temperature gradient. *Planta*, 150: 180–188, 1980.

U. Achenbach and K. E. Wohlfarth-Bottermann. Synchronization and signal transmission in protoplasmatic strands of *Physarum*. Effects of externally applied substances and mechanical influences. *Planta*, 151:574–583, 1981.

R. Adler. A study of locking phenomena in oscillators. *Proc. IRE*, 34:351–357, 1946. Reprinted in *Proc. IEEE*, 61(10):1380–1385, 1973.

V. S. Afraimovich, V. I. Nekorkin, G. V. Osipov, and V. D. Shalfeev. *Stability, Structures and Chaos in Nonlinear Synchronization Networks*. World Scientific, Singapore, 1994.

V. S. Afraimovich and L. P. Shilnikov. Invariant two-dimensional tori, their destroying and stochasticity. In *Methods of Qualitative Theory of Differential Equations*, pages 3–28. Gorki, 1983. In Russian; English translation *Amer. Math. Soc. Transl. Ser. 2* 149:201–212, 1991.

Y. Aizawa and T. Uezu. Global aspects of the dissipative dynamical systems. II. Periodic and chaotic responses in the forced Lorenz system. *Prog. Theor. Phys.*, 68(6):1864–1879, 1982.

M. K. Ali. Synchronization of a chaotic map in the presence of common noise. *Phys. Rev. E*, 55(4):4804–4805, 1997.

M. K. Ali and J.-Q. Fang. Synchronization of chaos and hyperchaos using linear and nonlinear feedback functions. *Phys. Rev. E*, 55(5):5285–5290, 1997.

K. T. Alligood, T. D. Sauer, and J. A. Yorke. *Chaos: An Introduction to Dynamical Systems*. Springer, New York, 1997.

P. Alstrøm, B. Christiansen, and M. T. Levinsen. Characterization of a simple class of modulated relaxation oscillators. *Phys. Rev. B*, 41(3): 1308–1319, 1990.

R. Andretic, S. Chaney, and J. Hirsh. Requirement of circadian genes for cocaine sensitization in *Drosophila*. *Science*, 285:1066–1068, 1999.

A. A. Andronov and A. A. Vitt. On mathematical theory of entrainment. *Zhurnal prikladnoi fiziki (J. Appl. Phys.)*, 7(4):3, 1930a (in Russian).

A. A. Andronov and A. A. Vitt. Zur Theorie des Mitnehmens von van der Pol. *Archiv für Elektrotechnik*, 24(1): 99–110, 1930b.

A. A. Andronov, A. A. Vitt, and S. E. Khaykin. *Theory of Oscillators*. Gostekhizdat, Moscow, 1937. (In Russian); English translation: Pergamon Press, Oxford, New York, Toronto, 1966.

V. S. Anishchenko. *Dynamical Chaos: Models and Experiments; Appearance Routes and Structure of Chaos in Simple Dynamical Systems*. World Scientific, Singapore; River Edge, NJ, 1995.

V. S. Anishchenko and V. V. Astakhov. Bifurcation phenomena in an autostochastic oscillator responding to a regular external signal. *Sov. Phys.*

*Tech. Phys.*, 28:1326, 1983.

V. S. Anishchenko, A. G. Balanov, N. B. Janson, N. B. Igosheva, and G. V. Bordyugov. Entrainment between heart rate and weak noninvasive forcing. *Int. J. Bifurc. Chaos*, 10:2339–2348, 2000.

V. S. Anishchenko, T. E. Vadivasova, D. E. Postnov, and M. A. Safonova. Forced and mutual synchronization of chaos. *Radioeng. Electron.*, 36:338–351, 1991.

V. S. Anischenko, T. E. Vadivasova, D. E. Postnov, and M. A. Safonova. Synchronization of chaos. *Int. J. Bifurc. Chaos*, 2(3):633–644, 1992.

V. V. Antyukhov, A. F. Glova, O. R. Kachurin, F. V. Lebedev, V. V. Likhanskii, A. P. Napartovich, and V. D. Pis'mennyi. Effective phase locking of an array of lasers. *JETP Lett.*, 44(2):78–81, 1986.

T. Aoyagi and Y. Kuramoto. Frequency order and wave patterns of mutual entrainment in two-dimensional oscillator lattices. *Phys. Lett. A*, 155 (6,7):410–414, 1991.

E. V. Appleton. The automatic synchronization of triode oscillator. *Proc. Cambridge Phil. Soc. (Math. and Phys. Sci.)*, 21:231–248, 1922.

J. Argyris, G. Faust, and M. Haase. *An Exploration of Chaos*. North-Holland, Amsterdam, 1994.

J. Arnhold, P. Grassberger, K. Lehnertz, and C. E. Elger. A robust method for detecting interdependences: Application to intracranially recorded EEG. *Physica D*, 134(4):419–430, 1999.

L. Arnold. *Random Dynamical Systems*. Springer, Berlin, 1998.

V. I. Arnold. Small denominators. I. Mappings of the circumference onto itself. *Izv. Akad. Nauk Ser. Mat.*, 25(1):

21–86, 1961. (In Russian); English translation: *AMS Transl. Ser.* 2, 46:213–284.

V. I. Arnold. Remarks on the perturbation problem for problems of Mathieu type. *Usp. Mat. Nauk*, 38(4):189–203, 1983. (In Russian); English translation: *Russ. Math. Surveys*, 38:215–233, 1983.

V. I. Arnold. Cardiac arrhythmias and circle mappings. *Chaos*, 1:20–24, 1991.

D. G. Aronson, M. A. Chory, G. R. Hall, and R. P. McGehee. Bifurcations from an invariant circle for two parameter families of maps of the plane: A computer-assisted study. *Commun. Math. Phys.*, 83:303–353, 1982.

D. G. Aronson, G. B. Ermentrout, and N. Koppel. Amplitude response of coupled oscillators. *Physica D*, 41: 403–449, 1990.

D. G. Aronson, R. P. McGehee, I. G. Kevrekidis, and R. Aris. Entrainment regions for periodically forced oscillations. *Phys. Rev. A*, 33(3): 2190–2192, 1986.

R. Artuso, E. Aurell, and P. Cvitanović. Recycling of strange sets. I. Cycle expansions. *Nonlinearity*, 3:325–359, 1990a.

R. Artuso, E. Aurell, and P. Cvitanović. Recycling of strange sets. II. Applications. *Nonlinearity*, 3: 361–386, 1990b.

J. Aschoff, S. Daan, and G. A. Groos. *Vertebrate Circadian Systems. Structure and Physiology.* Springer, Berlin, 1982.

P. Ashwin and P. J. Aston. Blowout bifurcations of codimension two. *Phys. Lett. A*, 244(4):261–270, 1998.

P. Ashwin, J. Buescu, and I. Stewart. Bubbling of attractors and synchronization of chaotic oscillators.

*Phys. Lett. A*, 193:126–139, 1994.

P. Ashwin, J. Buescu, and I. Stewart. From attractor to chaotic saddle: a tale of transverse instability. *Nonlinearity*, 9 (3):703–737, 1996.

P. Ashwin, J. R. Terry, K. S. Thornburg, and R. Roy. Blowout bifurcation in a system of coupled chaotic lasers. *Phys. Rev. E*, 58(6):7186–7189, 1998.

V. Astakhov, T. Kapitaniak, A. Shabunin, and V. Anishchenko. Non-bifurcational mechanism of loss of chaos synchronization in coupled non-identical systems. *Phys. Lett. A*, 258(2–3):99–102, 1999.

P. J. Aston and M. Dellnitz. Symmetry breaking bifurcations of chaotic attractors. *Int. J. Bifurc. Chaos*, 5(6): 1643–1676, 1995.

R. Badii, G. Broggi, B. Derighetti, M. Ravani S. Ciliberto, A. Politi, and M. A. Rubio. Dimension increase in filtered chaotic signals. *Phys. Rev. Lett.*, 60:979–982, 1988.

R. Badii and A. Politi. *Complexity. Hierarchical Structures and Scaling in Physics.* Cambridge University Press, Cambridge, 1997.

F. Bagnoli, L. Baroni, and P. Palmerini. Synchronization and directed percolation in coupled map lattices. *Phys. Rev. E*, 59(1):409–416, 1999.

F. Bagnoli and R. Rechtman. Synchronization and maximum Lyapunov exponents of cellular automata. *Phys. Rev. E*, 59(2): R1307–R1310, 1999.

G. L. Baker, J. A. Blackburn, and H. J. T. Smith. A stochastic model of synchronization for chaotic pendulums. *Phys. Lett. A*, 252(3–4): 191–197, 1999.

J. Baker and J. Gollub. *Chaotic Dynamics.* Cambridge University Press,

Cambridge, 1996.

N. Balmforth, C. Tresser, P. Worfolk, and C. W. Wu. Master–slave synchronization and the Lorenz equation. *Chaos*, 7(3):392–394, 1997.

N. J. Balmforth, A. Jacobson, and A. Provenzale. Synchronized family dynamics in globally coupled maps. *Chaos*, 9(3):738–754, 1999.

N. J. Balmforth and R. Sassi. A shocking display of synchrony. *Physica* D, 143 (1–4):21–55, 2000.

M. Banaji and P. Glendinning. Towards a quasi-periodic mean flow theory for globally coupled oscillators. *Phys. Lett.* A, 251:297–302, 1994.

A.-L. Barabási and H. E. Stanley. *Fractal Concepts in Surface Growth*. Cambridge University Press, Cambridge, 1995.

S. Barbay, G. Giacomelli, and F. Marin. Stochastic resonance in vertical cavity surface emitting lasers. *Phys. Rev. E.*, 61(1):157–166, 2000.

A. Barone and G. Paterno. *Physics and Applications of the Josephson Effect*. Wiley, New York, 1982.

C. Beck and F. Schlögl. *Thermodynamics of Chaotic Systems*. Cambridge University Press, Cambridge, 1997.

J. Benford, H. Sze, W. Woo, R. R. Smith, and B. Harteneck. Phase locking of relativistic magnetrons. *Phys. Rev. Lett.*, 62(8):969–971, 1989.

T. B. Benjamin and J. E. Feir. The disintegration of wave trains on deep water. *J. Fluid Mech.*, 27:417, 1967.

R. E. Best. *Phase-Locked Loops*. McGraw-Hill, New York, 1984.

L. Bezaeva, L. Kaptsov, and P. S. Landa. Synchronization threshold as the criterium of stochasticity in the generator with inertial nonlinearity. *Zhurnal Tekhnicheskoi Fiziki*, 32:
467–650, 1987 (in Russian).

B. P. Bezruchko. *Experimental study of nonstationary and chaotic effects in the distributed self-oscillating system electron beam–backward electromagnetic wave*. Ph.D. dissertation, Saratov Univ., Saratov, 1980 (in Russian).

B. P. Bezruchko, L. V. Bulgakova, S. P. Kuznetsov, and D. I. Trubetskov. Experimental and theoretical study of stochastic self-sustained oscillations in the backward wave oscillator. In *Lectures on High-Frequency Electronics (5th Winter School)*, Saratov 1981, pages 25–77, Saratov University Press, 1981 (in Russian).

B. P. Bezruchko, S. P. Kuznetsov, and D. I. Trubetskov. Experimental observation of stochastic self-oscillations in the electron beam–backscattered electromagnetic wave dynamic system. *JETP Lett.*, 29(3):162–165, 1979.

L. Billings, J. H. Curry, and E. Phipps. Lyapunov exponents, singularities, and a riddling bifurcation. *Phys. Rev. Lett.*, 79(6):1018–1021, 1997.

B. Blasius, A. Huppert, and L. Stone. Complex dynamics and phase synchronization in spatially extended ecological systems. *Nature*, 399: 354–359, 1999.

I. I. Blekhman. *Synchronization of Dynamical Systems*. Nauka, Moscow, 1971 (in Russian).

I. I. Blekhman. *Synchronization in Science and Technology*. Nauka, Moscow, 1981. In Russian; English translation: 1988, ASME Press, New York.

I. I. Blekhman, P. S. Landa, and M. G. Rosenblum. Synchronization and chaotization in interacting dynamical systems. *Appl. Mech. Rev.*, 48(11): 733–752, 1995.

B. Boashash. Estimating and interpreting the instantaneous frequency of a signal. *Proc. IEEE*, 80(4):520–568, 1992.

S. Boccaletti, J. Bragard, F. T. Arecchi, and H. Mancini. Synchronization in nonidentical extended systems. *Phys. Rev. Lett.*, 83(3):536–539, 1999.

S. Boccaletti, A. Farini, and F. T. Arecchi. Adaptive synchronization of chaos for secure communication. *Phys. Rev. E*, 55(5):4979–4981, 1997.

N. N. Bogoliubov and Yu. A. Mitropolsky. *Asymptotic Methods in the Theory of Nonlinear Oscillations*. Gordon and Breach, New York, 1961.

T. Bohr, M. H. Jensen, G. Paladin, and A. Vulpiani. *Dynamical Systems Approach to Turbulence*. Cambridge University Press, Cambridge, 1998.

R. Bonaccini and A. Politi. Chaotic-like behavior in chains of stable nonlinear oscillators. *Physica* D, 103:362, 1997.

A. V. Bondarenko, A. F. Glova, S. N. Kozlov, F. V. Lebedev, V. V. Likhanskii, A. P. Napartovich, V. D. Pismennyi, and V. P. Yartsev. Bifurcation and chaos in a system of optically coupled $CO_2$ lasers. *Sov. Phys. JETP*, 68(3):461–466, 1989.

L. L. Bonilla, J. M. Casado, and M. Morillo. Self-synchronization of populations of nonlinear oscillators in the thermodynamic limit. *J. Stat. Phys.*, 48(3/4):571–591, 1987.

L. L. Bonilla, J. C. Neu, and R. Spigler. Nonlinear stability of incoherence and collective synchronization in a population of coupled oscillators. *J. Stat. Phys.*, 67(1/2):313–330, 1992.

L. L. Bonilla, C. J. Pérez Vicente, F. Ritort, and J. Soler. Exactly solvable phase oscillator models with synchronization dynamics. *Phys. Rev. Lett.*, 81(17): 3643–3646, 1998.

L. L. Bonilla, C. J. Pérez Vicente, and J. M. Rubi. Glassy synchronization in a population of coupled oscillators. *J. Stat. Phys.*, 70(3/4):921–936, 1993.

S. Bottani. Pulse-coupled relaxation oscillators: From biological synchronization to self-organized criticality. *Phys. Rev. Lett.*, 74(21): 4189–4192, 1995.

S. Bottani. Synchronization of integrate and fire oscillators with global coupling. *Phys. Rev. E*, 54(3): 2334–2350, 1996.

C. Boyd. On the structure of the family of Cherry fields on the torus. *Ergod. Theor. Dynam. Syst.*, 5:27–46, 1985.

M. Bračič and A. Stefanovska. Synchronization and modulation in the human cardiorespiratory system. *Physica* A, 283:451–461, 2000.

Y. Braiman, W. L. Ditto, K. Wiesenfeld, and M. L. Spano. Disorder-enhanced synchronization. *Phys. Lett.* A, 206: 54–60, 1995.

D. M. Bramble and D. R. Carrier. Running and breathing in mammals. *Science*, 219:251–256, 1983.

H. Bremmer. The scientific work of Balthasar van der Pol. *Philips Tech. Rev.*, 22(2):36–52, 1960/61.

P. C. Bressloff and S. Coombes. Symmetry and phase-locking in a ring of pulse-coupled oscillators with distributed delays. *Physica* D, 126 (1–2):99–122, 1999.

J. Brindley and T. Kapitaniak. Analytic predictors for strange non-chaotic attractors. *Phys. Lett.* A, 155:361–364, 1991.

V. Ya. Brodsky. Ultradian rhythms in a population of cells. *Bull. Exp. Biol. Med.*, 124(12):604–609, 1997 (in Russian).

R. Brown and N. Rulkov. Designing a

coupling that guarantees synchronization between identical chaotic systems. *Phys. Rev. Lett.*, 78 (22):4189–4192, 1997a.

R. Brown and N. Rulkov. Synchronization of chaotic systems: Transverse stability of trajectories in invariant manifolds. *Chaos*, 7(3):395–413, 1997b.

L. Brunnet and H. Chaté. Phase coherence in chaotic oscillatory media. *Physica A*, 257:347–356, 1998.

L. Brunnet, H. Chaté, and P. Manneville. Long-range order with local chaos in lattices of diffusively coupled ODEs. *Physica D*, 78:141–154, 1994.

P. Bryant and C. Jeffries. The dynamics of phase locking and points of resonance in a forced magnetic resonator. *Physica D*, 25:196–232, 1987.

J. Buck and E. Buck. Mechanism of rhythmic synchronous flashing of fireflies. *Science*, 159:1319–1327, 1968.

J. Buck, E. Buck, F. E. Hanson, J. F. Case, L. Mets, and G. J. Atta. Control of flushing in fireflies. IV. Free run pacemaking in a synchronic *Pteroptyx*. *J. Comp. Physiol.*, 144:277–286, 1981.

C. J. Buczek, R. J. Freiberg, and M. L. Skolnick. Laser injection locking. *Proc. IEEE*, 61(10):1411–1431, 1973.

N. V. Butenin, Yu. I. Neimark, and N. A. Fufaev. *Introduction to the Theory of Nonlinear Oscillations*. Nauka, Moscow, 1987 (in Russian).

P. J. Butler and A. J. Woakes. Heart rate, respiratory frequency and wing beat frequency of free flying barnacle geese. *J. Exp. Biol.*, 85:213–226, 1980.

J. L. Cardy, Editor. *Finite-Size Scaling*. North-Holland, Amsterdam, 1988.

T. L. Carroll. Amplitude-independent chaotic synchronization. *Phys. Rev. E*, 53(4):3117–3122, 1996.

T. L. Carroll, J. F. Heagy, and L. M. Pecora. Transforming signals with chaotic synchronization. *Phys. Rev. E*, 54(5):4676–4680, 1996.

T. L. Carroll and G. A. Johnson. Synchronizing broadband chaotic systems to narrow-band signals. *Phys. Rev. E*, 57(2):1555–1558, 1998.

T. L. Carroll and L. M. Pecora. Cascading synchronized chaotic systems. *Physica D*, 67:126–140, 1993a.

T. L. Carroll and L. M. Pecora. Synchronizing nonautonomous chaotic circuits. *IEEE Trans. Circ. Syst.*, 40: 646, 1993b.

T. L. Carroll and L. M. Pecora. Synchronizing hyperchaotic volume-preserving maps and circuits. *IEEE Trans. Circ. Syst. I*, 45(6): 656–659, 1998.

M. L. Cartwright and J. E. Littlewood. On nonlinear differential equations of the second order. *J. London Math. Soc.*, 20:180–189, 1945.

A. Čenys, A. N. Anagnostopoulos, and G. L. Bleris. Distribution of laminar lengths for noisy on–off intermittency. *Phys. Lett. A*, 224(6):346–352, 1997a.

A. Čenys, A. N. Anagnostopoulos, and G. L. Bleris. Symmetry between laminar and burst phases for on–off intermittency. *Phys. Rev. E*, 56(3): 2592–2596, 1997b.

A. Čenys, A. Namajunas, A. Tamasevicius, and T. Schneider. On–off intermittency in chaotic synchronization experiment. *Phys. Lett. A*, 213:259–264, 1996.

J. Cernacek. Stabilography in neurology. *Agressologie D*, 21:25–29, 1980.

H. Chaté and P. Manneville. Collective behaviors in spatially extended systems with local interactions and synchronous updating. *Prog. Theor. Phys.*, 87(1):1–60, 1992.

H. Chaté and P. Manneville. Phase diagram of the two-dimensional complex Ginzburg-Landau equation. *Physica* A, 224:348, 1996.

H. Chaté, A. Pikovsky, and O. Rudzick. Forcing oscillatory media: Phase kinks vs. synchronization. *Physica* D, 131 (1–4):17–30, 1999.

Chia-Chu Chen. Threshold effects on synchronization of pulse-coupled oscillators. *Phys. Rev.* E, 49: 2668–2672, 1994.

M. Y. Choi, H. J. Kim, D. Kim, and H. Hong. Synchronization in a system of globally coupled oscillators with time delay. *Phys. Rev.* E, 61(1): 371–381, 2000.

B. Christiansen, P. Alstrøm, and M. T. Levinsen. Routes to chaos and complete phase locking in modulated relaxation oscillators. *Phys. Rev.* A, 42(4):1891–1900, 1990.

B. Christiansen, P. Alstrøm, and M. T. Levinsen. Collective dynamics of coupled modulated oscillators with random pinning. *Physica* D, 56:23–35, 1992.

J. J. Collins and I. N. Stewart. Coupled nonlinear oscillators and the symmetries of animal gaits. *J. Nonlinear Sci.*, 3:349–392, 1993.

S. Coombes. Liapunov exponents and mode-locked solutions for integrate-and-fire dynamical systems. *Phys. Lett.* A, 255(1–2):49–57, 1999.

S. Coombes and P. C. Bressloff. Mode locking and Arnold tongues in integrate-and-fire neural oscillators. *Phys. Rev.* E, 60(2):2086–2096, 1999.

I. P. Cornfeld, S. V. Fomin, and Ya. G. Sinai. *Ergodic Theory*. Springer, New York, 1982.

Á. Corral, C. J. Pérez, A. Díaz-Guilera, and A. Arenas. Self-organized criticality and synchronization in a lattice model of integrate-and-fire oscillators. *Phys. Rev. Lett.*, 74(1): 118–121, 1995a.

Á. Corral, C. J. Pérez, A. Díaz-Guilera, and A. Arenas. Synchronization in a lattice model of pulse-coupled oscillators. *Phys. Rev. Lett.*, 75(20): 3697–3700, 1995b.

D. Cortez and S. J. Elledge. Conducting the mitotic symphony. *Nature*, 406: 354–356, 2000.

P. Coullet and K. Emilsson. Pattern formation in the strong resonant forcing of spatially distributed oscillators. *Physica* A, 188:190–200, 1992a.

P. Coullet and K. Emilsson. Strong resonances of spatially distributed oscillators: A laboratory to study patterns and defects. *Physica* D, 61: 119–131, 1992b.

H. Crauel and F. Flandoli. Attractors for random dynamical systems. *Probab. Theory Relat. Fields*, 100:365–393, 1994.

J. D. Crawford. Amplitude expansions for instabilities in populations of globally-coupled oscillators. *J. Stat. Phys.*, 74(5/6):1047–1084, 1994.

J. D. Crawford. Scaling and singularities in the entrainment of globally coupled oscillators. *Phys. Rev. Lett.*, 74(21): 4341–4344, 1995.

J. D. Crawford and K. T. R. Davies. Synchronization of globally coupled phase oscillators: Singularities and scaling for general couplings. *Physica* D, 125(1–2):1–46, 1999.

A. Crisanti, M. Falconi, and A. Vulpiani. Broken ergodicity and glassy behavior in a deterministic chaotic map. *Phys. Rev. Lett.*, 76(4):612–615, 1996.

A. Crisanti, G. Paladin, and A. Vulpiani.

*Products of Random Matrices in Statistical Physics.* Springer, Berlin, 1993.

M. C. Cross and P. C Hohenberg. Pattern formation outside of equilibrium. *Rev. Mod. Phys.*, 65:851–1112, 1993.

J. P. Crutchfield and K. Kaneko. Are attractors relevant to turbulence? *Phys. Rev. Lett.*, 60:2715–2718, 1988.

K. M. Cuomo and A. V. Oppenheim. Circuit implementation of synchronized chaos with applications to communications. *Phys. Rev. Lett.*, 71(1):65–68, 1993.

K. M. Cuomo, A. V. Oppenheim, and S. H. Strogatz. Robustness and signal recovery in a synchronized chaotic system. *Int. J. Bifurc. Chaos*, 3(6): 1629–1638, 1993a.

K. M. Cuomo, A. V. Oppenheim, and S. H. Strogatz. Synchronization of Lorenz-based chaotic circuits with applications to communications. *IEEE Trans. Circ. Syst.*, 40:626–633, 1993b.

C. A. Czeisler, J. S. Allan, S. H. Strogatz, J. M. Ronda, R. Sánches, C. D. Ríos, W. O. Freitag, G. S. Richardson, and R. E. Kronauer. Bright light resets the human circadian pacemaker independent of the timing of the sleep–wake cycle. *Science*, 233: 667–671, 1986.

C. A. Czeisler, J. F. Duffy, T. L. Shanahan, E. N. Brown, J. F. Mitchell, D. W. Rimmer, J. M. Ronda, E. J. Silva, J. S. Allan, J. S. Emens, D.-J. Dijk, and R. E. Kronauer. Stability, precision, and near-24-hour period of the human circadian pacemaker. *Science*, 284: 2177–2181, 1999.

H. Daido. Intrinsic fluctuations and a phase transition in a class of large population of interacting oscillators. *J. Stat. Phys.*, 60(5/6):753–800, 1990.

H. Daido. Order function and macroscopic mutual entrainment in uniformly coupled limit-cycle oscillators. *Prog. Theor. Phys.*, 88(6):1213–1218, 1992a.

H. Daido. Quasientrainment and slow relaxation in a population of oscillators with random and frustrated interactions. *Phys. Rev. Lett.*, 68(7): 1073–1076, 1992b.

H. Daido. Critical conditions of macroscopic mutual entrainment in uniformly coupled limit-cycle oscillators. *Prog. Theor. Phys.*, 89(4): 929–934, 1993a.

H. Daido. A solvable model of coupled limit-cycle oscillators exhibiting perfect synchrony and novel frequency spectra. *Physica D*, 69:394–403, 1993b.

H. Daido. Multi-branch entrainment and multi-peaked order-functions in a phase model of limit-cycle oscillators with uniform all-to-all coupling. *J. Phys. A: Math. Gen.*, 28:L151–L157, 1995.

H. Daido. Onset of cooperative entrainment in limit-cycle oscillators with uniform all-to-all interactions: Bifurcation of the order function. *Physica D*, 91:24–66, 1996.

D. Dawson and J. Gärtner. Large deviations from the McKean–Vlasov limit for weakly interacting diffusions. *Stochastics*, 20:247–308, 1987.

M. de Sousa Vieira and A. J. Lichtenberg. Nonuniversality of weak synchronization in chaotic systems. *Phys. Rev. E*, 56(4):R3741–3744, 1997.

M. de Sousa Vieira, A. J. Lichtenberg, and M. A. Lieberman. Synchronization of regular and chaotic systems. *Phys. Rev. A*, 46(12):R7359–R7362, 1992.

M. de Sousa Vieira, A. J. Lichtenberg, and M. A. Lieberman. Self synchronization of many coupled oscillations. *Int. J. Bifurc. Chaos*, 4(6):1563–1577, 1994.

A. Denjoy. Sur les courbes définies par les équations différentielles à la surface du tore. *J. Math. Pure Appl.*, 11:333–375, 1932.

R. L. Devaney. *An Introduction to Chaotic Dynamical Systems*. Addison-Wesley, Reading MA, 1989.

N. E. Diamant and A. Bortoff. Nature of the intestinal slow-wave frequency. *Am. J. Physiol.*, 216(2):301–307, 1969.

A. Díaz-Guilera, C. J. Pérez, and A. Arenas. Mechanisms of synchronization and pattern formation in a lattice of pulse-coupled oscillators. *Phys. Rev. E*, 57(4):3820–3828, 1998.

M. Ding, C. Grebogi, and E. Ott. Evolution of attractors in quasiperiodically forced systems: From quasiperiodic to strange nonchaotic to chaotic. *Phys. Rev. A*, 39 (5):2593–2598, 1989.

M. Ding and W. Yang. Stability of synchronous chaos and on–off intermittency in coupled map lattices. *Phys. Rev. E*, 56(4):4009–4016, 1997.

W. L. Ditto, M. L. Spano, H. T. Savage, S. N. Rauseo, J. Heagy, and E. Ott. Experimental observation of a strange nonchaotic attractor. *Phys. Rev. Lett.*, 65:533, 1990.

M. Dolnik and I. R. Epstein. Coupled chaotic chemical oscillators. *Phys. Rev. E*, 54(4):3361–3368, 1996.

B. Drossel. Self-organized criticality and synchronization in a forest-fire-model. *Phys. Rev. Lett.*, 76(6):936–939, 1996.

G. S. Duane. Synchronized chaos in extended systems and meteorological teleconnections. *Phys. Rev. E*, 56(6): 6475–6493, 1997.

J. Dudel and W. Trautwein. Der Mechanismus der automatischen rhythmischen Impulsbildung der Herzmuscelfaser. *Pflügers Arch.*, 313: 553, 1958.

G. I. Dykman, P. S. Landa, and Yu. I. Neymark. Synchronizing the chaotic oscillations by external force. *Chaos, Solit. Fract.*, 1(4):339–353, 1991.

W. H. Eccles and J. H. Vincent. British Patent Spec. clxiii p.462, 1920. Application date 17.02.1920.

R. J. Elble and W. C. Koller. *Tremor*. John Hopkins University Press, Baltimore, 1990.

C. Elphick, A. Hagberg, and E. Meron. Multiphase patterns in periodically forced oscillatory systems. *Phys. Rev. E*, 59(5):5285–5291, 1999.

R. C. Elson, A. I. Selverston, R. Huerta, N. F. Rulkov, M. I. Rabinovich, and H. D. I. Abarbanel. Synchronous behavior of two coupled biological neurons. *Phys. Rev. Lett.*, 81(25): 5692–5695, 1998.

C. Elton and M. Nicholson. The ten-year cycle in numbers of the lynx in Canada. *J. Anim. Ecol.*, 11:215–244, 1942.

J. Engel and T. A. Pedley. *Epilepsy: A Comprehensive Textbook*. Lippincott-Raven, Philadelphia, 1975.

G. B. Ermentrout. The behavior of rings of coupled oscillators. *J. Math. Biol.*, 23: 55–74, 1985.

G. B. Ermentrout. Oscillator death in populations of "all to all" coupled nonlinear oscillators. *Physica D*, 41: 219–231, 1990.

G. B. Ermentrout and N. Kopell. Frequency plateaus in a chain of weakly coupled oscillators, I. *SIAM J. Math. Anal.*, 15(2):215–237, 1984.

G. B. Ermentrout and J. Rinzel. Beyond a pacemaker's entrainment limit: Phase walk-through. *Am. J. Physiol.*, 246:

R102–106, 1984.

U. Ernst, K. Pawelzik, and T. Geisel. Synchronization induced by temporal delays in pulse-coupled oscillators. *Phys. Rev. Lett.*, 74(9):1570–1573, 1995.

U. Ernst, K. Pawelzik, and T. Geisel. Delay-induced multistable synchronization of biological oscillators. *Phys. Rev. E*, 57(2): 2150–2162, 1998.

M. S. Feldman. Non-linear system vibration analysis using Hilbert transform I. Free vibration analysis method "FREEVIB". *Mech. Syst. Signal Proc.*, 8(2):119–127, 1994.

W. Feller. *An Introduction to Probability Theory and its Applications*. Wiley, New York, 1974.

R. Femat and G. Solis-Perales. On the chaos synchronization phenomena. *Phys. Lett.* A, 262(1):50–60, 1999.

M. Franz and M. Zhang. Supression and creation of chaos in a periodically forced Lorenz system. *Phys. Rev. E*, 52 (4):3558–3565, 1995.

H.-J. Freund. Motor unit and muscle activity in voluntary motor control. *Physiol. Rev.*, 63(2):387–436, 1983.

H. Fujigaki, M. Nishi, and T. Shimada. Synchronization of nonlinear systems with distinct parameters: Phase synchronization and metamorphosis. *Phys. Rev. E*, 53(4):3192–3197, 1996.

H. Fujigaki and T. Shimada. Phase synchronization and nonlinearity decision in the network of chaotic flows. *Phys. Rev. E*, 55(3):2426–2433, 1997.

H. Fujisaka, H. Ishii, M. Inoue, and T. Yamada. Intermittency caused by chaotic modulation II. *Prog. Theor. Phys.*, 76(6):1198–1209, 1986.

H. Fujisaka, S. Matsushita, and T. Yamada.

Fluctuation-controlled transient below the on–off intermittency transition. *J. Phys. A: Math. Gen.*, 30(16): 5697–5707, 1997.

H. Fujisaka, K. Ouchi, H. Hata, B. Masaoka, and S. Miyazaki. On-off intermittency in oscillatory media. *Physica* D, 114(3–4):237–250, 1998.

H. Fujisaka and T. Yamada. Stability theory of synchronized motion in coupled-oscillator systems. *Prog. Theor. Phys.*, 69(1):32–47, 1983.

H. Fujisaka and T. Yamada. A new intermittency in coupled dynamical systems. *Prog. Theor. Phys.*, 74(4): 918–921, 1985.

H. Fujisaka and T. Yamada. Stability theory of synchronized motion in coupled-oscillator systems. IV. Instability of synchronized chaos and new intermittency. *Prog. Theor. Phys.*, 75(5):1087–1104, 1986.

H. Fujisaka and T. Yamada. Intermittency caused by chaotic modulation. III. Self-similarity and higher order correlation functions. *Prog. Theor. Phys.*, 77(5):1045–1056, 1987.

J. M. Furman. Posturography: Uses and limitations. In *Bailliére's Clinical Neurology*, volume 3, pages 501–513. Bailliére Tindall, London, 1994.

D. Gabor. Theory of communication. *J. IEE (London)*, 93(3):429–457, 1946.

P. M. Gade. Synchronization of oscillators with random nonlocal connectivity. *Phys. Rev. E*, 54(1):64–70, 1996.

J. A. C. Gallas, P. Grassberger, H. J. Herremann, and P. Ueberholz. Noisy collective behaviour in deterministic cellular automata. *Physica* A, 180: 19–41, 1992.

L. Gammaitoni, P. Hänggi, P. Jung, and F. Marchesoni. Stochastic resonance. *Rev. Mod. Phys.*, 70:223–288, 1998.

C. W. Gardiner. *Handbook of Stochastic Methods*. Springer, Berlin, 1990.

D. J. Gauthier and J. C. Bienfang. Intermittent loss of synchronization in coupled chaotic oscillators: Toward a new criterion for high-qualitiy synchronization. *Phys. Rev. Lett.*, 77 (9):1751–1754, 1996.

I. M. Gel'fand, S. A. Kovalev, and L. M. Chailahyan. Intracellular stimulation of different regions of a frog's heart. *Doklady Akademii Nauk SSSR*, 148(4): 973–976, 1963 (in Russian).

W. Gerstner. Time structure of the activity in neural network models. *Phys. Rev. E*, 51(1):738–758, 1995.

L. Glass. Cardiac arrhythmias and circle maps. *Chaos*, 1:13–19, 1991.

L. Glass. Synchronization and rhythmic processes in physiology. *Nature*, 410:277–284, 2001.

L. Glass and M. C. Mackey. *From Clocks to Chaos: The Rhythms of Life*. Princeton University Press, Princeton, NJ, 1988.

L. Glass and A. Shrier. Low-dimensional dynamics in the heart. In L. Glass, P. Hunter, and A. McCulloch, Editors, *Theory of Heart*, pages 289–312. Springer, New York, 1991.

P. Glendinning. *Stability, Instability and Chaos*. Cambridge University Press, Cambridge, 1994.

P. Glendinning. The stability boundary of synchronized states in globally coupled dynamical systems. *Phys. Lett. A*, 259: 129–134, 1999.

P. Glendinning, U. Feudel, A. Pikovsky, and J. Stark. The structure of mode-locking regions in quasi-periodically forced circle maps. *Physica* D, 140(1):227–243, 2000.

P. Glendinning and M. Proctor. Travelling waves with spatially resonant forcing:

Bifurcation of a modified Landau equation. *Int. J. Bifurc. Chaos*, 3(6): 1447–1455, 1993.

N. R. J. Glossop, L. C. Lyons, and P. E. Hardin. Interlocked feedback loops within the *Drosophila* circadian oscillator. *Science*, 286:766–768, 1999.

A. F. Glova, S. Yu. Kurchatov, V. V. Likhanskii, A. Yu. Lysikov, and A. P. Napartovich. Coherent emission of a linear array of $CO_2$ waveguide lasers with a spatial filter. *Quant. Electron.*, 26(6):500–502, 1996.

L. Goldberg, H. F. Taylor, and J. F. Weller. Injection locking of coupled-stripe diode laser arrays. *Appl. Phys. Lett.*, 46 (3):236–238, 1985.

D. Golomb, D. Hansel, B. Shraiman, and H. Sompolinsky. Clustering in globally coupled phase oscillators. *Phys. Rev. A*, 45(6):3516, 1992.

D. L. Gonzalez and O. Piro. Symmetric kicked self-oscillators: Iterated maps, strange attractors, and symmetry of the phase-locking Farey hierarchy. *Phys. Rev. Lett.*, 55(1):17–20, 1985.

J. M. González-Miranda. Chaotic systems with a null conditional Lyapunov exponent under nonlinear driving. *Phys. Rev. E*, 53(1):R5–R8, 1996a.

J. M. González-Miranda. Synchronization of symmetric chaotic systems. *Phys. Rev. E*, 53(6):5656–5669, 1996b.

J. M. Gonzalez-Miranda. Communications by synchronization of spatially symmetric chaotic systems. *Phys. Lett. A*, 251(2):115–120, 1999.

M. Gorman, P. J. Widmann, and K. A. Robbins. Chaotic flow regimes in a convective loop. *Phys. Rev. Lett*, 52 (25):2241–2244, 1984.

M. Gorman, P. J. Widmann, and K. A. Robbins. Nonlinear dynamics of a convection loop: a quantitative comparison of experiment with theory. *Physica* D, 19(2):255–267, 1986.

A. Goryachev, H. Chaté, and R. Kapral. Synchronization defects and broken symmetry in spiral waves. *Phys. Rev. Lett.*, 80(4):873–876, 1998.

A. Goryachev and R. Kapral. Spiral waves in chaotic systems. *Phys. Rev. Lett.*, 76 (10):1619–1622, 1996.

P. Grassberger. Are damage spreading transitions generically in the universality class of directed percolation? *J. Stat. Phys.*, 79(1–2):13, 1995.

P. Grassberger. Synchronization of coupled systems with spatiotemporal chaos. *Phys. Rev.* E, 59(3): R2520–R2522, 1999.

C. Graves, L. Glass, D. Laporta, R. Meloche, and A. Grassino. Respiratory phase locking during mechanical ventilation in anesthetized human subjects. *Am. J. Physiol.*, 250: R902–R909, 1986.

C. M. Gray, P. König, A. K. Engel, and W. Singer. Oscillatory responses in cat visual cortex exhibit inter-columnar synchronization which reflects global stimulus properties. *Nature*, 338: 334–337, 1989.

C. Grebogi, E. Ott, S. Pelikan, and J. A. Yorke. Strange attractors that are not chaotic. *Physica* D, 13:261–268, 1984.

G. Grinstein, D. Mukamel, R. Seidin, and C. H. Bennett. Temporally periodic phases and kinetic roughening. *Phys. Rev. Lett.*, 70(23):3607–3610, 1993.

J. Guckenheimer. Isochrons and phaseless sets. *J. Math. Biol.*, 1:259–273, 1975.

J. Guckenheimer and P. Holmes. *Nonlinear Oscillations, Dynamical Systems, and Bifurcations of Vector Fields*. Springer, New York, 1986.

J. Güemez, C. Martín, and M. A. Matías. Approach to the chaotic synchronized state of some driving methods. *Phys. Rev.* E, 55(1):124–134, 1997.

J. Güemez and M. A. Matías. Synchronization in small assemblies of chaotic systems. *Phys. Rev.* E, 53(4): 3059–3067, 1996.

M. R. Guevara. Iteration of the human atrioventricular (AV) nodal recovery curve predicts many rhythms of AV block. In L. Glass, P. Hunter, and A. McCulloch, Editors, *Theory of Heart*, pages 313–358. Springer, New York, 1991.

M. R. Guevara, L. Glass, and A. Shrier. Phase-locking, period-doubling bifurcations and irregular dynamics in periodically stimulated cardiac cells. *Science*, 214:1350–1353, 1981.

M. R. Guevara, A. Shrier, and L. Glass. Phase-locked rhythms in periodically stimulated cardiac cells. *Am. J. Physiol.*, 254:H1–10, 1989.

N. Gupte and R. E. Amritkar. Synchronization of chaotic orbits: The influence of unstable periodic orbits. *Phys. Rev.* E, 48(3):R1620–R1623, 1993.

V. S. Gurfinkel, Ya. M. Kots, and M. L. Shik. *Regulation of Posture in Humans*. Nauka, Moscow, 1965 (in Russian).

H. Gutowitz, Editor. *Cellular Automata: Theory and Experiment*. North-Holland, Amsterdam, 1990.

R. Guttman, S. Lewis, and J. Rinzel. Control of repetitive firing in squid axon membrane as a model for a neuron oscillator. *J. Physiol. (London)*, 305:377–395, 1980.

H. Haken. *Information and Self-Organization. A Macroscopic Approach to Complex Systems*.

Springer, Berlin, 1988; 2nd edition, 1999.

H. Haken. *Advanced Synergetics: Instability Hierarchies of Self-Organizing Systems.* Springer, Berlin, 1993.

H. Haken, J. A. S. Kelso, and H. Bunz. A theoretical model of phase transitions in human hand movements. *Biol. Cybern.*, 51:347–356, 1985.

V. Hakim and W. J. Rappel. Dynamics of the globally coupled complex Ginzburg–Landau equation. *Phys. Rev. A*, 46(12):R7347–R7350, 1992.

J. K. Hale and H. Koçak. *Dynamics and Bifurcations.* Springer, New York, 1991.

G. M. Hall, S. Bahar, and D. J. Gauthier. Prevalence of rate-dependent behaviors in cardiac muscle. *Phys. Rev. Lett.*, 82 (14):2995–2998, 1999.

T. Halpin-Healy and Y.-C. Zhang. Kinetic roughening phenomena, stochastic growth, directed polymers and all that. *Physics Reports*, 254:215–414, 1995.

M. Hämäläinen, R. Hari, R. J. Ilmoniemi, J. Knuutila, and O. V. Lounasmaa. Magnetoencephalography – Theory, instrumentation, and applications to noninvasive studies of the working human brain. *Rev. Mod. Phys.*, 65: 413–497, 1993.

P. W. Hammer, N. Platt, S. M. Hammel, J. F. Heagy, and B. D. Lee. Experimental observation of on–off intermittency. *Phys. Rev. Lett.*, 73(8): 1095–1098, 1994.

S. K. Han, C. Kurrer, and Y. Kuramoto. Dephasing and bursting in coupled neural oscillators. *Phys. Rev. Lett.*, 75: 3190–3193, 1995.

S. K. Han, Ch. Kurrer, and Y. Kuramoto. Diffusive interaction leading to dephasing of coupled neural

oscillators. *Int. J. Bifurc. Chaos*, 7(4): 869–876, 1997.

D. Hansel, G. Mato, and C. Meunier. Clustering and slow switching in globally coupled phase oscillators. *Phys. Rev. E*, 48(5):3470–3477, 1993.

M. Hasler, Yu. Maistrenko, and O. Popovych. Simple example of partial synchronization of chaotic systems. *Phys. Rev. E*, 58(5): 6843–6846, 1998.

C. Hayashi. *Nonlinear Oscillations in Physical Systems.* McGraw-Hill, New York, 1964.

R. He and P. G. Vaidya. Time delayed chaotic systems and their synchronization. *Phys. Rev. E*, 59(4): 4048–4051, 1999.

J. F. Heagy and T. L. Carroll. Chaotic synchronization in Hamiltonian systems. *Chaos*, 4:385, 1994.

J. F. Heagy, T. L. Carroll, and L. M. Pecora. Experimental and numerical evidence for riddled basins in coupled chaotic systems. *Phys. Rev. Lett.*, 73 (26):3528, 1994a.

J. F. Heagy, T. L. Carroll, and L. M. Pecora. Synchronous chaos in coupled oscillator systems. *Phys. Rev. E*, 50(3): 1874–1884, 1994b.

J. F. Heagy, T. L. Carroll, and L. M. Pecora. Desynchronization by periodic orbits. *Phys. Rev. E*, 52(2): R1253–R1256, 1995.

J. F. Heagy and S. M. Hammel. The birth of strange nonchaotic attractors. *Physica* D, 70:140–153, 1994.

J. F. Heagy, N. Platt, and S. M. Hammel. Characterization of on–off intermittency. *Phys. Rev. E*, 49(2): 1140–1150, 1994c.

M. R. Herman. Une méthode pour minorer les exposants de Lyapunov et quelques exemples montrant le caractère local d'un théorème d'Arnold et de Moser sur le tori de dimension 2. *Comment. Math. Helvetici*, 58:453, 1983.

A. V. Herz and J. J. Hopfield. Earthquake cycles and neural reverberations: Collective oscillations in systems with pulse-coupled threshold elements. *Phys. Rev. Lett.*, 75(6):1222–1225, 1995.

H. Herzel and J. Freund. Chaos, noise, and synchronization reconsidered. *Phys. Rev. E*, 52(3):3238, 1995.

R. C. Hilborn. *Chaos and Nonlinear Dynamics: An Introduction for Scientists and Engineers*. Oxford University Press, Oxford, New York, 1994.

P. Holmes and D. R. Rand. Bifurcations of the forced van der Pol oscillator. *Quart. Appl. Math.*, 35:495–509, 1978.

H. Hong, M. Y. Choi, K. Park, B. G. Yoon, and K. S. Soh. Synchronization and resonance in a driven system of coupled oscillators. *Phys. Rev. E*, 60(4):4014–4020, 1999a.

H. Hong, M. Y. Choi, J. Yi, and K.-S. Soh. Inertia effects on periodic synchronization in a system of coupled oscillators. *Phys. Rev. E*, 59(1):353–363, 1999b.

H. Hong, M. Y. Choi, B.-G. Yoon, K. Park, and K.-S. Soh. Noise effects on synchronization in systems of coupled oscillators. *J. Phys. A: Math. Gen.*, 32:L9–L15, 1999c.

J. J. Hopfield. Neurons, dynamics and computation. *Phys. Today*, pages 40–46, February 1994.

J. J. Hopfield and A. V. M. Herz. Rapid local synchronization of action potentials: Toward computation with coupled integrate-and-fire neurons. *Proc. Nat. Acad. Sci. USA*, 92:6655–6662, 1995.

F. C. Hoppensteadt and E. M. Izhikevich. *Weakly Connected Neural Networks*. Springer, Berlin, 1997.

F. C. Hoppensteadt and E. M. Izhikevich. Oscillatory neurocomputers with dynamic connectivity. *Phys. Rev. Lett.*, 82(14):2983–2986, 1999.

W. Horsthemke and R. Lefever. *Noise Induced Transitions: Theory and Applications in Physics, Chemistry and Biology*. Springer, Berlin, 1989.

D. Hoyer, O. Hader, and U. Zwiener. Relative and intermittent cardiorespiratory coordination. *IEEE Eng. Med. Biol.*, 16(6):97–104, 1997.

G. Hu, J. Xiao, J. Yang, F. Xie, and Zh. Qu. Synchronization of spatiotemporal chaos and its applications. *Phys. Rev. E*, 56(3):2738–2746, 1997.

B. R. Hunt, E. Ott, and J. A. Yorke. Differentiable generalized synchronization of chaos. *Phys. Rev. E*, 55(4):4029–4034, 1997.

Ch. Huygens (Hugenii). *Horologium Oscillatorium*. Apud F. Muguet, Parisiis, France, 1673. English translation: *The Pendulum Clock*, Iowa State University Press, Ames, 1986.

Ch. Huygens. *Œvres Complètes*, volume 15. Swets & Zeitlinger B. V., Amsterdam, 1967a.

Ch. Huygens. *Œvres Complètes*, volume 17. Swets & Zeitlinger B. V., Amsterdam, 1967b.

J. M. Hyman, B. Nicolaenko, and S. Zaleski. Order and complexity in the Kuramoto–Sivashinsky model of weakly turbulent interfaces. *Physica D*, 23(1–3):265–292, 1986.

G. Iooss and D. D. Joseph. *Elementary Stability and Bifurcation Theory*. Springer, New York, 1980.

A. K. Jain, K. K. Likharev, J. E. Lukens, and J. E. Sauvageau. Mutual phase-locking in Josephson junction arrays. *Phys. Reports*, 109(6): 309–426, 1984.

J. Jalife and C. Antzelevitch. Phase resetting and annihilation of pacemaker activity in cardiac tissue. *Science*, 206:695–697, 1979.

M. H. Jensen, P. Bak, and T. Bohr. Complete devil's staircase, fractal dimension and universality of mode-locking structure in the circle map. *Phys. Rev. Lett*, 50(21): 1637–1639, 1983.

M. H. Jensen, P. Bak, and T. Bohr. Transition to chaos by interaction of resonances in dissipative systems. I. Circle maps. *Phys. Rev. A*, 30(4): 1960–1969, 1984.

Y. Jiang and P. Parmananda. Synchronization of spatiotemporal chaos in asymmetrically coupled map lattices. *Phys. Rev. E*, 57(4): 4135–4139, 1998.

G. A. Johnson, D. J. Mar, T. L. Carroll, and L. M. Pecora. Synchronization and imposed bifurcations in the presence of large parameter mismatch. *Phys. Rev. Lett.*, 80(18):3956–3959, 1998.

L. Junge and U. Parlitz. Phase synchronization of coupled Ginzburg–Landau equations. *Phys. Rev. E*, 62(1):438–441, 2000.

E. Kaempfer. *The History of Japan (With a Description of the Kingdom of Siam)*. Sloane, London, 1727. Posthumous translation; or reprint by McLehose, Glasgow, 1906.

P. B. Kahn. *Mathematical Methods for Scientists and Engineers: Linear and Nonlinear Systems*. Wiley, New York, 1990.

K. Kaneko. Clustering, coding, switching, hierarchical ordering and control in network of chaotic elements. *Physica D*, 41:137–172, 1990.

K. Kaneko. Globally coupled circle maps. *Physica D*, 54(1):5–19, 1991.

K. Kaneko, Editor. *Theory and Applications of Coupled Map Lattices*. Wiley, Chichester, 1993.

K. Kaneko. Dominance of Milnor attractors and noise-induced selection in a multiattractor system. *Phys. Rev. Lett.*, 78(14):2736–2739, 1997.

K. Kaneko. On the strength of attractors in a high-dimensional system: Milnor attractor network, robust global attraction, and noise-induced selection. *Physica D*, 124:322–344, 1998.

H. Kantz and T. Schreiber. *Nonlinear Time Series Analysis*. Cambridge University Press, Cambridge, 1997.

T. Kapitaniak, Y. Maistrenko, A. Stefanski, and J. Brindley. Bifurcations from locally to globally riddled basins. *Phys. Rev. E*, 57(6):R6253–R6256, 1998.

T. Kapitaniak and Yu. L. Maistrenko. Chaos synchronization and riddled basins in two coupled one-dimensional maps. *Chaos Solit. Fract.*, 9(1–2): 271–282, 1998.

T. Kapitaniak, J. Wojewoda, and J. Brindley. Synchronization and desynchronization in quasi-hyperbolic chaotic systems. *Phys. Lett. A*, 210: 283–289, 1996.

D. Kaplan and L. Glass. *Understanding Nonlinear Dynamics*. Springer, New York, 1995.

J. L. Kaplan, J. Mallet-Paret, and J. A. Yorke. The Lyapunov dimension of a nowhere differentiable attracting torus. *Ergod. Theor. Dynam. Syst.*, 4: 261–281, 1984.

J. L. Kaplan and J. A. Yorke. Chaotic behavior of multidimensional difference equations. In H. O. Walter and H.-O. Peitgen, Editors, *Functional Differential Equations and Approximation of Fixed Points*, volume 730 of *Lecture Notes in Mathematics*, pages 204–227. Springer, Berlin, 1979.

R. Kapral and K. Showalter, Editors. *Chemical Waves and Patterns*. Kluwer, Dordrecht, 1995.

A. Katok and B. Hasselblatt. *Introduction to the Modern Theory of Dynamical Systems*. Cambridge University Press, Cambridge, 1995.

G. Keller. A note on strange nonchaotic attractors. *Fund. Math.*, 151:139–148, 1996.

T. Kenner, H. Pessenhofer, and G. Schwaberger. Method for the analysis of the entrainment between heart rate and ventilation rate. *Pflügers Arch.*, 363:263–265, 1976.

A. A. Kharkevich. *Basics of Radio Engineering*. Svyazizdat, Moscow, 1962 (in Russian).

P. Khoury, M. A. Lieberman, and A. J. Lichtenberg. Degree of synchronization of noisy maps on the circle. *Phys. Rev. E*, 54(4):3377–3388, 1996.

P. Khoury, M. A. Lieberman, and A. J. Lichtenberg. Experimental measurement of the degree of chaotic synchronization using a distribution exponent. *Phys. Rev. E*, 57(5): 5448–5466, 1998.

C.-M. Kim. Mechanism of chaos synchronization and on–off intermittency. *Phys. Rev. E*, 56(3): 3697–3700, 1997.

V. Kirk and E. Stone. Effect of a refractory period on the entrainment of pulse-coupled integrate-and-fire oscillators. *Phys. Lett. A*, 232:70–76, 1997.

L. Kocarev and U. Parlitz. Generalized synchronization, predictability, and equivalence of unidirectionally coupled dynamical systems. *Phys. Rev. Lett.*, 76(11):1816–1819, 1996.

L. Kocarev, Z. Tasev, and U. Parlitz. Synchronizing spatiotemporal chaos of partial differential equations. *Phys. Rev. Lett.*, 79(1):51–54, 1997.

H. P. Koepchen. Physiology of rhythms and control systems: An integrative approach. In H. Haken and H. P. Koepchen, Editors, *Rhythms in Physiological Systems*, volume 55 of *Springer Series in Synergetics*, pages 3–20, Springer, Berlin, 1991.

V. G. Kolin'ko, T. A. Arhangel'skaja, and Yu. M. Romanovsky. Protoplasm motion in slime mold *Physarum* under varying temperature. *Stud. Biophys.*, 106(3):215–222, 1985 (in Russian).

R. Konnur. Equivalence of synchronization and control of chaotic systems. *Phys. Rev. Lett.*, 77(14): 2937–2940, 1996.

N. Kopell and G. B. Ermentrout. Symmetry and phase locking in chains of weakly coupled oscillators. *Comm. Pure Appl. Math.*, 39:623–660, 1986.

N. Koshiya and J. C. Smith. Neuronal pacemaker for breathing visualized *in vitro*. *Nature*, 400:360–363, 1999.

I. O. Kulik and I. K. Yanson. *Josephson Effect in Superconductive Tunnel Structures*. Nauka, Moscow, 1970 (in Russian).

Y. Kuramoto. Self-entrainment of a population of coupled nonlinear oscillators. In H. Araki, Editor, *International Symposium on Mathematical Problems in Theoretical Physics*, volume 39 of *Springer Lecture Notes in Physics*, page 420, Springer, New York, 1975.

Y. Kuramoto. *Chemical Oscillations,*

*Waves and Turbulence.* Springer, Berlin, 1984.

Y. Kuramoto, T. Aoyagi, I. Nishikawa, T. Chawanya, and K. Okuda. Neural network model carrying phase information with application to collective dynamics. *Prog. Theor. Phys.*, 87(5):1119–1126, 1992.

Y. Kuramoto and T. Tsuzuki. Persistent propagation of concentration waves in dissipative media far from thermal equilibrium. *Prog. Theor. Phys.*, 55: 356, 1976.

K. Kurokawa. Injection locking of microwave solid-state oscillators. *Proc. IEEE*, 61(10):1386–1410, 1973.

C. Kurrer. Synchronization and desynchronization of weakly coupled oscillators. *Phys. Rev. E*, 56(4): 3799–3802, 1997.

J. Kurths, Editor. A focus issue on phase synchronization in chaotic systems. *Int. J. Bifurc. Chaos*, 10:2289–2667, 2000.

J. Kurths and A. S. Pikovsky. Symmetry breaking in distributed systems and modulational spatio-temporal intermittency. *Chaos Solit. Fract.*, 5 (10):1893–1899, 1995.

S. Kuznetsov, A. Pikovsky, and U. Feudel. Birth of a strange nonchaotic attractor: A renormalization group analysis. *Phys. Rev. E*, 51(3):R1629–R1632, 1995.

S. P. Kuznetsov and A. S. Pikovsky. Transition from a symmetric to a nonsymmetric regime under conditions of randomness dynamics in a system of dissipatively coupled recurrence mappings. *Radiophys. Quant. Electron.*, 32(1):41–45, 1989.

Yu. Kuznetsov, P. S. Landa, A. Ol'khovoi, and S. Perminov. Relationship between the amplitude threshold of

synchronization and the entropy in stochastic self-excited systems. *Sov. Phys. Dokl.*, 30(3):221–222, 1985.

Y.-Ch. Lai. Symmetry-breaking bifurcation with on–off intermittency in chaotic dynamical systems. *Phys. Rev. E*, 53(5):4267–4270, 1996a.

Y.-Ch. Lai. Transition from strange nonchaotic to strange chaotic attractors. *Phys. Rev. E*, 53(1):57–65, 1996b.

Y.-Ch. Lai. Scaling laws for noise-induced temporal riddling in chaotic systems. *Phys. Rev. E*, 56(4):3897–3908, 1997.

Y.-Ch. Lai and C. Grebogi. Noise-induced riddling in chaotic systems. *Phys. Rev. Lett.*, 77(25):5047–5050, 1996.

Y.-Ch. Lai, C. Grebogi, J. A. Yorke, and S. C. Venkataramani. Riddling bifurcation in chaotic dynamical systems. *Phys. Rev. Lett.*, 77(1):55–58, 1996.

B. C. Lampkin, T. Nagao, and A. M. Mauer. Synchronization of the mitotic cycle in acute leukaemia. *Nature*, 222: 1274–1275, 1969.

P. S. Landa. *Self–Oscillations in Systems with Finite Number of Degrees of Freedom.* Nauka, Moscow, 1980 (in Russian).

P. S. Landa. *Nonlinear Oscillations and Waves in Dynamical Systems.* Kluwer, Dordrecht, 1996.

P. S. Landa and S. M. Perminov. Interaction of periodic and stochastic self-oscillations. *Electronics*, 28(4): 285–287, 1985.

P. S. Landa and S. M. Perminov. Synchronizing the chaotic oscillations in the Mackey–Glass system. *Radiofizika*, 30:437–439, 1987 (in Russian).

P. S. Landa, Y. S. Rendel, and V. A. Sher. Synchronization of oscillations in the

Lorenz system. *Radiofizika*, 32: 1172–1174, 1989 (in Russian).

P. S. Landa and M. G. Rosenblum. Synchronization of random self-oscillating systems. *Sov. Phys. Dokl.*, 37(5):237–239, 1992.

P. S. Landa and M. G. Rosenblum. Synchronization and chaotization of oscillations in coupled self-oscillating systems. *Appl. Mech. Rev.*, 46(7): 414–426, 1993.

P. S. Landa and N. D. Tarankova. Synchronization of a generator with modulated natural frequency. *Radio Eng. Electron. Phys.*, 21(2):34–38, 1976.

K. J. Lee, Y. Kwak, and T. K. Lim. Phase jumps near a phase synchronization transition in systems of two coupled chaotic oscillators. *Phys. Rev. Lett.*, 81 (2):321–324, 1998.

G. V. Levina and A. A. Nepomnyaschiy. Analysis of an amplitude equation for autovibrational flow regimes at resonance external forces. *A. Angew. Math. Mech.*, 66(6):241–246, 1986.

A. J. Lichtenberg and M. A. Lieberman. *Regular and Chaotic Dynamics*. Springer, New York, 1992.

E. M. Lifshitz and L. P. Pitaevskii. *Physical Kinetics*. Pergamon Press, Cambridge, 1981.

K. K. Likharev. *Dynamics of Josephson Junctions and Circuits*. Gordon and Breach, Philadelphia, 1991.

A. L. Lin, M. Bertram, K. Martinez, H. L. Swinney, A. Ardelea, and G. F. Carey. Resonant phase patterns in a reaction–diffusion system. *Phys. Rev. Lett.*, 84(18):4240–4243, 2000.

W. C. Lindsey and C. M. Chie, Editors. *Phase-Locked Loops*. IEEE Press, New York, 1985.

M. Lipp and N. S. Longridge.

Computerized dynamic posturography: Its place in the evaluation of patients with dizziness and imbalance. *J. Otolaryngology*, 23(3):177–183, 1994.

J. N. Little and L. Shure. *Signal Processing Toolbox for Use with MATLAB. User's Guide*. Mathworks, Natick, MA, 1992.

Z. Liu and S. Chen. Symbolic analysis of generalized synchronization of chaos. *Phys. Rev. E*, 56(6):7297–7300, 1997.

Z. H. Liu, S. G. Chen, and B. B. Hu. Coupled synchronization of spatiotemporal chaos. *Phys. Rev. E*, 59 (3):2817–2821, 1999.

L. Longa, S. P. Dias, and E. M. F. Curado. Lyapunov exponents and coalescence of chaotic trajectories. *Phys. Rev. E*, 56 (1):259–263, 1997.

A. Longtin, A. Bulsara, D. Pierson, and F. Moss. Bistability and the dynamics of periodically forced sensory neurons. *Biol. Cybernetics*, 70:569–578, 1994.

A. Longtin and D. R. Chialvo. Stochastic and deterministic resonances for excitable systems. *Phys, Rev. Lett.*, 81(18):4012–4015, 1998.

R. Lopez-Ruiz and Y. Pomeau. Transition between two oscillation modes. *Phys. Rev. E*, 55(4):R3820–R3823, 1997.

E. N. Lorenz. Deterministic nonperiodic flow. *J. Atmos. Sci.*, 20:130–141, 1963.

E. N. Lorenz. *The Essence of Chaos*. University of Washington Press, Seattle, 1993.

M. N. Lorenzo, I. P. Mariño, V. Pérez-Muñuzuri, M. A. Matías, and V. Pérez-Villar. Synchronization waves in arrays of driven chaotic systems. *Phys. Rev. E*, 54(4):3094–3097, 1996.

C. Ludwig. Beiträge zur Kenntnis des Einflusses der Respirationsbewegung auf den Blutlauf im Aortensystem. *Arch. Anat. Physiol.*, 13:242–302,

1847.

W. A. MacKay. Synchronized neuronal oscillations and their role in motor processes. *Trends Cogn. Sci.*, 1: 176–183, 1997.

R. Mainieri and J. Rehacek. Projective synchronization in three-dimensional chaotic systems. *Phys. Rev. Lett.*, 82 (15):3042–3045, 1999.

Y. Maistrenko and T. Kapitaniak. Different types of chaos synchronization in two coupled piecewise linear maps. *Phys. Rev. E*, 54(4):3285–3292, 1996.

Y. Maistrenko, T. Kapitaniak, and P. Szuminski. Locally and globally riddled basins in two coupled piecewise-linear maps. *Phys. Rev. E*, 56(6):6393–6399, 1997.

Y. L. Maistrenko, V. L. Maistrenko, O. Popovych, and E. Mosekilde. Desynchronization of chaos in coupled logistic maps. *Phys. Rev. E*, 60(3): 2817–2830, 1999a.

Yu. L. Maistrenko, V. L. Maistrenko, O. Popovych, and E. Mosekilde. Transverse instability and riddled basins in a system of two coupled logistic maps. *Phys. Rev. E*, 57(3): 2713–2724, 1998.

Yu. L. Maistrenko, O. Popovych, and M. Hasler. On strong and weak chaotic partial synchronization. *Int. J. Bifurc. Chaos*, 10(1):179–204, 2000.

A. N. Malakhov. *Fluctuations in Self-Oscillatory Systems*. Nauka, Moscow, 1968 (in Russian).

I. G. Malkin. *Some Problems in Nonlinear Oscillation Theory*. Gostechizdat, Moscow, 1956 (in Russian).

S. C. Manrubia and A. S. Mikhailov. Mutual synchronization and clustering in randomly coupled chaotic dynamical networks. *Phys. Rev. E*, 60 (2):1579–1589, 1999.

M. Manscher, M. Nordham, E. Mosekilde, and Yu. L. Maistrenko. Riddled basins of attraction for synchronized type-I intermittency. *Phys. Lett. A*, 238: 358–364, 1998.

A. Maritan and J. R. Banavar. Chaos, noise, and synchronization. *Phys. Rev. Lett.*, 72:1451–1454, 1994.

S. Martin and W. Martienssen. Circle maps and mode locking in the driven electrical conductivity of barium sodium niobate crystals. *Phys. Rev. Lett.*, 56(15):1522–1525, 1986.

G. Matsumoto, K. Aihara, Y. Hanyu, N. Takahashi, S. Yoshizawa, and J. Nagumo. Chaos and phase locking in normal squid axons. *Phys. Lett. A*, 123:162–166, 1987.

K. Matsumoto and I. Tsuda. Noise-induced order. *J. Stat. Phys.*, 31 (1):87–106, 1983.

T. Matsumoto and M. Nishi. Subsystem decreasing for exponential synchronization of chaotic systems. *Phys. Rev. E*, 59(2):1711–1718, 1999.

P. C. Matthews, R. E. Mirollo, and S. H. Strogatz. Dynamics of a large system of coupled nonlinear oscillators. *Physica D*, 52:293–331, 1991.

P. C. Matthews and S. H. Strogatz. Phase diagram for the collective behavior of limit-cycle oscillators. *Phys. Rev. Lett.*, 65(14):1701–1704, 1990.

M. K. McClintock. Menstrual synchrony and suppression. *Nature*, 229: 244–245, 1971.

R. V. Mendes. Clustering and synchronization with positive Lyapunov exponents. *Phys. Lett. A*, 257:132–138, 1999.

R. Mettin, U. Parlitz, and W. Lauterborn. Bifurcation structure of the driven van der Pol oscillator. *Int. J. Bifurc. Chaos*, 3(6):1529–1555, 1993.

A. S. Mikhailov. *Foundations of Synergetics 1. Distributed Active Systems.* Springer, Berlin, 1994.

M. Milan, S. Campuzano, and A. Garcia-Bellido. Cell cycling and patterned cell proliferation in the *Drosophila* wing during metamorphosis. *Proc. Natl. Acad. Sci. USA*, 93(21):11687–11692, 1996.

J. Milnor. On the concept of attractor. *Commun. Math. Phys.*, 99:177–195, 1985.

A. A. Minai and T. Anand. Chaos-induced synchronization in discrete-time oscillators driven by a random input. *Phys. Rev. E*, 57(2):1559–1562, 1998.

A. A. Minai and T. Anand. Synchronization of chaotic maps through a noisy coupling channel with application to digital communication. *Phys. Rev. E*, 59(1):312–320, 1999a.

A. A. Minai and T. Anand. Synchronizing multiple chaotic maps with a randomized scalar coupling. *Physica D*, 125(3-4):241–259, 1999b.

N. Minorsky. *Nonlinear Oscillations.* Van Nostrand, Princeton, NJ, 1962.

R. Mirollo and S. Strogatz. Amplitude death in an array of limit-cycle oscillators. *J. Stat. Phys.*, 60:245–262, 1990a.

R. Mirollo and S. Strogatz. Synchronization of pulse-coupled biological oscillators. *SIAM J. Appl. Math.*, 50:1645–1662, 1990b.

V. M. Mitjushin, L. L. Litinskaja, and L. B. Kaminir. On synchronous variation of cell nuclei. In *Oscillatory Processes in Biological and Chemical Systems*, pages 325–331, Nauka, Moscow, 1967 (in Russian).

F. Mitschke. Acausal filters for chaotic signals. *Phys. Rev. A*, 41:1169–1171, 1990.

F. Mitschke, M. Möller, and H. W. Lange. Measuring filtered chaotic signals. *Phys. Rev. A*, 37:4518–4521, 1988.

S. Miyazaki and H. Hata. Universal scaling law of the power spectrum in the on–off intermittency. *Phys. Rev. E*, 58(6):7172–7175, 1998.

F. C. Moon. *Chaotic Vibration. An Introduction for Applied Scientists and Engineers.* Wiley, New York, 1987.

R. Y. Moore. A clock for the ages. *Science*, 284:2102–2103, 1999.

L. G. Morelli and D. H. Zanette. Synchronization of stochastically coupled cellular automata. *Phys. Rev. E*, 58(1):R8–R11, 1998.

O. Morgül. Necessary condition for observer-based chaos synchronization. *Phys. Rev. Lett.*, 82(1):77–80, 1999.

O. Morgül and M. Feki. A chaotic masking scheme by using synchronized chaotic systems. *Phys. Lett. A*, 251(3):169–176, 1999.

F. Mormann, K. Lehnertz, P. David, and C. E. Elger. Mean phase coherence as a measure for phase synchronization and its application to the EEG of epilepsy patients. *Physica D*, 144 (3–4):358–369, 2000.

F. Moss, A. Bulsara, and M. Shlesinger, Editors. The proceedings of the NATO advanced research workshop: Stochastic resonance in physics and biology. *J. Stat. Phys*, 70:1–512, 1993.

F. Moss, D. Pierson, and D. O'Gorman. Stochastic resonance: Tutorial and update. *Int. J. Bifurc. Chaos*, 4(6): 1383–1397, 1994.

N. Mousseau. Synchronization by disorder in coupled systems. *Phys. Rev. Lett.*, 77:968–971, 1996.

R. Mrowka, A. Patzak, and M. G. Rosenblum. Quantitative analysis of cardiorespiratory synchronization in infants. *Int. J. Bifurc. Chaos*, 10(11):2479–2518, 2000.

Y. Nagai and Y.-Ch. Lai. Characterization of blowout bifurcation by unstable periodic orbits. *Phys. Rev. E*, 55(2): 1251–1254, 1997a.

Y. Nagai and Y.-Ch. Lai. Periodic-orbit theory of the blowout bifurcation. *Phys. Rev. E*, 56(4):4031–4041, 1997b.

N. Nakagawa and Y. Kuramoto. Collective chaos in a population of globally coupled oscillators. *Prog. Theor. Phys.*, 89(2):313–323, 1993.

N. Nakagawa and Y. Kuramoto. From collective oscillations to collective chaos in a globally coupled oscillator system. *Physica* D, 75:74–80, 1994.

N. Nakagawa and Y. Kuramoto. Anomalous Lyapunov spectrum in globally coupled oscillators. *Physica* D, 80:307–316, 1995.

H. Nakao. Asymptotic power law of moments in a random multiplicative process with weak additive noise. *Phys. Rev. E*, 58(2):1591–1600, 1998.

S. Nakata, T. Miyata, N. Ojima, and K. Yoshikawa. Self-synchronization in coupled salt-water oscillators. *Physica* D, 115:313–320, 1998.

A. H. Nayfeh and D. T. Mook. *Nonlinear Oscillations*. Wiley, New York, 1979.

Z. Néda, E. Ravasz, Y. Brechet, T. Vicsek, and A.-L. Barabási. Tumultuous applause can transform itself into waves of synchronized clapping. *Nature*, 403(6772):849–850, 2000.

A. Neiman, X. Pei, D. F. Russell, W. Wojtenek, L. Wilkens, F. Moss, H. A. Braun, M. T. Huber, and K. Voigt. Synchronization of the noisy electrosensitive cells in the paddlefish. *Phys. Rev. Lett.*, 82(3):660–663, 1999a.

A. Neiman, L. Schimansky-Geier, A. Cornell-Bell, and F. Moss.

Noise-enhanced phase synchronization in excitable media. *Phys. Rev. Lett.*, 83 (23):4896–4899, 1999b.

A. Neiman, L. Schimansky-Geier, F. Moss, B. Shulgin, and J. J. Collins. Synchronization of noisy systems by stochastic signals. *Phys. Rev. E*, 60(1): 284–292, 1999c.

A. Neiman, A. Silchenko, V. S. Anishchenko, and L. Schimansky-Geier. Stochastic resonance: Noise-enhanced phase coherence. *Phys. Rev. E*, 58(6): 7118–7125, 1998.

A. B. Neiman, D. F. Russell, X. Pei, W. Wojtenek, J. Twitty, E. Simonotto, B. A. Wettring, E. Wagner, L. A. Wilkens, and F. Moss. Stochastic synchronization of electroreceptors in the paddlefish. *Int. J. Bifurc. Chaos*, 10(11):2499–2518, 2000.

Yu. I. Neimark and P. S. Landa. *Stochastic and Chaotic Oscillations*. Kluwer, Dordrecht, 1992.

A. Nepomnyashchy. Stability of the wavy regimes in the film flowing down an inclined plane. *Izv. AN SSSR, Mekh. Zhidk. i Gaza*, (3):28–34, 1974. English translation: *Fluid Dyn.*, 9:354–359, 1974.

A. C. Newell. Envelope equations. In *Lectures in Applied Mathematics*, volume 15, page 157, American Mathematical Society, Providence, RI, 1974.

S. Nichols and K. Wiesenfeld. Ubiquitous neutral stability of splay-phase states. *Phys. Rev. A*, 45:8430–8435, 1992.

K. Niizeki, K. Kawahara, and Y. Miyamoto. Interaction among cardiac, respiratory, and locomotor rhythms during cardiolocomotor synchronization. *J. Appl. Physiol.*, 75 (4):1815–1821, 1993.

T. Nishikawa and K. Kaneko. Fractalization of torus revisited as a strange nonchaotic attrator. *Phys. Rev. E*, 54(6):6114–6124, 1996.

E. P. Odum. *Fundamentals of Ecology*. Saunders, Philadelphia, 1953.

K. Okuda. Variety and generality of clustering in globally coupled oscillators. *Physica D*, 63:424–436, 1993.

F. Ollendorf and W. Peters. Schwingungsstabilität parallelarbeitender Synchromaschinen. In *Wissenschaftliche Veröffentlichungen aus dem Siemens–Konzern*, volume 6, pages 7–26. Springer, Berlin, 1925–1926.

G. Osipov, A. Pikovsky, M. Rosenblum, and J. Kurths. Phase synchronization effects in a lattice of nonidentical Rössler oscillators. *Phys. Rev. E*, 55 (3):2353–2361, 1997.

S. Ostlund, D. Rand, J. Sethna, and E. Siggia. Universal properties of the transition from quasi-periodicity to chaos in dissipative systems. *Physica D*, 8:303–342, 1983.

E. Ott. *Chaos in Dynamical Systems*. Cambridge University Press, Cambridge, 1992.

G. Paladin and A. Vulpiani. Anomalous scaling laws in multifractal objects. *Phys. Rep.*, 156:147–225, 1987.

M. Palus. Detecting phase synchronization in noisy systems. *Phys. Lett. A*, 227: 301–308, 1997.

Ya. G. Panovko and I. I. Gubanova. *Stability and Oscillation of Elastic Systems*. Nauka, Moscow, 1964 (in Russian).

P. Panter. *Modulation, Noise, and Spectral Analysis*. McGraw-Hill, New York, 1965.

P. Paoli, A. Politi, and R. Badii. Long-range order in the scaling behaviour of hyperbolic dynamical systems. *Physica D*, 36(6):263–286, 1989a.

P. Paoli, A. Politi, G. Broggi, M. Ravani, and R. Badii. Phase transitions in filtered chaotic signals. *Phys. Rev. Lett.*, 62(21):2429–2432, 1989b.

N. Parekh, V. R. Kumar, and B. D. Kulkarni. Analysis and characterization of complex spatio-temporal patterns in nonlinear reaction–diffusion systems. *Physica A*, 224(1–2):369–381, 1996.

E.-H. Park, M. A. Zaks, and J. Kurths. Phase synchronization in the forced Lorenz system. *Phys. Rev. E*, 60(6): 6627–6638, 1999.

K. Park, S. W. Rhee, and M. Y. Choi. Glass synchronization in the network of oscillators with random phase shifts. *Phys. Rev. E*, 57(5):5030–5035, 1998.

S. H. Park, S. Kim, H.-B. Pyo, and S. Lee. Effects of time-delayed interactions on dynamic patterns in a coupled oscillator system. *Phys. Rev. E*, 60(4): 4962–4965, 1999a.

S. H. Park, S. Kim, H.-B. Pyo, and S. Lee. Multistability analysis of phase locking patterns in an excitatory coupled neural system. *Phys. Rev. E*, 60(2): 2177–2181, 1999b.

U. Parlitz, L. Junge, and L. Kocarev. Subharmonic entrainment of unstable periodic orbits and generalized synchronization. *Phys. Rev. Lett.*, 79 (17):3158–3161, 1997.

U. Parlitz, L. Junge, W. Lauterborn, and L. Kocarev. Experimental observation of phase synchronization. *Phys. Rev. E*, 54(2):2115–2118, 1996.

U. Parlitz and L. Kocarev. Synchronization of chaotic systems. In H. Schuster, Editor, *Handbook of Chaos Control*, pages 271–303. Wiley-VCH, Weinheim, 1999.

V. Parlitz and W. Lauterborn. Period-doubling cascades and devil's staircases of the driven van der Pol oscillator. *Phys. Rev.* A, 36(3): 1428–1434, 1987.

F. Pasemann. Synchronized chaos and other coherent states for two coupled neurons. *Physica* D, 128(2–4): 236–249, 1999.

I. Pastor-Diáz and A. López-Fraguas. Dynamics of two coupled van der Pol oscillators. *Phys. Rev.* E, 52(2): 1480–1489, 1995.

I. Pastor-Diáz, V. Perez-Garcia, F. Encinas-Sanz, and J. M. Guerra. Ordered and chaotic behavior of two coupled van der Pol oscillators. *Phys. Rev.* E, 48(1):171–182, 1993.

T. Pavlidis. Populations of interacting oscillators and circadian rhythms. *J. Theor. Biol.*, 22:418–436, 1969.

L. Pecora, Editor. A focus issue on synchronization in chaotic systems. *Chaos*, 7(4):509–687, 1997.

L. M. Pecora and T. L. Carroll. Synchronization in chaotic systems. *Phys. Rev. Lett.*, 64:821–824, 1990.

L. M. Pecora and T. L. Carroll. Driving systems with chaotic signals. *Phys. Rev.* A, 44:2374–2383, 1991.

L. M. Pecora and T. L. Carroll. Discontinuous and nondifferentiable functions and dimension increase induced by filtering chaotic data. *Chaos*, 6(3):432–439, 1996.

L. M. Pecora and T. L. Carroll. Master stability functions for synchronized coupled systems. *Phys. Rev. Lett.*, 80 (10):2109–2112, 1998.

L. M. Pecora, T. L. Carroll, and J. F. Heagy. Statistics for continuity and differentiability: An application to attractor reconstruction from time series. In C. D. Cutler and D. T. Kaplan, Editors, *Nonlinear Dynamics and Time Series*, volume 11 of *Fields Inst. Communications*, pages 49–62. American Mathematical Society, Providence, RI, 1997a.

L. M. Pecora, T. L. Carroll, G. Johnson, and D. Mar. Volume-preserving and volume-expanding synchronized chaotic systems. *Phys. Rev.* E, 56(5): 5090–5100, 1997b.

L. M. Pecora, T. L. Carroll, G. A. Johnson, D. J. Mar, and J. F. Heagy. Fundamentals of synchronization in chaotic systems, concepts, and applications. *Chaos*, 7(4):520–543, 1997c.

J. Peinke, R. Richter, and J. Parisi. Spatial coherence of nonlinear dynamics in a semiconductor experiment. *Phys. Rev.* B, 47(1):115–124, 1993.

H.-O. Peitgen, H. Jürgens, and D. Saupe. *Chaos and Fractals: New Frontiers of Science.* Springer, New York, 1992.

J. H. Peng, E. J. Ding, M. Ding, and W. Yang. Synchronizing hyperchaos with a scalar transmitted signal. *Phys. Rev. Lett.*, 76(6):904–907, 1996.

H. Pessenhofer and T. Kenner. Zur Methodik der kontinuierlichen Bestimmung der Phasenbeziehung zwischen Herzschlag und Atmung. *Pflügers Arch.*, 355:77–83, 1975.

D. W. Peterman, M. Ye, and P. E. Wigen. High frequency synchronization of chaos. *Phys. Rev. Lett.*, 74:1740–1742, 1995.

D. Petracchi, M. Barbi, S. Chillemi, E. Pantazelou, D. Pierson, C. Dames, L. Wilkens, and F. Moss. A test for a biological signal encoded by noise. *Int. J. Bifurc. Chaos*, 5(1):89–100, 1995.

V. Petrov, Q. Ouyang, and H. L. Swinney.

Resonant pattern formation in a chemical system. *Nature*, 388: 655–657, 1997.

A. Pikovsky and U. Feudel. Correlations and spectra of strange nonchaotic attractors. *J. Phys. A: Math., Gen.*, 27 (15):5209–5219, 1994.

A. Pikovsky, G. Osipov, M. Rosenblum, M. Zaks, and J. Kurths. Attractor–repeller collision and eyelet intermittency at the transition to phase synchronization. *Phys. Rev. Lett.*, 79: 47–50, 1997a.

A. Pikovsky, M. Rosenblum, and J. Kurths. Synchronization in a population of globally coupled chaotic oscillators. *Europhys. Lett.*, 34(3):165–170, 1996.

A. Pikovsky, M. Rosenblum, G. Osipov, and J. Kurths. Phase synchronization of chaotic oscillators by external driving. *Physica D*, 104:219–238, 1997b.

A. Pikovsky and S. Ruffo. Finite-size effects in a polpulation of interacting oscillators. *Phys. Rev. E*, 59(2): 1633–1636, 1999.

A. Pikovsky, M. Zaks, M. Rosenblum, G. Osipov, and J. Kurths. Phase synchronization of chaotic oscillations in terms of periodic orbits. *Chaos*, 7 (4):680–687, 1997c.

A. S. Pikovsky. On the interaction of strange attractors. *Z. Physik* B, 55(2): 149–154, 1984a.

A. S. Pikovsky. Synchronization and stochastization of nonlinear oscillations by external noise. In R. Z. Sagdeev, Editor, *Nonlinear and Turbulent Processes in Physics*, pages 1601–1604, Harwood, Singapore, 1984b.

A. S. Pikovsky. Synchronization and stochastization of the ensemble of autogenerators by external noise.

*Radiophys. Quant. Electron.*, 27(5): 576–581, 1984c.

A. S. Pikovsky. Phase synchronization of chaotic oscillations by a periodic external field. *Sov. J. Commun. Technol. Electron.*, 30:85, 1985.

A. S. Pikovsky. Statistics of trajectory separation in noisy dynamical systems. *Phys. Lett.* A, 165:33, 1992.

A. S. Pikovsky. Comment on "Chaos, Noise, and Synchronization". *Phys. Rev. Lett.*, 73(21):2931, 1994.

A. S. Pikovsky and U. Feudel. Characterizing strange nonchaotic attractors. *Chaos*, 5(1):253–260, 1995.

A. S. Pikovsky and P. Grassberger. Symmetry breaking bifurcation for coupled chaotic attractors. *J. Phys. A: Math., Gen.*, 24(19):4587–4597, 1991.

A. S. Pikovsky and J. Kurths. Roughening interfaces in the dynamics of perturbations of spatiotemporal chaos. *Phys. Rev. E*, 49(1):898–901, 1994.

A. S. Pikovsky, M. G. Rosenblum, and J. Kurths. Phase synchronization in regular and chaotic systems. *Int. J. Bifurc. Chaos*, 10(1):2291–2306, 2000.

N. Platt, S. M. Hammel, and J. F. Heagy. Effects of additive noise on on–off intermittency. *Phys. Rev. Lett.*, 72: 3498–3501, 1994.

N. Platt, E. A. Spiegel, and C. Tresser. On–off intermittency: A mechanism for bursting. *Phys. Rev. Lett.*, 70: 279–282, 1989.

A. A. Polezhaev and E. I. Volkov. On the possible mechanism of cell cycle synchronization. *Biol. Cybern.*, 41: 81–89, 1981.

A. Politi, R. Livi, G.-L. Oppo, and R. Kapral. Unpredictable behavior of stable systems. *Europhys. Lett.*, 22: 571, 1993.

B. Pompe. Measuring statistical

dependencies in a time series. *J. Stat. Phys.*, 73:587–610, 1993.

D. Postnov, S. K. Han, and H. Kook. Synchronization of diffusively coupled oscillators near the homoclinic bifurcation. *Phys. Rev. E*, 60(3): 2799–2807, 1999a.

D. E. Postnov, T. E. Vadivasova, O. V. Sosnovtseva, A. G. Balanov, V. S. Anishchenko, and E. Mosekilde. Role of multistability in the transition to chaotic phase synchronization. *Chaos*, 9(1):227–232, 1999b.

A. Prasad, V. Mehra, and R. Ramaswamy. Strange nonchaotic attractors in the quasiperiodically forced logistic map. *Phys. Rev. E*, 57(2):1576–1584, 1998.

W. H. Press, S. T. Teukolsky, W. T. Vetterling, and B. P. Flannery. *Numerical Recipes in C: the Art of Scientific Computing*. Cambridge University Press, Cambridge, second edition, 1992.

K. Pyragas. Weak and strong synchronization of chaos. *Phys. Rev. E*, 54(5):R4508–R4511, 1996.

K. Pyragas. Conditional Lyapunov exponents from time series. *Phys. Rev. E*, 56(5):5183–5188, 1997.

L. R. Rabiner and B. Gold. *Theory and Application of Digital Signal Processing*. Prentice–Hall, Englewood Cliffs, NJ, 1975.

M. I. Rabinovich and D. I. Trubetskov. *Oscillations and Waves in Linear and Nonlinear Systems*. Kluwer, Dordrecht, 1989.

W. J. Rappel and A. Karma. Noise-induced coherence in neural networks. *Phys. Rev. Lett.*, 77(15): 3256–3259, 1996.

J. Rayleigh. *The Theory of Sound*. Dover Publishers, New York, 1945.

D. V. R. Reddy, A. Sen, and G. L.

Johnston. Time delay effects on coupled limit cycle oscillators at Hopf bifurcation. *Physica D*, 129(1–2): 15–34, 1999.

C. Reichhardt and F. Nori. Phase locking, devil's staircase, Farey trees, and Arnold tongues in driven vortex lattices with periodic pinning. *Phys. Rev. Lett.*, 82(2):414–417, 1999.

P. Reimann, C. Van den Broeck, and P. Kawai. Nonequilibrium noise in coupled phase oscillators. *Phys. Rev. E*, 60(6):6402–6406, 1999.

L. Ren and B. Ermentrout. Phase locking in chains of multiple-coupled oscillators. *Physica D*, 143(1–4): 56–73, 2000.

A. Rényi. *Probability Theory*. Akadémiai Kiadó, Budapest, 1970.

P. Richard, B. M. Bakker, B. Teusink, K. Van Dam, and H. V. Westerhoff. Acetaldehyde mediates the synchronization of sustained glycolytic oscillations in population of yeast cells. *Eur. J. Biochem.*, 235:238–241, 1996.

H. Z. Risken. *The Fokker–Planck Equation*. Springer, Berlin, 1989.

E. Rodriguez, N. George, J.-P. Lachaux, J. Martinerie, B. Renault, and F. J. Varela. Perception's shadow: Long distance synchronization of human brain activity. *Nature*, 397(4):430–433, 1999.

J. L. Rogers and L. T. Wille. Phase transitions in nonlinear oscillator chains. *Phys. Rev. E*, 54(3): R2193–R2196, 1996.

Yu. M. Romanovsky, N. V. Stepanova, and D. S. Chernavsky. *Mathematical Modelling in Biophysics*. Nauka, Moscow, 1975 (in Russian).

Yu. M. Romanovsky, N. V. Stepanova, and D. S. Chernavsky. *Mathematical Biophysics*. Nauka, Moscow, 1984 (in

Russian).

F. J. Romeiras, A. Bondeson, E. Edward Ott, Th. M. Antonsen Jr., and C. Grebogi. Quasiperiodically forced dynamical systems with strange nonchaotic attractors. *Physica* D, 26: 277–294, 1987.

E. Rosa Jr., E. Ott, and M. H. Hess. Transition to phase synchronization of chaos. *Phys. Rev. Lett.*, 80(8): 1642–1645, 1998.

E. Rosa Jr., W. B. Pardo, C. M. Ticos, J. A. Walkenstein, and M. Monti. Phase synchronization of chaos in a plasma discharge tube. *Int. J. Bifurc. Chaos*, 10(11):2551–2564, 2000.

J. E. Rose, J. F. Brugge, D. J. Anderson, and J. E. Hind. Phase-locked response to low-frequency tones in single auditory nerve fibers of the squirrel monkey. *J. Neurophysiol.*, 30:769–793, 1967.

M. Rosenblum, A. Pikovsky, and J. Kurths. Phase synchronization of chaotic oscillators. *Phys. Rev. Lett.*, 76: 1804, 1996.

M. Rosenblum, A. Pikovsky, and J. Kurths. Effect of phase synchronization in driven chaotic oscillators. *IEEE Trans. CAS-I*, 44(10):874–881, 1997a.

M. Rosenblum, A. Pikovsky, and J. Kurths. From phase to lag synchronization in coupled chaotic oscillators. *Phys. Rev. Lett.*, 78:4193–4196, 1997b.

M. G. Rosenblum. A characteristic frequency of chaotic dynamical system. *Chaos, Solit. Fract.*, 3(6): 617–626, 1993.

M. G. Rosenblum, G. I. Firsov, R. A. Kuuz, and B. Pompe. Human postural control: Force plate experiments and modelling. In H. Kantz, J. Kurths, and G. Mayer-Kress, Editors, *Nonlinear Analysis of Physiological Data*, pages

283–306. Springer, Berlin, 1998.

M. G. Rosenblum and J. Kurths. Analysing synchronization phenomena from bivariate data by means of the Hilbert transform. In H. Kantz, J. Kurths, and G. Mayer-Kress, Editors, *Nonlinear Analysis of Physiological Data*, pages 91–99. Springer, Berlin, 1998.

M. G. Rosenblum, A. S. Pikovsky, J. Kurths, C. Schäfer, and P. A. Tass. Phase synchronization: From theory to data analysis. In F. Moss and S. Gielen, Editors, *Handbook of Biological Physics. Vol. 4, Neuro-informatics*, pages 279–321. Elsevier, Amsterdam, 2001.

M. G. Rosenblum, P. A. Tass, and J. Kurths. Estimation of synchronization from noisy data with application to human brain activity. In J. A. Freund and T. Pöschel, Editors, *Stochastic Processes in Physics, Chemistry, and Biology*, Lecture Notes in Physics, LNP 557, pages 202–211. Springer, Berlin, 2000.

O. E. Rössler. An equation for continuous chaos. *Phys. Lett.* A, 57(5):397, 1976.

R. Roy and K. S. Thornburg. Experimental synchronization of chaotic lasers. *Phys. Rev. Lett.*, 72:2009–2012, 1994.

N. F. Rulkov. Images of synchronized chaos: Experiments with circuits. *Chaos*, 6(3):262–279, 1996.

N. F. Rulkov and M. M. Suschik. Experimental observation of synchronized chaos with frequency ratio 1:2. *Phys. Lett.* A, 214:145–150, 1996.

N. F. Rulkov, M. M. Sushchik, L. S. Tsimring, and H. D. I. Abarbanel. Generalized synchronization of chaos in directionally coupled chaotic systems. *Phys. Rev. E*, 51(2):980–994,

1995.

D. F. Russell, L. A. Wilkens, and F. Moss. Use of behavioral stochastic resonance by paddle fish for feeding. *Nature*, 402:291–294, 1999.

S. M. Rytov, Yu. A. Kravtsov, and V. I. Tatarskii. *Principles of Statistical Radiophysics. Volume 2: Correlation Theory of Random Processes*. Springer, Berlin, 1988.

T. Saitoh and T. Nishino. Phase locking in a double junction of Josephson weak links. *Phys. Rev.* B, 44(13):7070–7073, 1991.

H. Sakaguchi and Y. Kuramoto. A solvable active rotator model showing phase transition via mutual entrainment. *Prog. Theor. Phys.*, 76(3): 576–581, 1986.

H. Sakaguchi, S. Shinomoto, and Y. Kuramoto. Mutual entrainment in oscillator lattices with nonvariational type interaction. *Prog. Theor. Phys.*, 79 (5):1069–1079, 1988a.

H. Sakaguchi, S. Shinomoto, and Y. Kuramoto. Phase transitions and their bifurcation analysis in a large population of active rotators with mean-field coupling. *Prog. Theor. Phys.*, 79(3):600–607, 1988b.

E. Sánchez, M. A. Matías, and V. Pérez-Muñuzuri. Analysis of synchronization of chaotic systems by noise: An experimental study. *Phys. Rev.* E, 56(4):4068–4071, 1997.

G. Santoboni, A. Varone, and S. R. Bishop. Spatial distribution of chaotic transients in unidirectional synchronisation. *Phys. Lett.* A, 257 (3–4):175–181, 1999.

P. Sassone-Corsi. Molecular clocks: Mastering time by gene regulation. *Nature*, 392:871–874, 1999.

M. Sauer and F. Kaiser. Synchronized spatiotemporal chaos and spatiotemporal on–off intermittency in a nonlinear ring cavity. *Phys. Rev. E*, 54(3):2468–2473, 1996.

J. P. Saul. Cardiorespiratory variability: Fractals, white noise, nonlinear oscillators, and linear modeling. What's to be learned? In H. Haken and H. P. Koepchen, Editors, *Rhythms in Physiological Systems*, volume 55 of *Springer Series in Synergetics*, pages 115–126. Springer, Berlin, 1991.

C. Schäfer, M. G. Rosenblum, H.-H. Abel, and J. Kurths. Synchronization in the human cardiorespiratory system. *Phys. Rev. E*, 60:857–870, 1999.

C. Schäfer, M. G. Rosenblum, J. Kurths, and H.-H. Abel. Heartbeat synchronized with ventilation. *Nature*, 392(6673):239–240, March 1998.

M. Schiek, F. R. Drepper, R. Engbert, H.-H. Abel, and K. Suder. Cardiorespiratory synchronization. In H. Kantz, J. Kurths, and G. Mayer-Kress, Editors, *Nonlinear Analysis of Physiological Data*, pages 191–209. Springer, Berlin, 1998.

S. J. Schiff, P. So, T. Chang, R. E. Burke, and T. Sauer. Detecting dynamical interdependence and generalized synchrony through mutual prediction in a neural ensemble. *Phys. Rev. E*, 54 (6):6708–6724, 1996.

G. Schmidt and A. A. Chernikov. General form of coupling leading to synchronization of oscillating dynamical systems. *Phys. Rev. E*, 60 (3):2767–2770, 1999.

R. F. Schmidt and G. Thews. *Human Physiology*. Springer, New York, 1983.

A. Schrader, M. Braune, and H. Engel. Dynamics of spiral waves in excitable media subjected to external periodic forcing. *Phys. Rev. E*, 52(1):98–108,

1995.

H. G. Schuster. *Deterministic Chaos, An Introduction*. VCH, Weinheim, 1988.

H. G. Schuster, Editor. *Handbook of Chaos Control*. Wiley-VCH, Weinheim, 1999.

H. G. Schuster, S. Martin, and W. Martienssen. A new method for determining the largest Lyapunov exponent in simple nonlinear systems. *Phys. Rev.* A, 33:3547, 1986.

D. M. Scolnick and T. D. Halazonetis. *Chfr* defines a mitotic stress checkpoint that delays entry into metaphase. *Nature*, 406:354–356, 2000.

H. Seidel and H.-P. Herzel. Analyzing entrainment of heartbeat and respiration with surrogates. *IEEE Eng. Med. Biol.*, 17(6):54–57, 1998.

S. Shapiro. Josephson current in superconducting tunneling: The effect of microwaves and other observations. *Phys. Rev. Lett.*, 11(2):80–82, 1963.

S. Shinomoto and Y. Kuramoto. Cooperative phenomena in two-dimensional active rotator systems. *Prog. Theor. Phys.*, 75(6): 1319–1327, 1986.

B. I. Shraiman, A. Pumir, W. van Saarlos, P. C. Hohenberg, H. Chaté, and M. Holen. Spatiotemporal chaos in the one-dimensional Ginzburg-Landau equation. *Physica* D, 57:241–248, 1992.

J. W. Shuai and K. W. Wong. Noise and synchronization in chaotic neural networks. *Phys. Rev.* E, 57(6): 7002–7007, 1998.

B. Shulgin, A. Neiman, and V. Anishchenko. Mean switching frequency locking in stochastic bistable systems driven by a periodic force. *Phys. Rev. Lett.*, 75(23): 4157–4160, 1995.

A. E. Siegman. *Lasers*. University Science Books, Mill Valley, CA, 1986.

A. Simon and A. Libchaber. Escape and synchronization of a Brownian particle. *Phys. Rev. Lett.*, 68:3375, 1992.

J. Simonet, M. Warden, and E. Brun. Locking and Arnold tongues in an infinite-dimensional system: The nuclear magnetic resonance laser with delayed feedback. *Phys. Rev.* E, 50: 3383–3391, 1994.

W. Singer. Striving for coherence. *Nature*, 397(4):391–393, 1999.

W. Singer and C. M. Gray. Visual feature integration and the temporal correlation hypothesis. *Annu. Rev. Neurosci.*, 18:555–586, 1995.

G. Sivashinsky. Self-turbulence in the motion of a free particle. *Found. Phys.*, 8(9–10):735–744, 1978.

S. Smale. *The Mathematics of Time*. Springer, New York, 1980.

M. J. T. Smith and R. M. Mersereau. *Introduction to Digital Signal Processing. A Computer Laboratory Textbook*. Wiley, New York, 1992.

M. Smolensky. Chronobiology and chronotherapeutics: Applications to cardiovascular medicine. In P. C. Deedwania, Editor, *Circadian Rhythms of Cardiovascular Disorders*, pages 173–206. Futura, Armonk, NY, 1997.

Y. Soen, N. Cohen, D. Lipson, and E. Braun. Emergence of spontaneous rhythm disorders in self-assembled networks of heart cells. *Phys. Rev. Lett.*, 82(17):3556–3559, 1999.

O. V. Sosnovtseva, A. G. Balanov, T. E. Vadivasova, V. V. Astakhov, and E. Mosekilde. Loss of lag synchronization in coupled chaotic systems. *Phys. Rev.* E, 60(6): 6560–6565, 1999.

P. Stange, A. S. Mikhailov, and B. Hess. Mutual synchronization of molecular turnover cycles in allosteric enzymes. *J. Phys. Chem.* B, 102(32):6273–6289, 1998.

P. Stange, A. S. Mikhailov, and B. Hess. Mutual synchronization of molecular turnover cycles in allosteric enzymes II. *J. Phys. Chem.* B, 103(29):6111–6120, 1999.

J. Stark. Invariant graphs for forced systems. *Physica* D, 109(1-2): 163–179, 1997.

O. Steinbock, V. Zykov, and S. Müller. Control of spiral-wave dynamics in active media by periodic modulation of excitability. *Nature*, 366:322–324, 1993.

P. N. Steinmetz, A. Roy, P. J. Fitzgerald, S. S. Hsiao, K. O. Johnson, and E. Niebur. Attention modulates synchronized neuronal firing in primate somatosensory cortex. *Nature*, 404(9):187–190, 2000.

E. A. Stern, D. Jaeger, and C. J. Wilson. Membrane potential synchrony of simultaneously recorded striatal spiny neurons *in vivo*. *Nature*, 394:475–478, 1998.

E. F. Stone. Frequency entrainment of a phase coherent attractor. *Phys. Lett.* A, 163:367–374, 1992.

M. Stopfer, S. Bhagavan, B. H. Smith, and G. Laurent. Impaired odour descrimination on desynchronization of odour-encoding neural assemblies. *Nature*, 390(6):70–74, 1997.

R. L. Stratonovich. *Topics in the Theory of Random Noise*. Gordon and Breach, New York, 1963.

S. H. Strogatz. *Nonlinear Dynamics and Chaos: With Applications to Physics, Biology, Chemistry, and Engineering*. Addison-Wesley, Reading, MA, 1994.

S. H. Strogatz. From Kuramoto to Crawford: Exploring the onset of synchronization in populations of coupled oscillators. *Physica* D, 143 (1–4):1–20, 2000.

S. H. Strogatz, C. M. Marcus, R. M. Westervelt, and R. E. Mirollo. Collective dynamics of coupled oscillators with random pinning. *Physica* D, 36:23–50, 1989.

S. H. Strogatz and R. E. Mirollo. Stability of incoherence in a population of coupled oscillators. *J. Stat. Phys.*, 63 (3/4):613–635, 1991.

S. H. Strogatz and R. E. Mirollo. Splay states in globally coupled Josephson arrays: Analytical prediction of Floquet multipliers. *Phys. Rev.* E, 47 (1):220–227, 1993.

S. H. Strogatz, R. E. Mirollo, and P. C. Matthews. Coupled nonlinear oscillators below the synchronization threshold: Relaxation by generalized Landau damping. *Phys. Rev. Lett.*, 68 (18):2730–2733, 1992.

S. H. Strogatz and I. Stewart. Coupled oscillators and biological synchronization. *Sci. Am.*, 12:68–75, 1993.

J. Sturis, C. Knudsen, N. M. O'Meara, J. S. Thomsen, E. Mosekilde, E. Van Cauter, and K. S. Polonsky. Phase-locking regions in a forced model of slow insulin and glucose oscillations. *Chaos*, 5(1):193–199, 1995.

J. Sturis, E. Van Cauter, J. Blackman, and K. S. Polonsky. Entrainment of pulsatile insulin secretion by oscillatory glucose infusion. *J. Clin. Invest.*, 87:439–445, 1991.

K. H. Stutte and G. Hildebrandt. Untersuchungen über die Koordination von Herzschlag und Atmung. *Pflügers*

*Arch.*, 289:R47, 1966.

T. Sugawara, M. Tachikawa, T. Tsukamoto, and T. Shimizu. Observation of synchronization in laser chaos. *Phys. Rev. Lett.*, 72:3502–3505, 1994.

J. W. Swift, S. H. Strogatz, and K. Wiesenfeld. Averaging of globally coupled oscillators. *Physica D*, 55: 239–250, 1992.

S. Taherion and Y.-C. Lai. Observability of lag synchronization of coupled chaotic oscillators. *Phys. Rev. E*, 59(6): R6247–R6250, 1999.

F. Takens. Detecting strange attractors in turbulence. In *Dynamical Systems and Turbulence*, volume 898 of *Springer Lecture Notes in Mathematics*, pages 366–381. Springer, New York, 1981.

A. Tamasevicius and A. Čenys. Synchronizing hyperchaos with a single variable. *Phys. Rev. E*, 55(1): 297–299, 1997.

T. Tamura, N. Inaba, and J. Miyamichi. Mechanism for taming chaos by weak harmonic perturbations. *Phys. Rev. Lett.*, 83(19):3824–3827, 1999.

H. Tanaka, A. Lichtenberg, and S. Oishi. First order phase transition resulting from finite inertia in coupled oscillator systems. *Phys. Rev. Lett.*, 78(11): 2104–2107, 1997a.

H.-A. Tanaka, A. J. Lichtenberg, and Sh. Oishi. Self-synchronization of coupled oscillators with hysteretic responses. *Physica D*, 100:279–300, 1997b.

D. Y. Tang, R. Dykstra, M. W. Hamilton, and N. R. Heckenberg. Experimental evidence of frequency entrainment between coupled chaotic oscillations. *Phys. Rev. E*, 57(3):3649–3651, 1998a.

D. Y. Tang, R. Dykstra, M. W. Hamilton, and N. R. Heckenberg. Observation of generalized synchronization of chaos

in a driven chaotic system. *Phys. Rev. E*, 57(5):5247–5251, 1998b.

D. Y. Tang, R. Dykstra, M. W. Hamilton, and N. R. Heckenberg. Stages of chaotic synchronization. *Chaos*, 8(3): 697–701, 1998c.

P. Tass. Phase and frequency shifts of two nonlinearly coupled oscillators. *Z. Physik B*, 99:111–121, 1995.

P. Tass. Phase and frequency shifts in a population of phase oscillators. *Phys. Rev. E*, 56(2):2043–2060, 1997.

P. Tass and H. Haken. Synchronization in networks of limit cycle oscillators. *Z. Physik B*, 100:303–320, 1996.

P. Tass, J. Kurths, M. G. Rosenblum, J. Weule, A. S. Pikovsky, J. Volkmann, A. Schnitzler, and H.-J. Freund. Complex phase synchronization in neurophysiological data. In C. Uhl, Editor, *Analysis of Neurophysiological Brain Functioning, Springer Series in Synergetics*, pages 252–273. Springer, Berlin, 1999.

P. Tass, M. G. Rosenblum, J. Weule, J. Kurths, A. S. Pikovsky, J. Volkmann, A. Schnitzler, and H.-J. Freund. Detection of $n : m$ phase locking from noisy data: Application to magnetoencephalography. *Phys. Rev. Lett.*, 81(15):3291–3294, 1998.

P. A. Tass. *Phase Resetting in Medicine and Biology. Stochastic Modelling and Data Analysis.* Springer, Berlin, 1999.

M. L. Tcetlin. *Studies on the Theory of Automata and Modelling of Biological Systems.* Nauka, Moscow, 1969 (in Russian).

K. F. Teodorchik. *Self-Oscillatory Systems.* Gostekhizdat, Moscow, 1952 (in Russian).

J. R. Terry, K. S. Thornbug, D. J. DeShazer, G. D. VanWiggeren, S. Q. Zhu, P. Ashwin, and R. Roy.

Synchronization of chaos in an array of three lasers. *Phys. Rev. E*, 59(4): 4036–4043, 1999.

K. S. Thornburg, M. Möller, R. Roy, T. W. Carr, R.-D. Li, and T. Erneux. Chaos and coherence in coupled lasers. *Phys. Rev. E*, 55(4):3865–3869, 1997.

C. M. Ticos, E. Rosa Jr., W. B. Pardo, J. A. Walkenstein, and M. Monti. Experimental real-time phase synchronization of a paced chaotic plasma discharge. *Phys. Rev. Lett.*, 85 (14):2929–2932, 2000.

E. Toledo, M. G. Rosenblum, J. Kurths, and S. Akselrod. Cardiorespiratory synchronization: Is it a real phenomenon? In A. Murray and S. Swiryn, Editors, *Computers in Cardiology*, pages 237–240. IEEE Computer Society Press, Hannover, 1999.

E. Toledo, M. G. Rosenblum, C. Schäfer, J. Kurths, and S. Akselrod. Quantification of cardiorespiratory synchronization in normal and heart transplant subjects. In *Proc. of Int. Symposium on Nonlinear Theory and its Applications*, volume 1, pages 171–174. Crans-Montana, Switzerland, Sept. 14–17, 1998. Presses Polytechniques et Universitaires Romandes, 1998.

A. Torcini, P. Grassberger, and A. Politi. Error propagation in extended systems. *J. Phys. A: Math. Gen.*, 27:4533, 1995.

C. Tresser, P. Worfolk, and C. W. Wu. Master–slave synchronization from the point of view of global dynamics. *Chaos*, 5(4):693–699, 1995.

K. Y. Tsang, R. E. Mirollo, S. H. Strogatz, and K. Wiesenfeld. Dynamics of globally coupled oscillator array. *Physica* D, 48:102–112, 1991a.

K. Y. Tsang, S. H. Strogatz, and

K. Wiesenfeld. Reversibility and noise sensitivity in globally coupled oscillators. *Phys. Rev. Lett.*, 66: 1094–1097, 1991b.

M. Tsodyks, I. Mitkov, and H. Sompolinsky. Pattern of synchrony in inhomogeneous networks of oscillators with pulse interactions. *Phys. Rev. Lett.*, 71(8):1280–1283, 1993.

T. Tsukamoto, M. Tachikawa, T. Hirano, T. Kuga, and T. Shimizu. Synchronization of a chaotic laser pulsation with its prerecorded history. *Phys. Rev. E*, 54(4):4476–4479, 1996.

T. Tsukamoto, M. Tachikawa, T. Tohei, T. Hirano, T. Kuga, and T. Shimizu. Synchronization of a laser system to a modulation signal artificially constructed from its strange attractor. *Phys. Rev. E*, 56(6):6564–6568, 1997.

N. Tufillaro, T. R. Abbott, and J. Reilly. *An Experimental Approach to Nonlinear Dynamics and Chaos*. Addison-Wesley, Reading, MA,1992.

J. Urias, G. Salazar, and E. Ugalde. Synchronization of cellular automation pairs. *Chaos*, 8(4):814–818, 1998.

T. E. Vadivasova, A. G. Balanov, O. V. Sosnovtseva, D. E. Postnov, and E. Mosekilde. Synchronization in driven chaotic systems: Diagnostics and bifurcations. *Phys. Lett.* A, 253: 66–74, 1999.

T. P. Valkering, C. L. A. Hooijer, and M. F. Kroon. Dynamics of two capacitively coupled Josephson junctions in the overdamped limit. *Physica* D, 135 (1–2):137–153, 2000.

B. van der Pol. A theory of the amplitude of free and forced triode vibration. *Radio Rev.*, 1:701, 1920.

B. van der Pol. On relaxation oscillation. *Phil. Mag.*, 2:978–992, 1926.

B. van der Pol. Forced oscillations in a circuit with non-linear resistance. (Reception with reactive triode). *Phil. Mag.*, 3:64–80, 1927.

B. van der Pol and J. van der Mark. Frequency demultiplication. *Nature*, 120(3019):363–364, 1927.

B. van der Pol and J. van der Mark. The heartbeat considered as a relaxation oscillation, and an electrical model of the heart. *Phil. Mag.*, 6:763–775, 1928.

J. L. van Hemmen and W. F. Wreszinski. Lyapunov function for the Kuramoto model of nonlinearly coupled oscillators. *J. Stat. Phys.*, 72(1/2): 145–166, 1993.

N. G. van Kampen. *Stochastic Processes in Physics and Chemistry*, 2nd edition. North Holland, Amsterdam, 1992.

C. van Vreeswijk. Partial synchronization in populations of pulse-coupled oscillators. *Phys. Rev.* E, 54(5): 5522–5537, 1996.

V. K. Vanag, L. Yang, M. Dolnik, A. M. Zhabotinsky, and I. R. Epshtein. Oscillatory cluster patterns in a homogeneous chemical system with global feedback. *Nature*, 406:389–391, 2000.

S. R. S. Varadhan. *Large Deviations and Applications*. SIAM, Philadelphia, 1984.

Ju. M. Vasiliev, I. M. Gelfand, V. I. Guelstein, and A. G. Malenkov. Interrelationships of contacting cells in the cell complexes of mouse ascites hepatoma. *Int. J. Cancer*, 1:451–462, 1966.

J. J. P. Veerman. Irrational rotation numbers. *Nonlinearity*, 2:419–428, 1989.

S. C. Venkataramani, T. M. Antonsen, E. Ott, and J. C. Sommerer. On–off intermittency: Power spectrum and fractal properties of time series. *Physica* D, 96(1–4):66–99, 1996.

Sh. C. Venkataramani, Th. M. Antonsen Jr., E. Ott, and J. C. Sommerer. Characterization of on-off intermittent time series. *Phys. Lett.* A, 207: 173–179, 1995.

M. D. Vieira. Chaos and synchronized chaos in an earthquake model. *Phys. Rev. Lett.*, 82(1):201–204, 1999.

J. Volkmann, M. Joliot, A. Mogilner, A. A. Ioannides, F. Lado, E. Fazzini, U. Ribary, and R. Llinás. Central motor loop oscillations in Parkinsonian resting tremor revealed by magnetoencephalography. *Neurology*, 46:1359–1370, 1996.

E. I. Volkov and V. A. Romanov. Bifurcations in the system of two identical diffusively coupled brusselators. *Phys. Scr.*, 51(1):19–28, 1994.

H. Voss and J. Kurths. Reconstruction of nonlinear time delay models from data by the use of optimal transformations. *Phys. Lett.* A, 234:336–344, 1997.

H. U. Voss. Anticipating chaotic synchronization. *Phys. Rev.* E, 61(5): 5115–5119, 2000.

D. Walgraef. *Spatio-Temporal Pattern Formation*. Springer, New York, 1997.

T. J. Walker. Acoustic synchrony: Two mechanisms in the snowy tree cricket. *Science*, 166:891–894, 1969.

I. Waller and R. Kapral. Synchronization and chaos in coupled nonlinear oscillators. *Phys. Lett.* A, 105: 163–168, 1984.

W. Wang, I. Z. Kiss, and J. L. Hudson. Experiments on arrays of globally coupled chaotic electrochemical oscillators: Synchronization and clustering. *Chaos*, 10(1):248–256, 2000a.

W. Wang, G. Perez, and H. A. Cerdeira. Dynamical behavior of the firings in a coupled neuronal system. *Phys. Rev. E*, 47(4):2893–2898, 1993.

Y. Wang, D. T. W. Chik, and Z. D. Wang. Coherence resonance and noise-induced synchronization in globally coupled Hodgkin-Huxley neurons. *Phys. Rev. E*, 61(1):740–746, 2000b.

S. Watanabe and S. H. Strogatz. Integrability of a globally coupled oscillator array. *Phys. Rev. Lett.*, 70 (16):2391–2394, 1993.

S. Watanabe and S. H. Strogatz. Constants of motion for superconducting Josephson arrays. *Physica* D, 74: 197–253, 1994.

D. Whitmore, N. S. Foulkes, and P. Sassone-Corsi. Light acts directly on organs and cells in culture to set the vertebrate circadian clock. *Nature*, 404:87–91, 2000.

K. Wiesenfeld. Noise, coherence, and reversibility in Josephson arrays. *Phys. Rev. B*, 45(1):431–435, 1992.

K. Wiesenfeld, P. Colet, and S. H. Strogatz. Synchronization transition in a disordered Josephson series array. *Phys. Rev. Lett.*, 76(3):404–407, 1996.

K. Wiesenfeld and F. Moss. Stochastic resonance: From ice ages to crayfish and SQUIDs. *Nature*, 373:33–36, 1995.

K. Wiesenfeld and J. W. Swift. Averaged equations for Josephson junction series arrays. *Phys. Rev. E*, 51(2):1020–1025, 1995.

S. Wiggins. *Global Bifurcations and Chaos (Analytical Methods)*. Springer, New York, 1988.

S. Wiggins. *Introduction to Applied Dynamical Systems and Chaos*. Springer, New York, 1990.

L. A. Wilkens, D. F. Russell, X. Pei, and C. Gurgens. The paddlefish rostrum functions as an electrosensory antenna in plankton feeding. *Proc. Roy. Soc. Lond.* B, 264:1723–1729, 1997.

A. T. Winfree. Biological rhythms and the behavior of populations of coupled oscillators. *J. Theor. Biol.*, 16:15–42, 1967.

A. T. Winfree. *The Geometry of Biological Time*. Springer, New York, 1980.

H. G. Winful and L. Rahman. Synchronized chaos and spatiotemporal chaos in arrays of coupled lasers. *Phys. Rev. Lett.*, 65: 1575–1578, 1990.

F. Xie, G. Hu, and Z. Qu. On–off intermittency in a coupled-map lattice system. *Phys. Rev. E*, 52(2):1265, 1995.

V. Yakhot. Large-scale properties of unstable systems governed by the Kuramoto–Sivashinski equation. *Phys. Rev. A*, 24:642, 1981.

T. Yalcinkaya and Y.-Ch. Lai. Bifurcation to strange nonchaotic attractors. *Phys. Rev. E*, 56(2):1623–1630, 1997.

T. Yamada and H. Fujisaka. Stability theory of synchronized motion in coupled-oscillator systems. II. The mapping approach. *Prog. Theor. Phys.*, 70(5):1240–1248, 1983.

T. Yamada and H. Fujisaka. Stability theory of synchronized motion in coupled-oscillator systems. III. Mapping model for continuous system. *Prog. Theor. Phys.*, 72(5):885–894, 1984.

T. Yamada and H. Fujisaka. Intermittency caused by chaotic modulation. I. Analysis with a multiplicative noise model. *Prog. Theor. Phys.*, 76(3): 582–591, 1986.

T. Yamada and H. Fujisaka. Effect of

inhomogenety on intermittent chaos in a coupled system. *Phys. Lett.* A, 124 (8):421–425, 1987.

Y. Yamaguchi and H. Shimizu. Theory of self-synchronization in the presence of native frequency distribution and external noises. *Physica* D, 11: 212–226, 1984.

H. L. Yang and E. J. Ding. Synchronization of chaotic systems and on–off intermittency. *Phys. Rev.* E, 54(2):1361–1365, 1996.

T. Yang and K. Bilimgut. Experimental results of strange nonchaotic phenomenon in a second-order quasi-periodic forced electronic circuit. *Phys. Lett.* A, 236(5–6):494–504, 1997.

A. R. Yehia, D. Jeandupeux, F. Alonso, and M. R. Guevara. Hysteresis and bistability in the direct transition from 1:1 to 2:1 rhythm in periodically driven ventricular cells. *Chaos*, 9(4):916–931, 1999.

M. K. Stephen Yeung and S. H. Strogatz. Time delay in the Kuramoto model of coupled oscillators. *Phys. Rev. Lett.*, 82 (3):648–651, 1999.

K.-P. Yip and N.-H. Holstein-Rathlou. Chaos and non-linear phenomena in renal vascular control. *Cardiovasc. Res.*, 31:359–370, 1996.

L. Yu, E. Ott, and Q. Chen. Transition to chaos for random dynamical systems. *Phys. Rev. Lett.*, 65:2935–2938, 1990.

Y. H. Yu, K. Kwak, and T. K. Lim. On–off intermittency in an experimental synchronization process. *Phys. Lett.* A, 198(1):34–38, 1995.

M. A. Zaks, E.-H. Park, M. G. Rosenblum, and J. Kurths. Alternating locking ratios in imperfect phase synchronization. *Phys. Rev. Lett.*, 82: 4228–4231, 1999.

D. Zanette and A. S. Mikhailov. Condensation in globally coupled populations of chaotic dynamical systems. *Phys. Rev.* E, 57(1):276–281, 1998a.

D. Zanette and A. S. Mikhailov. Mutual synchronization in ensembles of globally coupled neural networks. *Phys. Rev.* E, 58(1):872–875, 1998b.

G. M. Zaslavsky. The simplest case of a strange attractor. *Phys. Lett.* A, 69(3): 145–147, 1978.

W.-Z. Zeng, M. Courtemanche, L. Sehn, A. Shrier, and L. Glass. Theoretical computation of phase locking in embrionic atrial heart cell aggregates. *J. Theoretical Biology*, 145:225–244, 1990.

Z. G. Zheng, G. Hu, and B. Hu. Phase slips and phase synchronization of coupled oscillators. *Phys. Rev. Lett.*, 81 (24):5318–5321, 1998.

T. Zhou, F. Moss, and A. Bulsara. Simulations of a strange non-chaotic attractor in a SQUID. In S. Vohra, M. Spano, M. Schlesinger, L. Pecora, and W. Ditto, Editors, *Proc. of the 1st Experiment. Chaos Conf., Arlington VA, Oct. 1–3, 1991*, pages 303–314. World Scientific, Singapore, 1992.

Zh. Zhu and Zh. Liu. Strange nonchaotic attractors of Chua's circuit with quasiperiodic excitation. *J. Bifurc. Chaos*, 7(1):227–238, 1997.

L. Zonghua and C. Shigang. General method of synchronization. *Phys. Rev.* E, 55(6):6651–6655, 1997a.

L. Zonghua and C. Shigang. Synchronization of a conservative map. *Phys. Rev.* E, 56(2):1585–1589, 1997b.

# Index